INTRODUCTION TO
NANOTECHNOLOGY

INTRODUCTION TO NANOTECHNOLOGY

Charles P. Poole, Jr.

Frank J. Owens

WILEY-
INTERSCIENCE

A JOHN WILEY & SONS, INC., PUBLICATION

For general information on our other products and services please contact our Customer Care Department
within the U.S. at 877-762-2974, outside the U.S. at 317-572-3993 or fax 317-572-4002.

Wiley also publishes its books in a variety of electronic formats. Some content that appears in print,
however, may not be available in electronic format.

Library of Congress Cataloging-in-Publication Data:

Poole, Charles P.
    Introduction to nanotechnology/Charles P. Poole, Jr., Frank J. Owens.
       p.cm.
    "A Wiley-Interscience publication."
    Includes bibliographical references and index.
    ISBN 0-471-07935-9 (cloth)
    1. Nanotechnology, I. Owens, Frank J. II. Title

    T174.7. P66 2003
    620′.5–dc21   2002191031

Printed in the United States of America

10 9 8 7 6 5 4 3 2 1

CONTENTS

PREFACE

In recent years nanotechnology has become one of the most important and exciting forefront fields in Physics, Chemistry, Engineering and Biology. It shows great promise for providing us in the near future with many breakthroughs that will change the direction of technological advances in a wide range of applications. To facilitate the timely widespread utilization of this new technology it is important to have available an overall summary and commentary on this subject which is sufficiently detailed to provide a broad coverage and insight into the area, and at the same time is sufficiently readable and thorough so that it can reach a wide audience of those who have a need to know the nature and prospects for the field. The present book hopes to achieve these two aims.

The current widespread interest in nanotechnology dates back to the years 1996 to 1998 when a panel under the auspices of the World Technology Evaluation Center (WTEC), funded by the National Science Foundation and other federal agencies, undertook a world-wide study of research and development in the area of nano-technology, with the purpose of assessing its potential for technological innovation. Nanotechnology is based on the recognition that particles less than the size of 100 nanometers (a nanometer is a billionth of a meter) impart to nanostructures built from them new properties and behavior. This happens because particles which are smaller than the characteristic lengths associated with particular phenomena often display new chemistry and physics, leading to new behavior which depends on the size. So, for example, the electronic structure, conductivity, reactivity, melting temperature, and mechanical properties have all been observed to change when particles become smaller than a critical size. The dependence of the behavior on the particle sizes can allow one to engineer their properties. The WTEC study concluded that this technology has enormous potential to contribute to significant advances over a wide and diverse range of technological areas ranging from producing stronger and lighter materials, to shortening the delivery time of nano structured pharmaceuticals to the body's circulatory system, increasing the storage capacity of magnetic tapes, and providing faster switches for computers. Recommendations made by this and subsequent panels have led to the appropriation of very high levels of funding in recent years. The research area of nanotechnology is interdisciplinary, covering a wide variety of subjects ranging from the chemistry of the catalysis of nanoparticles, to the physics of the quantum dot laser. As a result researchers in any one particular area need to reach beyond their expertise in order to appreciate the broader implications of nanotechnology, and learn how to contribute to this exciting new field. Technical managers, evaluators, and those who must make funding decisions will need to understand a wide variety of disciplines. Although this book was originally intended to be an introduction to nanotechnology, due to the nature of the field it has developed into an introduction to selected topics in nanotechnology, which are thought to be representative of the overall field. Because of the rapid pace of development of the subject, and its interdisciplinary nature, a truly comprehensive coverage does not seem feasible. The topics presented here were chosen based on the maturity of understanding of the subjects, their potential for applications, or the number of already existing applications. Many of the chapters discuss present and future possibilities. General references are included for those who wish to pursue further some of the areas in which this technology is moving ahead.

We have attempted to provide an introduction to the subject of nanotechnology written at a level such that researchers in different areas can obtain an appreciation of developments outside their present areas of expertise, and so that technical administrators and managers can obtain an overview of the subject. It is possible that such a book could be used as a text for a graduate course on nanotechnology. Many of the chapters contain introductions to the basic physical and chemical principles of the subject area under discussion, hence the various chapters are self contained, and may be read independently of each other. Thus Chapter 2 begins with a brief overview of the properties of bulk materials that need to be understood if one is to appreciate how, and why, changes occur in these materials when their sizes approach a billionth of a meter. An important impetus that caused nanotechnology to advance so rapidly has been the development of instrumentation such as the scanning tunneling microscope that allows the visualization of the surfaces of nanometer sized materials. Hence Chapter 3 presents descriptions of important instrumentation systems, and provides illustrations of measurements on nano materials. The remaining chapters cover various aspects of the field.

One of us (CPP) would like to thank his son Michael for drawing several dozen of the figures that appear throughout the book, and his grandson Jude Jackson for helping with several of these figures. We appreciate the comments of Prof. Austin Hughes on the biology chapter. We have greatly benefited from the information found in the five volume Handbook of Nanostructured Materials and Nanotechnology edited by H. S. Nalwa, and in the book Advanced Catalysis and Nanostructured Materials edited by W. R. Moser, both from Academic Press.

1

INTRODUCTION

The prefix *nano* in the word *nanotechnology* means a billionth (1×10^{-9}). Nanotechnology deals with various structures of matter having dimensions of the order of a billionth of a meter. While the word *nanotechnology* is relatively new, the existence of functional devices and structures of nanometer dimensions is not new, and in fact such structures have existed on Earth as long as life itself. The abalone, a mollusk, constructs very strong shells having iridescent inner surfaces by organizing calcium carbonate into strong nanostructured bricks held together by a glue made of a carbohydrate–protein mix. Cracks initiated on the outside are unable to move through the shell because of the nanostructured bricks. The shells represent a natural demonstration that a structure fabricated from nanoparticles can be much stronger. We will discuss how and why nanostructuring leads to stronger materials in Chapter 6.

It is not clear when humans first began to take advantage of nanosized materials. It is known that in the fourth-century A.D. Roman glassmakers were fabricating glasses containing nanosized metals. An artifact from this period called the Lycurgus cup resides in the British Museum in London. The cup, which depicts the death of King Lycurgus, is made from soda lime glass containing sliver and gold nanoparticles. The color of the cup changes from green to a deep red when a light source is placed inside it. The great varieties of beautiful colors of the windows of medieval cathedrals are due to the presence of metal nanoparticles in the glass.

Introduction to Nanotechnology, by Charles P. Poole Jr. and Frank J. Owens.
ISBN 0-471-07935-9. Copyright © 2003 John Wiley & Sons, Inc.

The potential importance of clusters was recognized by the Irish-born chemist Robert Boyle in his *Sceptical Chymist* published in 1661. In it Boyle criticizes Aristotle's belief that matter is composed of four elements: earth, fire, water, and air. Instead, he suggests that tiny particles of matter combine in various ways to form what he calls *corpuscles*. He refers to "minute masses or clusters that were not easily dissipable into such particles that composed them."

Photography is an advanced and mature technology, developed in the eighteenth and nineteenth centuries, which depends on production of silver nanoparticles sensitive to light. Photographic film is an emulsion, a thin layer of gelatin containing silver halides, such as silver bromide, and a base of transparent cellulose acetate. The light decomposes the silver halides, producing nanoparticles of silver, which are the pixels of the image. In the late eighteenth century the British scientists Thomas Wedgewood and Sir Humprey Davy were able to produce images using silver nitrate and chloride, but their images were not permanent. A number of French and British researchers worked on the problem in the nineteenth century. Such names as Daguerre, Niecpce, Talbot, Archer, and Kennet were involved. Interestingly James Clark Maxwell, whose major contributions were to electromagnetic theory, produced the first color photograph in 1861. Around 1883 the American inventor George Eastman, who would later found the Kodak Corporation, produced a film consisting of a long paper strip coated with an emulsion containing silver halides. He later developed this into a flexible film that could be rolled, which made photography accessible to many. So technology based on nanosized materials is really not that new.

In 1857 Michael Faraday published a paper in the *Philosophical Transactions* of *the Royal Society*, which attempted to explain how metal particles affect the color of church windows. Gustav Mie was the first to provide an explanation of the dependence of the color of the glasses on metal size and kind. His paper was published in the German Journal *Annalen der Physik* (Leipzig) in 1908.

Richard Feynman was awarded the Nobel Prize in physics in 1965 for his contributions to quantum electrodynamics, a subject far removed from nanotechnology. Feynman was also a very gifted and flamboyant teacher and lecturer on science, and is regarded as one of the great theoretical physicists of his time. He had a wide range of interests beyond science from playing bongo drums to attempting to interpret Mayan hieroglyphics. The range of his interests and wit can be appreciated by reading his lighthearted autobiographical book *Surely You're Joking, Mr. Feynman*. In 1960 he presented a visionary and prophetic lecture at a meeting of the American Physical Society, entitled "There is Plenty of Room at the Bottom," where he speculated on the possibility and potential of nanosized materials. He envisioned etching lines a few atoms wide with beams of electrons, effectively predicting the existence of electron-beam lithography, which is used today to make silicon chips. He proposed manipulating individual atoms to make new small structures having very different properties. Indeed, this has now been accomplished using a scanning tunneling microscope, discussed in Chapter 3. He envisioned building circuits on the scale of nanometers that can be used as elements in more powerful computers. Like many of present-day nanotechnology researchers, he recognized the existence of nanostructures in biological systems. Many of Feynman's speculations have become reality. However, his thinking did not resonate with scientists at the time. Perhaps because of his reputation for wit, the reaction of many in the audience could best be described by the title of his later book *Surely You're Joking, Mr. Feynman*. Of course, the lecture is now legendary among present-day nanotechnology researchers, but as one scientist has commented, "it was so visionary that it did not connect with people until the technology caught up with it."

There were other visionaries. Ralph Landauer, a theoretical physicist working for IBM in 1957, had ideas on nanoscale electronics and realized the importance that quantum-mechanical effects would play in such devices.

Although Feynman presented his visionary lecture in 1960, there was experimental activity in the 1950s and 1960s on small metal particles. It was not called *nanotechnology* at that time, and there was not much of it. Uhlir reported the first observation of porous silicon in 1956, but it was not until 1990 when room-temperature fluorescence was observed in this material that interest grew. The properties of porous silicon are discussed in Chapter 6. Other work in this era involved making alkali metal nanoparticles by vaporizing sodium or potassium metal and then condensing them on cooler materials called *substrates*. Magnetic fluids called *ferrofluids* were developed in the 1960s. They consist of nanosized magnetic particles dispersed in liquids. The particles were made by ballmilling in the presence of a surface-active agent (surfactant) and liquid carrier. They have a number of interesting properties and applications, which are discussed in Chapter 7. Another area of activity in the 1960s involved electron paramagnetic resonance (EPR) of conduction electrons in metal particles of nanodimensions referred to as *colloids* in those days. The particles were produced by thermal decomposition and irradiation of solids having positive metal ions, and negative molecular ions such as sodium and potassium azide. In fact, decomposing these kinds of solids by heat is one way to make nanometal particles, and we discuss this subject in Chapter 4. Structural features of metal nanoparticles such as the existence of magic numbers were revealed in the 1970s using mass spectroscopic studies of sodium metal beams. Herman and co-workers measured the ionization potential of sodium clusters in 1978 and observed that it depended on the size of the cluster, which led to the development of the jellium model of clusters discussed in Chapter 4.

Groups at Bell Laboratories and IBM fabricated the first two-dimensional quantum wells in the early 1970s. They were made by thin-film (epitaxial) growth techniques that build a semiconductor layer one atom at a time. The work was the beginning of the development of the zero-dimensional quantum dot, which is now one of the more mature nanotechnologies with commercial applications. The quantum dot and its applications are discussed in Chapter 9.

However, it was not until the 1980s with the emergence of appropriate methods of fabrication of nanostructures that a notable increase in research activity occurred, and a number of significant developments resulted. In 1981, a method was developed to make metal clusters using a high-powered focused laser to vaporize metals into a hot plasma. This is discussed in Chapter 4. A gust of helium cools the vapor, condensing the metal atoms into clusters of various sizes. In 1985, this method was used to synthesize the fullerene (C_{60}). In 1982, two Russian scientists, Ekimov and Omushchenko, reported the first observation of quantum confinement, which is discussed in Chapter 9. The scanning tunneling microscope was developed during this decade by G. K. Binnig and H. Roher of the IBM Research Laboratory in Zürich, and they were awarded the Nobel Prize in 1986 for this. The invention of the scanning tunneling microscope (STM) and the atomic force microscope (AFM), which are described in Chapter 3, provided new important tools for viewing, characterizing, and atomic manipulation of nanostructures. In 1987, B. J. van Wees and H. van Houten of the Netherlands observed steps in the current–voltage curves of small point contacts. Similar steps were observed by D. Wharam and M. Pepper of Cambridge University. This represented the first observation of the quantization of conductance. At the same time T. A. Fulton and G. J. Dolan of Bell Laboratories made a single-electron transistor and observed the Coulomb blockade, which is explained in Chapter 9. This period was marked by development of methods of fabrication such as electron-beam lithography, which are capable of producing 10-nm structures. Also in this decade layered alternating metal magnetic and nonmagnetic materials, which displayed the fascinating property of giant magnetoresistance, were fabricated. The layers were a nanometer thick, and the materials have an important application in magnetic storage devices in computers. This subject is discussed in Chapter 7.

Although the concept of photonic crystals was theoretically formulated in the late 1980s, the first three-dimensional periodic photonic crystal possessing a complete bandgap was fabricated by Yablonovitch in 1991. Photonic crystals are discussed in Chapter 6. In the 1990s, Iijima made carbon nanotubes, and superconductivity and ferromagnetism were found in C_{60} structures. Efforts also began to make molecular switches and measure the electrical conductivity of molecules. A field-effect transistor based on carbon nanotubes was demonstrated. All of these subjects are discussed in this book. The study of self-assembly of molecules on metal surfaces intensified. *Self-assembly* refers to the spontaneous bonding of molecules to metal surfaces, forming an organized array of molecules on the surface. Self-assembly of thiol and disulfide compounds on gold has been most widely studied, and the work is presented in Chapter 10.

In 1996, a number of government agencies led by the National Science Foundation commissioned a study to assess the current worldwide status of trends, research, and development in nanoscience and nanotechnology. The detailed recommendations led to a commitment by the government to provide major funding and establish a national nanotechnology initiative. Figure 1.1 shows the growth of U.S. government funding for nanotechnology and the projected increase due to the national nanotechnology initiative. Two general findings emerged from the study.

The first observation was that materials have been and can be nanostructured for new properties and novel performance. The underlying basis for this, which we discuss in more detail in later chapters, is that every property of a material has a characteristic or critical length associated with it. For example, the resistance of a material that results from the conduction electrons being scattered out of the direction of flow by collisions with vibrating atoms and impurities, can be

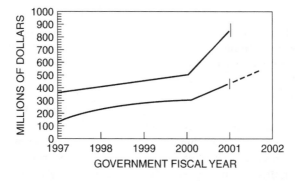

Figure 1.1. Funding for nanotechnology research by year. The top tracing indicates spending by foreign governments; bottom, U.S. government spending. The dashed line represents proposed spending for 2002. (From U.S. Senate Briefing on nanotechnology, May 24, 2001 and National Science Foundation.)

characterized by a length called the *scattering length*. This length is the average distance an electron travels before being deflected. The fundamental physics and chemistry changes when the dimensions of a solid become comparable to one or more of these characteristic lengths, many of which are in the nanometer range. One of the most important examples of this is what happens when the size of a semiconducting material is in the order of the wavelength of the electrons or holes that carry current. As we discuss in Chapter 9, the electronic structure of the system completely changes. This is the basis of the quantum dot, which is a relatively mature application of nanotechnology resulting in the quantum-dot laser presently used to read compact disks (CDs). However, as we shall see in Chapter 9, the electron structure is strongly influenced by the number of dimensions that are nanosized.

If only one length of a three-dimensional nanostructure is of a nanodimension, the structure is known as a *quantum well*, and the electronic structure is quite different from the arrangement where two sides are of nanometer length, constituting what is referred to as a *quantum wire*. A quantum dot has all three dimensions in the nanorange. Chapter 9 discusses in detail the effect of dimension on the electronic properties of nanostructures. The changes in electronic properties with size result in major changes in the optical properties of nanosized materials, which is discussed in Chapter 8, along with the effects of reduced size on the vibrational properties of materials.

The second general observation of the U.S. government study was a recognition of the broad range of disciplines that are contributing to developments in the field. Work in nanotechnology can be found in university departments of physics, chemistry, and environmental science, as well as electrical, mechanical, and chemical engineering. The interdisciplinary nature of the field makes it somewhat difficult for researchers in one field to understand and draw on developments in another area. As Feynman correctly pointed out, biological systems have been making nanometer functional devices since the beginning of life, and there is much to learn from biology about how to build nanostructured devices. How, then, can a solid-state physicist who is involved in building nanostructures but who does not know the difference between an amino acid and a protein learn from biological systems? It is this issue that motivated the writing of the present book. The book attempts to present important selected topics in nanotechnology in various disciplines in such a way that workers in one field can understand developments in other fields. In order to accomplish this, it is necessary to include in each chapter some introductory material. Thus, the chapter on the effect of nanostructuring on ferromagnetism (Chapter 7) starts with a brief introduction to the theory and properties of ferromagnets. As we have mentioned above, the driving force behind nanotechnology is the recognition that nanostructured materials can have chemistry and physics different from those of bulk materials, and a major objective of this book is to explain these differences and the reasons for them. In order to do that, one has to understand the basic chemistry and physics of the bulk solid state. Thus, Chapter 2 provides an introduction to the theory of bulk solids. Chapter 3 is devoted to describing the various experimental methods used to characterize nanostructures. Many of the experimental methods described such as the scanning tunneling microscope have been developed quite recently, as was mentioned above, and without their existence, the field of nanotechnology would not have made the progress it has. The remaining chapters deal with selected topics in nanotechnology. The field of nanotechnology is simply too vast, too interdisciplinary, and too rapidly changing to cover exhaustively. We have therefore selected a number of topics to present. The criteria for selection of subjects is the maturity of the field, the degree of understanding of the phenomena, and existing and potential applications. Thus most of the chapters describe examples of existing applications and potential new ones. The applications potential of nanostructured materials is certainly a cause of the intense interest in the subject, and there are many applications already in the commercial world. Giant magnetoresistivity of nanostructured materials has been introduced into commercial use, and some examples are given in Chapter 7. The effect of nanostructuring to increase the storage capacity of magnetic tape devices is an active area of research, which we will examine in some detail in Chapter 7. Another area of intense activity is the use of nanotechnology to make smaller switches, which are the basic elements of computers. The potential use of carbon nanotubes as the basic elements of computer switches is described in Chapter 5. In Chapter 13 we discuss how nanosized molecular switching devices are subjects of research activity. Another area of potential application is the role of nanostructuring and its effect on the mechanical properties of materials. In Chapter 6, we discuss how consolidated materials made of nanosized grains can have significantly different mechanical properties such as enhanced yield strength. Also discussed in most of the chapters are methods of fabrication of the various nanostructure types under discussion. Development of large-scale inexpensive methods of fabrication is a major challenge for nanoscience if it is to have an impact on technology. As we discuss in Chapter 5, single-walled carbon nanotubes have enormous application potential ranging from gas sensors to switching elements in fast computers. However, methods of manufacturing large quantities of the tubes will have to be developed before they will have an impact on technology. Michael Roukes, who is working on development of nanoelectromechanical devices, has pointed out some other challenges in the September 2001 issue of *Scientific American*. One major challenge deals with communication between the nanoworld and the macroworld. For example, the resonant vibrational frequency of a rigid beam increases as the size of the beam decreases. In the nanoregime the frequencies can be as high 10^{10} Hertz, and the amplitudes of vibration in the picometer (10^{-12}) to femtometer (10^{-15}) range. The sensor must be able to detect these small displacements and high frequencies. Optical deflection schemes, such as those used in scanning tunneling microscopy discussed in Chapter 3, may not work because of the diffraction limit, which becomes a problem when the wavelength of the light is in the order of the size of the object from which the light is to be reflected. Another obstacle to be overcome is the effect of surface on nanostructures. A silicon beam 10 nm wide and 100 nm long has almost 10% of its atoms at or near the surface. The surface atoms will affect the mechanical behavior (strength, flexibility, etc.) of the beam, albeit in a way that is not yet understood.

2

INTRODUCTION TO PHYSICS OF THE SOLID STATE

In this book we will be discussing various types of nanostructures. The materials used to form these structures generally have bulk properties that become modified when their sizes are reduced to the nanorange, and the present chapter presents background material on bulk properties of this type. Much of what is discussed here can be found in a standard text on solid-state physics [e.g., Burns (1985); Kittel (1996); see also Yu and Cardona (2001)].

2.1. STRUCTURE

2.1.1. Size Dependence of Properties

Many properties of solids depend on the size range over which they are measured. Microscopic details become averaged when investigating bulk materials. At the macro- or large-scale range ordinarily studied in traditional fields of physics such as mechanics, electricity and magnetism, and optics, the sizes of the objects under study range from millimeters to kilometers. The properties that we associate with these materials are averaged properties, such as the density and elastic moduli in

Introduction to Nanotechnology, by Charles P. Poole Jr. and Frank J. Owens.
ISBN 0-471-07935-9. Copyright © 2003 John Wiley & Sons, Inc.

mechanics, the resistivity and magnetization in electricity and magnetism, and the dielectric constant in optics. When measurements are made in the micrometer or nanometer range, many properties of materials change, such as mechanical, ferroelectric, and ferromagnetic properties. The aim of the present book is to examine characteristics of solids at the next lower level of size, namely, the nanoscale level, perhaps from 1 to 100 nm. Below this there is the atomic scale near 0.1 nm, followed by the nuclear scale near a femtometer (10^{-15} m). In order to understand properties at the nanoscale level it is necessary to know something about the corresponding properties at the macroscopic and mesoscopic levels, and the present chapter aims to provide some of this background.

Many important nanostructures are composed of the group IV elements Si or Ge, type III–V semiconducting compounds such as GaAs, or type II–VI semiconducting materials such as CdS, so these semiconductor materials will be used to illustrate some of the bulk properties that become modified with incorporation into nanostructures. The Roman numerals IV, III, V, and so on, refer to columns of the periodic table. Appendix B provides tabulations of various properties of these semiconductors.

2.1.2. Crystal Structures

Most solids are crystalline with their atoms arranged in a regular manner. They have what is called *long-range order* because the regularity can extend throughout the crystal. In contrast to this, amorphous materials such as glass and wax lack long-range order, but they have what is called *short-range order* so the local environment of each atom is similar to that of other equivalent atoms, but this regularity does not persist over appreciable distances. Liquids also have short-range order, but lack long-range order. Gases lack both long-range and short-range order.

Figure 2.1 shows the five regular arrangements of lattice points that can occur in two dimensions: the square (a), primitive rectangular (b), centered rectangular (c), hexagonal (d), and oblique (e) types. These arrangements are called *Bravais lattices*. The general or oblique Bravais lattice has two unequal lattice constants $a \neq b$ and an arbitrary angle θ between them. For the perpendicular case when $\theta = 90°$, the lattice becomes the rectangular type. For the special case $a = b$ and $\theta = 60°$, the lattice is the hexagonal type formed from equilateral triangles. Each lattice has a unit cell, indicated in the figures, which can replicate throughout the plane and generate the lattice.

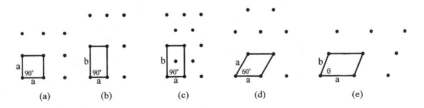

Figure 2.1. The five Bravais lattices that occur in two dimensions, with the unit cells indicated: (a) square; (b) primitive rectangular; (c) centered rectangular; (d) hexagonal; (e) oblique.

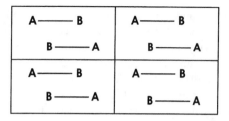

Figure 2.2. Sketch of a two-dimensional crystal structure based on a primitive rectangular lattice containing two diatomic molecules A–B in each unit cell.

A crystal structure is formed by associating with a lattice a regular arrangement of atoms or molecules. Figure 2.2 presents a two-dimensional crystal structure based on a primitive rectangular lattice containing two diatomic molecules A–B in each unit cell. A single unit cell can generate the overall lattice.

In three dimensions there are three lattice constants, a, b, and c, and three angles: α between b and c; β between a and c, and γ between lattice constants a and b. There are 14 Bravais lattices, ranging from the lowest-symmetry triclinic type in which all three lattice constants and all three angles differ from each other ($a \neq b \neq c$ and $\alpha \neq \beta \neq \gamma$), to the highest-symmetry cubic case in which all the lattice constants are equal and all the angles are 90° ($a = b = c$ and $\alpha = \beta = \gamma = 90°$). There are three Bravais lattices in the cubic system, namely, a primitive or simple cubic (SC) lattice in which the atoms occupy the eight apices of the cubic unit cell, as shown in Fig. 2.3a, a body-centered cubic (BCC) lattice with lattice points occupied at the apices and in the center of the unit cell, as indicated in Fig. 2.3b, and a face-centered cubic (FCC) Bravais lattice with atoms at the apices and in the centers of the faces, as shown in Fig. 2.3c.

In two dimensions the most efficient way to pack identical circles (or spheres) is the equilateral triangle arrangement shown in Fig. 2.4a, corresponding to the hexagonal Bravais lattice of Fig. 2.1d. A second hexagonal layer of spheres can be placed on top of the first to form the most efficient packing of two layers, as shown in Fig. 2.4b. For efficient packing, the third layer can be placed either above

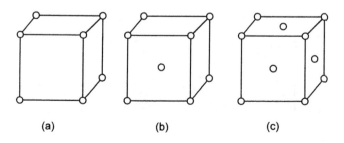

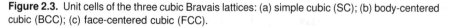

Figure 2.3. Unit cells of the three cubic Bravais lattices: (a) simple cubic (SC); (b) body-centered cubic (BCC); (c) face-centered cubic (FCC).

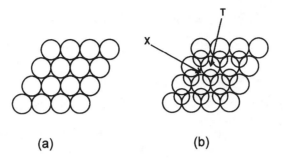

Figure 2.4. Close packing of spheres on a flat surface: (a) for a monolayer; (b) with a second layer added. The circles of the second layer are drawn smaller for clarity. The location of an octahedral site is indicated by x, and the position of a tetrahedral site is designated by T on panel (b).

the first layer with an atom at the location indicated by T or in the third possible arrangement with an atom above the position marked by X on the figure. In the first case a hexagonal lattice with a hexagonal close-packed (HCP) structure is generated, and in the second case a face-centered cubic lattice results. The former is easy to visualize, but the latter is not so easy to picture.

In the three-dimensional case of close-packed spheres there are spaces or sites between the spheres where smaller atoms can reside. The point marked by X on Fig. 2.4b, called an *octahedral site*, is equidistant from the three spheres O below it, and from the three spheres O above it. An atom A at this site has the local coordination AO_6. The radius a_{oct}, of this octahedral site is

$$a_{oct} = \tfrac{1}{4}(2 - \sqrt{2})a = (\sqrt{2} - 1)a_0 = 0.41421a_0 \qquad (2.1)$$

where a is the lattice constant and a_0 is the radius of the spheres. The number of octahedral sites is equal to the number of spheres. There are also smaller sites, called *tetrahedral sites*, labeled T in the figure that are equidistant from four nearest-neighbor spheres, one below and three above, corresponding to AO_4 for the local coordination. This is a smaller site since its radius a_T is

$$a_T = \tfrac{1}{4}(\sqrt{3} - \sqrt{2})a = [(\tfrac{3}{2})^{1/2} - 1]a_0 = 0.2247a_0 \qquad (2.2)$$

There are twice as many tetrahedral sites as there are spheres in the structure. Many diatomic oxides and sulfides such as MgO, MgS, MnO, and MnS have their larger oxygen or sulfur anions in a perfect FCC arrangement with the smaller metal cations located at octahedral sites. This is called the *NaCl lattice type*, where we use the term *anion* for a negative ion (e.g., Cl^-) and cation for a positive ion (e.g., Na^+). The mineral spinel $MgAl_2O_4$ has a face-centered arrangement of divalent oxygens O^{2-} (radius 0.132 nm) with the Al^{3+} ions (radius 0.051 nm) occupying one-half of the octahedral sites and Mg^{2+} (radius 0.066 nm) located in one-eighth of the tetrahedral sites in a regular manner.

2.1.3. Face-Centered Cubic Nanoparticles

Most metals in the solid state form close-packed lattices; thus Ag, Al, Au, Co, Cu, Pb, Pt, and Rh, as well as the rare gases Ne, Ar, Kr, and Xe, are face-centered cubic (FCC), and Mg, Nd, Os, Re, Ru, Y, and Zn, are hexagonal close-packed (HCP). A number of other metallic atoms crystallize in the not so closely packed body-centered cubic (BCC) lattice, and a few such as Cr, Li, and Sr crystallize in all three structure types, depending on the temperature. An atom in each of the two close-packed lattices has 12 nearest neighbors. Figure 2.5 shows the 12 neighbors that surround an atom (darkened circle) located in the center of a cube for a FCC lattice. Figure 10.18 (of Chapter 10) presents another perspective of the 12 nearest neighbors. These 13 atoms constitute the smallest theoretical nanoparticle for an FCC lattice. Figure 2.6 shows the 14-sided polyhedron, called a *dekatessarahedron*, that is generated by connecting the atoms with planar faces. Sugano and Koizumi (1998) call this polyhedron a *cuboctahedron*. The three open circles at the upper right of Fig. 2.6 are the three atoms in the top layer of Fig. 10.18, the six darkened circles plus an atom in the center of the cube of Fig. 2.6 constitute the middle layer of that figure, and the open circle at the lower left of Fig. 2.5 is one of the three obscured atoms in the plane below the cluster pictured in Fig. 10.18. This 14-sided polyhedron has six square faces and eight equilateral triangle faces.

If another layer of 42 atoms is layed down around the 13-atom nanoparticle, one obtains a 55-atom nanoparticle with the same dekatessarahedron shape. Larger

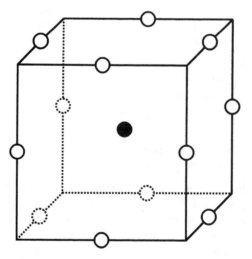

Figure 2.5. Face-centered cubic unit cell showing the 12 nearest-neighbor atoms that surround the atom (darkened circle) in the center.

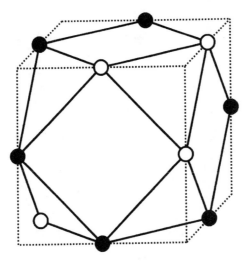

Figure 2.6. Thirteen-atom nanoparticle set in its FCC unit cell, showing the shape of the 14-sided polyhedron associated with the nanocluster. The three open circles at the upper right correspond to the atoms of the top layer of the nanoparticle sketched in Fig. 10.8 (of Chapter 10), the six solid circles plus the atom (not pictured) in the center of the cube constitute the middle hexagonal layer of that figure, and the open circle at the lower left corner of the cube is one of the three atoms at the bottom of the cluster of Fig. 10.18.

nanoparticles with the same polyhedral shape are obtained by adding more layers, and the sequence of numbers in the resulting particles, $N = 1, 13, 55, 147, 309, 561, \ldots$, which are listed in Table 2.1, are called *structural magic numbers*. For n layers the number of atoms N in this FCC nanoparticle is given by the formula

$$N = \tfrac{1}{3}[10n^3 - 15n^2 + 11n - 3] \qquad (2.3)$$

and the number of atoms on the surface N_{surf} is

$$N_{\text{surf}} = 10n^2 - 20n + 12 \qquad (2.4)$$

For each value of n, Table 2.1 lists the number of atoms on the surface, as well as the percentage of atoms on the surface. The table also lists the diameter of each nanoparticle, which is given by the expression $(2n - 1)d$, where d is the distance between the centers of nearest-neighbor atoms, and $d = a/\sqrt{2}$, where a is the lattice constant. If the same procedure is used to construct nanoparticles with the hexagonal close-packed structure that was discussed in the previous section, a slightly different set of structural magic numbers is obtained, namely, $1, 13, 57, 153, 321, 581, \ldots$.

Purely metallic FCC nanoparticles such as Au_{55} tend to be very reactive and have short lifetimes. They can be ligand-stabilized by adding atomic groups between their atoms and on their surfaces. The Au_{55} nanoparticle has been studied in the ligand-

Table 2.1. Number of atoms (structural magic numbers) in rare gas or metallic nanoparticles with face-centered cubic close-packed structure[a]

Shell Number	Diameter	Number of FCC Nanoparticle Atoms		
		Total	On Surface	% Surface
1	$1d$	1	1	100
2	$3d$	13	12	92.3
3	$5d$	55	42	76.4
4	$7d$	147	92	62.6
5	$9d$	309	162	52.4
6	$11d$	561	252	44.9
7	$13d$	923	362	39.2
8	$15d$	1415	492	34.8
9	$17d$	2057	642	31.2
10	$19d$	2869	812	28.3
11	$21d$	3871	1002	25.9
12	$23d$	5083	1212	23.8
25	$49d$	4.90×10^4	5.76×10^3	11.7
50	$99d$	4.04×10^5	2.40×10^4	5.9
75	$149d$	1.38×10^6	5.48×10^4	4.0
100	$199d$	3.28×10^6	9.80×10^4	3.0

[a]The diameters d in nanometers for some representative FCC atoms are Al 0.286, Ar 0.376, Au 0.288, Cu 0.256, Fe 0.248, Kr 0.400, Pb 0.350, and Pd 0.275.

stabilized form $Au_{55}(PPh_3)_{12}Cl_6$ which has the diameter of ~ 1.4 nm, where PPh_3 is an organic group. Further examples are the magic number nanoparticles $Pt_{309}(1,10\text{-phenanthroline})_{36}O_{30}$ and $Pd_{561}(1,10\text{-phenanthroline})_{36}O_{200}$.

The magic numbers that we have been discussing are called structural magic numbers because they arise from minimum-volume, maximum-density nano-particles that approximate a spherical shape, and have close-packed structures characteristic of a bulk solid. These magic numbers take no account of the electronic structure of the consitituent atoms in the nanoparticle. Sometimes the dominant factor in determining the minimum-energy structure of small nanoparticles is the interactions of the valence electrons of the constituent atoms with an averaged molecular potential, so that the electrons occupy orbital levels associated with this potential. Atomic cluster configurations in which these electrons fill closed shells are especially stable, and constitute electronic magic numbers. Their atomic structures differ from the FCC arrangement, as will be discussed in Sections 4.2.1 and 4.2.2 (of Chapter 4).

When mass spectra were recorded for sodium nanoparticles Na_N, it was found that mass peaks corresponding to the first 15 electronic magic numbers $N = 3, 9, 20, 36, 61, \ldots$ were observed for cluster sizes up to $N = 1220$ atoms ($n = 15$), and FCC structural magic numbers starting with $N = 1415$ for $n = 8$ were observed for larger sizes [Martin et al. (1990); see also Sugano and Koizumi (1998), p. 90]. The mass spectral data are plotted versus the cube root of the number of atoms $N^{1/3}$ in Fig. 2.7, and it is clear that the lines from both sets of magic numbers are

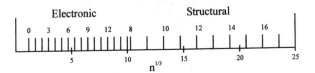

Figure 2.7. Dependence of the observed mass spectra lines from Na_N nanoparticles on the cube root $N^{1/3}$ of the number of atoms N in the cluster. The lines are labeled with the index n of their electronic and structural magic numbers obtained from Martin et al. (1990).

approximately equally spaced, with the spacing between the structural magic numbers about 2.6 times that between the electronic ones. This result provides evidence that small clusters tend to satisfy electronic criteria and large structures tend to be structurally determined.

2.1.4. Tetrahedrally Bonded Semiconductor Structures

The type III–V and type II–VI binary semiconducting compounds, such as GaAs and ZnS, respectively, crystallize with one atom situated on a FCC sublattice at the positions 000, $\frac{1}{2}\frac{1}{2}0$, $\frac{1}{2}0\frac{1}{2}$ and $0\frac{1}{2}\frac{1}{2}$, and the other atom on a second FCC sublattice displaced from the first by the amount $\frac{1}{4}\frac{1}{4}\frac{1}{4}$ along the body diagonal, as shown in Fig. 2.8b. This is called the *zinc blende* or *ZnS structure*. It is clear from the figure that each Zn atom (white sphere) is centered in a tetrahedron of S atoms (black spheres), and likewise each S has four similarly situated Zn nearest neighbors. The small half-sized, dashed-line cube delineates one such tetrahedron. The same structure would result if the Zn and S atoms were interchanged.

The elements Si and Ge crystallize in this same structure by having the Si (or Ge) atoms occupying all the sites on the two sublattices, so there are eight identical atoms in the unit cell. This atom arrangement, sketched in Fig. 2.8a, is called the *diamond structure*. Both Si and Ge have a valence of 4, so from bonding considerations, it is appropriate for each to be bound to four other atoms in the shape of a regular tetrahedron.

In Appendix B we see that Table B.1 lists the lattice constants a for various compounds with the zinc blende structure, and Table B.2 provides the crystal radii of their monatomic lattices in which the atoms are uncharged, as well as the ionic radii for ionic compounds in which the atoms are charged. We see from Table B.2 that the negative anions are considerably larger than the positive cations, in accordance with the sketch of the unit cell presented in Fig. 2.9, and this size differential is greater for the III–V compounds than for the II–VI compounds. However, these size changes for the negative and positive ions tend to balance each other so that the III–V compounds have the same range of lattice constants as the II–VI compounds, with Si and Ge also in this range. Table B.4 gives the molecular masses, and Table B.5 gives the densities of these semiconductors. The three tables B.1, B.4, and B.5 show a regular progression in the values as one goes from left to right in a particular row,

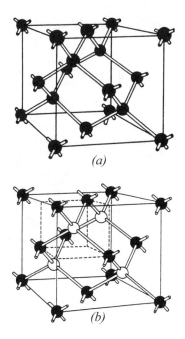

(a)

(b)

Figure 2.8. Unit cell of the diamond structure (a), which contains only one type of atom, and corresponding unit cell of the zinc blende (sphalerite) structure (b), which contains two atom types. The rods represent the tetrahedral bonds between nearest-neighbor atoms. The small dashed line cube in (b) delineates a tetrahedron. (From G. Burns, *Solid State Physics*, Academic Press, Boston, 1985, p.148.)

and as one goes from top to bottom in a particular column. This occurs because of the systematic increase in the size of the atoms in each group with increasing atomic number, as indicated in Table B.2.

There are two simple models for representing these AC binary compound structures. For an ionic model the lattice $A^{n-}C^{n+}$ consists of a FCC arrangement

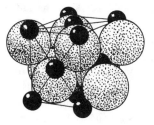

Figure 2.9. Packing of larger S atoms and smaller Zn atoms in the zinc blende (ZnS) structure. Each atom is centered in a tetrahedron of the other atom type. (From R. W. G. Wyckoff, *Crystal Structures*, Vol. 1, Wiley, New York, 1963, p. 109.)

of the large anions A^{n-} with the small cations C^{n+} located in the tetrahedral sites of the anion FCC lattice. If the anions touch each other, their radii have the value $a_0 = a/2\sqrt{2}$, where a is the lattice parameter, and the radius a_T of the tetrahedral site $a_T = 0.2247a_0$ is given by Eq. (2.2). This is the case for the very small Al^{3+} cation in the AlSb structure. In all other cases the cations in Table B.2 are too large to fit in the tetrahedral site so they push the larger anions further apart, and the latter no longer touch each other, in accordance with Fig. 2.9. In a covalent model for the structure consisting of neutral atoms A and C the atom sizes are comparable, as the data in Table B.2 indicate, and the structure resembles that of Si or Ge. To compare these two models, we note that the distance between atom A at lattice position $0\,0\,0$ and its nearest neighbor C at position $\frac{1}{4}\frac{1}{4}\frac{1}{4}$ is equal to $\frac{1}{4}\sqrt{3}a$, and in Table B.3 we compare this crystallographically evaluated distance with the sums of radii of ions A^{n-}, C^{n+} from the ionic model, and with the sums of radii of neutral atoms A and C of the covalent model using the data of Table B.2. We see from the results on Table B.3 that neither model fits the data in all cases, but the neutral atom covalent model is closer to agreement. For comparison purposes we also list corresponding data for several alkali halides and alkaline-earth chalcogenides that crystallize in the cubic rock salt or NaCl structure, and we see that all of these compounds fit the ionic model very well. In these compounds each atom type forms a FCC lattice, with the atoms of one FCC lattice located at octahedral sites of the other lattice. The octahedral site has the radius $a_{oct} = 0.41411a_0$ given by Eq. (2.1), which is larger than the tetrahedral one of Eq. (2.2).

Since the alkali halide and alkaline-earth chalcogenide compounds fit the ionic model so well, it is significant that neither model fits the structures of the semiconductor compounds. The extent to which the semiconductor crystals exhibit ionic or covalent bonding is not clear from crystallographic data. If the wavefunction describing the bonding is written in the form

$$\Psi = a_{cov}\psi_{cov} + a_{ion}\psi_{ion} \tag{2.5}$$

where the coefficients of the covalent and ionic wavefunction components are normalized

$$a_{cov}^2 + a_{ion}^2 = 1 \tag{2.6}$$

then a_{cov}^2 is the fractional covalency and a_{ion}^2 is the fractional ionicity of the bond. A chapter (Poole and Farach 2001) in a book by Karl Boer (2001) tabulates the effective charges e^* associated with various II–VI and III–V semiconducting compounds, and this effective charge is related to the fractional covalency by the expression

$$a_{cov}^2 = \frac{8 - N + e^*}{8} \tag{2.7}$$

where $N = 2$ for II–VI and $N = 3$ for III–V compounds. The fractional charges all lie in the range from 0.43 to 0.49 for the compounds under consideration. Using the e^* tabulations in the Boer book and Eq. (2.7), we obtain the fractional covalencies of $a_{cov}^2 \sim 0.81$ for all the II–VI compounds, and $a_{cov}^2 \sim 0.68$ for all the III–V compounds listed in Tables B.1, B.4, B.5, and so on. These values are consistent with the better fit of the covalent model to the crystallographic data for these compounds.

We conclude this section with some observations that will be of use in later chapters. Table B.1 shows that the typical compound GaAs has the lattice constant $a = 0.565$ nm, so the volume of its unit cell is 0.180 nm^3, corresponding to about 22 of each atom type per cubic nanometer. The distances between atomic layers in the 100, 110, and 111 directions are, respectively, 0.565 nm, 0.400 nm, and 0.326 nm for GaAs. The various III–V semiconducting compounds under discussion form mixed crystals over broad concentration ranges, as do the group of II–VI compounds. In a mixed crystal of the type In$_x$Ga$_{1-x}$As it is ordinarily safe to assume that Vegard's law is valid, whereby the lattice constant a scales linearly with the concentration parameter x. As a result, we have the expressions

$$a(x) = a(\text{GaAs}) + [a(\text{InAs}) - a(\text{GaAs})]x$$
$$= 0.565 + 0.041x \qquad (2.8)$$

where $0 \le x \le 1$. In the corresponding expression for the mixed semiconductor Al$_x$Ga$_{1-x}$As the term $+0.001x$ replaces the term $+0.041x$, so the fraction of lattice mismatch $2|a_{\text{AlAs}} - a_{\text{GaAs}}|/(a_{\text{AlAs}} + a_{\text{GaAs}}) = 0.0018 = 0.18\%$ for this system is quite minimal compared to that $[2|a_{\text{InAs}} - a_{\text{GaAs}}|/(a_{\text{InAs}} + a_{\text{GaAs}}) = 0.070 = 7.0\%]$ of the In$_x$Ga$_{1-x}$As system, as calculated from Eq. (2.8) [see also Eq. (10-3).] Table B.1 gives the lattice constants a for various III–V and II–VI semiconductors with the zinc blende structure.

2.1.5. Lattice Vibrations

We have discussed atoms in a crystal as residing at particular lattice sites, but in reality they undergo continuous fluctuations in the neighborhood of their regular positions in the lattice. These fluctuations arise from the heat or thermal energy in the lattice, and become more pronounced at higher temperatures. Since the atoms are bound together by chemical bonds, the movement of one atom about its site causes the neighboring atoms to respond to this motion. The chemical bonds act like springs that stretch and compress repeatedly during oscillatory motion. The result is that many atoms vibrate in unison, and this collective motion spreads throughout the crystal. Every type of lattice has its own characteristic modes or frequencies of vibration called *normal modes*, and the overall collective vibrational motion of the lattice is a combination or superposition of many, many normal modes. For a diatomic lattice such as GaAs, there are low-frequency modes called *acoustic modes*, in which the heavy and light atoms tend to vibrate in phase or in unison with each other, and high-frequency modes called *optical modes*, in which they tend to vibrate out of phase.

A simple model for analyzing these vibratory modes is a linear chain of alternating atoms with a large mass M and a small mass m joined to each other by springs ($\sim$) as follows:

$$\sim m \sim M \sim m \sim M \sim m \sim M \sim m \sim M \sim$$

When one of the springs stretches or compresses by an amount Δx, a force is exerted on the adjacent masses with the magnitude $C \Delta x$, where C is the spring constant. As the various springs stretch and compress in step with each other, longitudinal modes of vibration take place in which the motion of each atom is along the string direction. Each such normal mode has a particular frequency ω and a wavevector $k = 2\pi/\lambda$, where λ is the wavelength, and the energy E, associated with the mode is given by $E = \hbar\omega$. There are also transverse normal modes in which the atoms vibrate back and forth in directions perpendicular to the line of atoms. Figure 2.10 shows the dependence of ω on k for the low-frequency acoustic and the high-frequency optical longitudinal modes. We see that the acoustic branch continually increases in frequency ω with increasing wavenumber k, and the optical branch continuously decreases in frequency. The two branches have respective limiting frequencies given by $(2C/M)^{1/2}$ and $(2C/m)^{1/2}$, with an energy gap between them at the edge of the

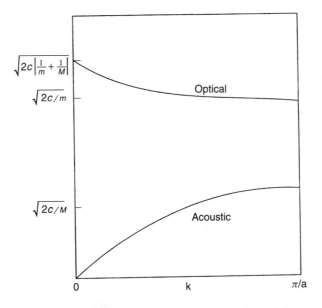

Figure 2.10. Dependence of the longitudinal normal-mode vibrational frequency ω on the wavenumber $k = 2\pi/\lambda$ for a linear diatomic chain of atoms with alternating masses $m < M$ having an equilibrium spacing a, and connected by bonds with spring constant C. (From C. P. Poole, Jr., *The Physics Handbook*, Wiley, New York, 1998, p. 53.)

Brillouin zone $k_{max} = \pi/a$, where a is the distance between atoms m and M at equilibrium. The *Brillouin zone* is a unit cell in wavenumber or reciprocal space, as will be explained later in this chapter. The optical branch vibrational frequencies are in the infrared region of the spectrum, generally with frequencies in the range from 10^{12} to 3×10^{14} Hz, and the acoustic branch frequencies are much lower. In three dimensions the situation is more complicated, and there are longitudinal acoustic (LA), transverse acoustic (TA), longitudinal optical (LO), and transverse optical (TO) modes.

The atoms in molecules also undergo vibratory motion, and a molecule containing N atoms has $3N - 6$ normal modes of vibration. Particular molecular groups such as hydroxyl —OH, amino —NH_2 and nitro —NO_2 have characteristic normal modes that can be used to detect their presence in molecules and solids.

The atomic vibrations that we have been discussing correspond to standing-wave types. This vibrational motion can also produce traveling waves in which localized regions of vibratory atomic motion travel through the lattice. Examples of such traveling waves are sound moving through the air, or seismic waves that start at the epicenter of an earthquake, and travel thousands of miles to reach a seismograph detector that records the earthquake event many minutes later. Localized traveling waves of atomic vibrations in solids, called *phonons*, are quantized with the energy $\hbar\omega = h\nu$, where $\nu = \omega/2\pi$ is the frequency of vibration of the wave. Phonons play an important role in the physics of the solid state.

2.2. ENERGY BANDS

2.2.1. Insulators, Semiconductors, and Conductors

When a solid is formed the energy levels of the atoms broaden and form bands with forbidden gaps between them. The electrons can have energy values that exist within one of the bands, but cannot have energies corresponding to values in the gaps between the bands. The lower energy bands due to the inner atomic levels are narrower and are all full of electrons, so they do not contribute to the electronic properties of a material. They are not shown in the figures. The outer or valence electrons that bond the crystal together occupy what is called a *valence band*. For an insulating material the valence band is full of electrons that cannot move since they are fixed in position in chemical bonds. There are no delocalized electrons to carry current, so the material is an insulator. The conduction band is far above the valence band in energy, as shown in Fig. 2.11a, so it is not thermally accessible, and remains essentially empty. In other words, the heat content of the insulating material at room temperature $T = 300$ K is not sufficient to raise an appreciable number of electrons from the valence band to the conduction band, so the number in the conduction band is negligible. Another way to express this is to say that the value of the gap energy E_g far exceeds the value $k_B T$ of the thermal energy, where k_B is Boltzmann's constant.

In the case of a semiconductor the gap between the valence and conduction bands is much less, as shown in Fig. 2.11b, so E_g is closer to the thermal energy $k_B T$, and

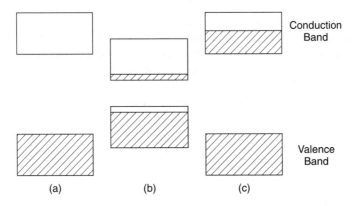

Figure 2.11. Energy bands of (a) an insulator, (b) an intrinsic semiconductor, and (c) a conductor. The cross-hatching indicates the presence of electrons in the bands.

the heat content of the material at room temperature can bring about the thermal excitation of some electrons from the valence band to the conduction band where they carry current. The density of electrons reaching the conduction band by this thermal excitation process is relatively low, but by no means negligible, so the electrical conductivity is small; hence the term *semiconducting*. A material of this type is called an *intrinsic semiconductor*. A semiconductor can be doped with donor atoms that give electrons to the conduction band where they can carry current. The material can also be doped with acceptor atoms that obtain electrons from the valence band and leave behind positive charges called *holes* that can also carry current. The energy levels of these donors and acceptors lie in the energy gap, as shown in Fig. 2.12. The former produces n-type, that is, negative-charge or electron, conductivity, and the latter produces p-type, that is, positive-charge or hole,

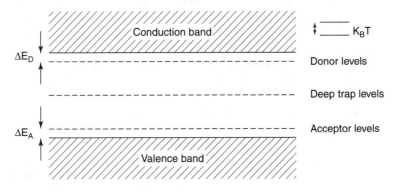

Figure 2.12. Sketch of the forbidden energy gap showing acceptor levels the typical distance Δ_A above the top of the valence band, donor levels the typical distance Δ_D below the bottom of the conduction band, and deep trap levels nearer to the center of the gap. The value of the thermal energy $k_B T$ is indicated on the right.

conductivity, as will be clarified in Section 2.3.1. These two types of conductivity in semiconductors are temperature-dependent, as is the intrinsic semiconductivity.

A conductor is a material with a full valence band, and a conduction band partly full with delocalized conduction electrons that are efficient in carrying electric current. The positively charged metal ions at the lattice sites have given up their electrons to the conduction band, and constitute a background of positive charge for the delocalized electrons. Figure 2.11c shows the energy bands for this case.

In actual crystals the energy bands are much more complicated than is suggested by the sketches of Fig. 2.11, with the bands depending on the direction in the lattice, as we shall see below.

2.2.2. Reciprocal Space

In Sections 2.1.2 and 2.1.3 we discussed the structures of different types of crystals in ordinary or coordinate space. These provided us with the positions of the atoms in the lattice. To treat the motion of conduction electrons, it is necessary to consider a different type of space that is mathematically called a *dual space* relative to the coordinate space. This dual or reciprocal space arises in quantum mechanics, and a brief qualitative description of it is presented here.

The basic relationship between the frequency $f = \omega/2\pi$, the wavelength λ, and the velocity v of a wave is $\lambda f = v$. It is convenient to define the wavevector $k = 2\pi/\lambda$ to give $f = (k/2\pi)v$. For a matter wave, or the wave associated with conduction electrons, the momentum $p = mv$ of an electron of mass m is given by $p = (h/2\pi)k$, where Planck's constant h is a universal constant of physics. Sometimes a reduced Planck's constant $\hbar = h/2\pi$ is used, where $p = \hbar k$. Thus for this simple case the momentum is proportional to the wavevector k, and k is inversely proportional to the wavelength with the units of reciprocal length, or reciprocal meters. We can define a reciprocal space called k *space* to describe the motion of electrons.

If a one-dimensional crystal has a lattice constant a and a length that we take to be $L = 10a$, then the atoms will be present along a line at positions $x = 0, a$, $2a, 3a, \ldots, 10a = L$. The corresponding wavevector k will assume the values $k = 2\pi/L, 4\pi/L, 6\pi/L, \ldots, 20\pi/L = 2\pi/a$. We see that the smallest value of k is $2\pi/L$, and the largest value is $2\pi/a$. The unit cell in this one-dimensional coordinate space has the length a, and the important characteristic cell in reciprocal space, called the *Brillouin zone*, has the value $2\pi/a$. The electron sites within the Brillouin zone are at the reciprocal lattice points $k = 2\pi n/L$, where for our example $n = 1, 2, 3, \ldots, 10$, and $k = 2\pi/a$ at the Brillouin zone boundary where $n = 10$.

For a rectangular direct lattice in two dimensions with coordinates x and y, and lattice constants a and b, the reciprocal space is also two-dimensional with the wavevectors k_x and k_y. By analogy with the direct lattice case, the Brillouin zone in this two-dimensional reciprocal space has the length $2\pi/a$ and width $2\pi/b$, as shown sketched in Fig. 2.13. The extension to three dimensions is straightforward. It is important to keep in mind that k_x is proportional to the momentum p_x of the conduction electron in the x direction, and similarly for the relationship between k_y and p_y.

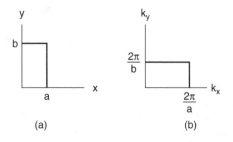

Figure 2.13. Sketch of (a) unit cell in two-dimensional x, y coordinate space and (b) corresponding Brillouin zone in reciprocal space k_x, k_y for a rectangular Bravais lattice.

2.2.3. Energy Bands and Gaps of Semiconductors

The electrical, optical, and other properties of semiconductors depend strongly on how the energy of the delocalized electrons involves the wavevector k in reciprocal or k space, with the electron momentum p given by $p = mv = \hbar k$, as explained above. We will consider three-dimensional crystals, and in particular we are interested in the properties of the III–V and the II–VI semiconducting compounds, which have a cubic structure, so their three lattice constants are the same: $a = b = c$. The electron motion expressed in the coordinates k_x, k_y, k_z of reciprocal space takes place in the Brillouin zone, and the shape of this zone for these cubic compounds is shown in Fig. 2.14. Points of high symmetry in the Brillouin zone are designated by capital Greek or Roman letters, as indicated.

The energy bands depend on the position in the Brillouin zone, and Fig. 2.15 presents these bands for the intrinsic (i.e., undoped) III–V compound GaAs. The

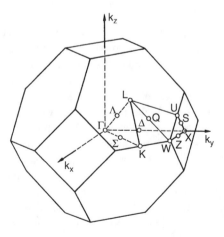

Figure 2.14. Brillouin zone of the gallium arsenide and zinc blende semiconductors showing the high-symmetry points Γ, K, L, U, W, and X and the high-symmetry lines $\Delta, \Lambda, \Sigma, Q, S$, and Z. (From G. Burns, *Solid State Physics*, Academic Press, Boston, 1985, p. 302.)

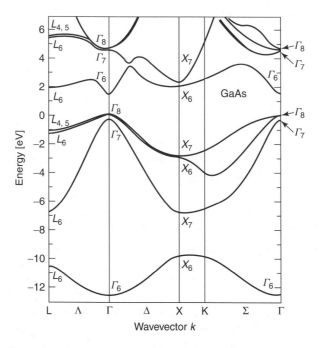

Figure 2.15. Band structure of the semiconductor GaAs calculated by the pseudopotential method. (From M. L. Cohen and J. Chelikowsky, *Electronic Structure and Electronic Properties of Semiconductors*, 2nd ed., Springer-Verlag; *Solid State Sci.* **75**, Springer, Berlin, 1989.)

figure plots energy versus the wavevector k in the following Brillouin zone directions: along Δ from point Γ to X, along Λ from Γ to L, along Σ from Γ to K, and along the path between points X and K. These points and paths are indicated in the sketch of the Brillouin zone in Fig. 2.14. We see from Fig. 2.15 that the various bands have prominent maxima and minima at the central point Γ of the Brillouin zone. The energy gap or region where no band appears extends from the zero of energy at point Γ_8 to the point Γ_6 directly above the gap at the energy $E_g = 1.35$ eV. The bands below point Γ_8 constitute the valence band, and those above point Γ_6 form the conduction band. Hence Γ_8 is the lowest energy point of the conduction band, and Γ_8 is the highest point of the valence band.

At absolute zero all the energy bands below the gap are filled with electrons, and all the bands above the gap are empty, so at absolute temperature 0 K the material is an insulator. At room temperature the gap is sufficiently small so that some electrons are thermally excited from the valence band to the conduction band, and these relatively few excited electrons gather in the region of the conduction band immediately above its minimum at Γ_6, a region that is referred to as a "valley." These electrons carry some electric current, hence the material is a semiconductor.

Gallium arsenide is called a *direct-bandgap semiconductor* because the top of the valence band and the bottom of the conduction band are both at the same center point (Γ) in the Brillouin zone, as is clear from Fig. 2.15. Electrons in the valence band at point Γ_8 can become thermally excited to point Γ_6 in the conduction band with no change in the wavevector k. The compounds GaAs, GaSb, InP, InAs, and InSb and all the II–VI compounds included in Table B.6 have direct gaps. In some semiconductors such as Si and Ge the top of the valence band is at a position in the Brillouin zone different from that for the bottom of the conduction band, and these are called *indirect-gap semiconductors*.

Figure 2.16 depicts the situation at point Γ of a direct-gap semiconductor on an expanded scale, at temperatures above absolute zero, with the energy bands approximated by parabolas. The conduction band valley at Γ_6 is shown occupied by electrons up to the Fermi level, which is defined as the energy of the highest occupied state. The excited electrons leave behind empty states near the top of the valence band, and these act like positive charges called "holes" in an otherwise full valence band. These hole levels exist above the energy $-E'_{F'}$, as indicated in Fig. 2.16. Since an intrinsic or undoped semiconductor has just as many holes in the valence band as it has electrons in the conduction band, the corresponding volumes filled with these electrons and holes in k space are equal to each other. These electrons and holes are the charge carriers of current, and the temperature dependence of their concentration in GaAs, Si, and Ge is given in Fig. 2.17.

In every semiconductor listed in Table B.6, including Si and Ge, the top of its valence band is at the center of the Brillouin zone, but the indirect-bandgap semiconductors Si, Ge, AlAs, AlSb, and GaP have the lowest valley of their conduction bands at a different location in k space than the point Γ. This is shown in Fig. 2.18 for the indirect-bandgap materials Si and Ge. We see from Fig. 2.18b that Ge has its conduction band minimum at the point L, which is in the middle of the hexagonal face of the Brillouin zone along the Λ or (1 1 1) direction

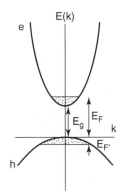

Figure 2.16. Sketch of lower valence band and upper conduction band of a semiconductor approximated by parabolas. The region in the valence band containing holes and that in the conduction band containing electrons are cross-hatched. The Fermi energies E_F and $E'_{F'}$ mark the highest occupied level of the conduction band and the lowest unoccupied level of the valence band, respectively. The zero of energy is taken as the top of the valence band, and the direct-band gap energy E_g is indicated.

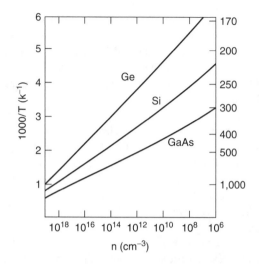

Figure 2.17. Temperature dependencies of the intrinsic carrier density of the semiconductors Ge, Si, and GaAs. Note the lack of linearity at the lower temperatures arising from the $T^{3/2}$ factor in Eq. (2.15). (From G. Burns, *Solid State Physics*, Academic Press, Boston, 1985, p. 315.)

depicted in Fig. 2.14. The Brillouin zone has eight such faces, with each point L shared by two zones, so the zone actually contains only four of these points proper to it, and we say that the valley degeneracy for Ge is 4. The semiconductor Si has its lowest conduction band minimum along the Δ or (001) direction about 85% of the way to the X point, as shown in Fig. 2.16. The compound GaP, not shown, also has its corresponding valley along Δ about 92% of the way to X. We see from Fig. 2.14 that there are six such Δ lines in the Brillouin zone, so this valley degeneracy is 6. Associated with each of the valleys that we have been discussing, at point L for Ge and along direction Δ for Si, there is a three-dimensional constant-energy surface in the shape of an ellipsoid that encloses the conduction electrons in the corresponding valleys, and these ellipsoids are sketched in Figs. 2.19a and 2.19b for Ge and Si, respectively.

Some interesting experiments such as cyclotron resonance have been carried out to map the configuration of these ellipsoid-type constant-energy surfaces. In a cyclotron resonance experiment conduction electrons are induced to move along constant energy surfaces at a velocity that always remains perpendicular to an applied magnetic field direction. By utilizing various orientations of applied magnetic fields relative to the ellipsoid the electrons at the surface execute a variety of orbits, and by measuring the trajectories of these orbits the shape of the energy surface can be delineated.

Table B.6 lists values of the energy gap E_g for the type III–V and II–VI semiconductors at room temperature (left-hand value) and in several cases it also lists the value at absolute zero temperature (right-hand value). Table B.7 gives the temperature and pressure dependencies, dE_g/dT and dE_g/dP, respectively, of the gap at room temperature.

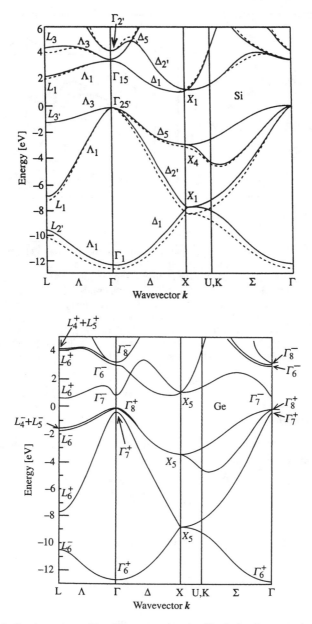

Figure 2.18. Band structure plots of the energy bands of the indirect-gap semiconductors Si and Ge. The bandgap (absence of bands) lies slightly above the Fermi energy $E = 0$ on both figures, with the conduction band above and the valence band below the gap. The figure shows that the lowest point or bottom of the conduction band of Ge is at the energy $E = 0.6$ at the symmetry point L (labeled by $L_6{}^+$), and for Si it is 85% of the way along the direction Δ_1 from Γ_{15} to X_1. It is clear from Fig. 2.15 that the bottom of the conduction band of the direct-gap semiconductor GaAs is at the symmetry point Γ_6. The top of the valence band is at the center point Γ of the Brillouin zone for both materials.

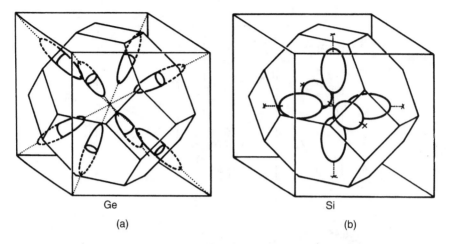

Ge   Si

(a)   (b)

Figure 2.19. Ellipsoidal constant-energy surfaces in the conduction band of germanium (left) and silicon (right). The constant energy surfaces of Ge are aligned along symmetry direction Λ and centered at symmetry point L. As a result, they lie half inside (solid lines) and half outside (dashed lines) the first Brillouin zone, so this zone contains the equivalent of four complete energy surfaces. The surfaces of Si lie along the six symmetry directions Δ (i.e., along $\pm k_x, \pm k_y, \pm k_z$), and are centered 85% of the way from the center point Γ to symmetry point X. All six of them lie entirely within the Brillouin zone, as shown. Figure 2.14 shows the positions of symmetry points Γ, L, and X, and of symmetry lines Δ and Λ, in the Brillouin zone. (From G. Burns, *Solid State Physics*, Academic Press, Boston, 1985, p. 313.)

2.2.4. Effective Masses

On a simple one-dimensional model the energy E of a conduction electron has a quadratic dependence on the wavevector k through the expression

$$E = \frac{\hbar^2 k^2}{2m^*}$$ (2.9)

The first derivative of this expression provides the velocity v

$$\frac{1}{\hbar}\frac{dE}{dk} = \frac{\hbar k}{m^*} = v$$ (2.10)

and the second derivative provides the effective mass m^*

$$\frac{1}{\hbar^2}\frac{d^2E}{dk^2} = \frac{1}{m^*}$$ (2.11)

which differs, in general, from the free-electron mass. These equations are rather trivial for the simple parabolic energy expression (2.9), but we see from the energy bands of Figs. 2.15 and 2.18 that the actual dependence of the energy E on k is much more complex than Eq. (2.9) indicates. Equation (2.11) provides a general definition of the effective mass, designated by the symbol m^*, and the wavevector k dependence of m^* can be evaluated from the band structure plots by carrying out the differentiations. We see from a comparison of the slopes near the conduction band minimum and the valence band maximum of GaAs at the Γ point on Fig. 2.15 (see also Fig. 2.16) that the upper electron bands have steeper slopes and hence lighter masses than do the lower hole bands with more gradual slopes.

2.2.5. Fermi Surfaces

At very low temperatures electrons fill the energy bands of solids up to an energy called the *Fermi energy* E_F, and the bands are empty for energies that exceed E_F. In three-dimensional k space the set of values of k_x, k_y, and k_z, which satisfy the equation $\hbar^2(k_x^2 + k_y^2 + k_z^2)/2m = E_F$, form a surface called the *Fermi surface*. All k_x, k_y, k_z energy states that lie below this surface are full, and the states above the surface are empty. The Fermi surface encloses all the electrons in the conduction band that carry electric current. In the good conductors copper and silver the conduction electron density is 8.5×10^{22} and 5.86×10^{22} electrons/cm^3, respectively. From another viewpoint, the Fermi surface of a good conductor can fill the entire Brillouin zone. In the intrinsic semiconductors GaAs, Si, and Ge the carrier density at room temperature from Fig. 2.17 is approximately $10^6, 10^{10}$, and 10^{13} carriers/cm^3, respectively, many orders of magnitude below that of metals, and semiconductors are seldom doped to concentrations above 10^{19} centers/cm^3. An intrinsic semiconductor is one with a full valence band and an empty conduction band at absolute zero of temperature. As we saw above, at ambient temperatures some electrons are thermally excited to the bottom of the conduction band, and an equal number of empty sites or holes are left behind near the top of the valence band. This means that only a small percentage of the Brillouin zone contains electrons in the conduction band, and the number of holes in the valence band is correspondingly small. In a one-dimensional representation this reflects the electron and hole occupancies depicted in Fig. 2.16.

If the conduction band minimum is at the Γ point in the center of the Brillouin zone, as is the case with GaAs, then it will be very close to a sphere since the symmetry is cubic, and to a good approximation we can assume a quadratic dependence of the energy on the wavevector k, corresponding to Eq. (2.9). Therefore the Fermi surface in k space is a small sphere given by the standard equation for a sphere

$$E_F = E_g + \frac{\hbar^2}{2m_e}(k_x^2 + k_y^2 + k_z^2)_F = E_g + \frac{\hbar^2}{2m_e}k_F^2 \qquad (2.12)$$

where E_F is the Fermi energy, and the electron mass m_e relative to the free-electron mass has the values given in Table B.8 for various direct-gap semiconductors.

Equations (2.12) are valid for direct-gap materials where the conduction band minimum is at point Γ. All the semiconductor materials under consideration have their valence band maxima at the Brillouin zone center point Γ.

The situation for the conduction electron Fermi surface is more complicated for the indirect-gap semiconductors. We mentioned above that Si and GaP have their conduction band minima along the Δ direction of the Brillouin zone, and the corresponding six ellipsoidal energy surfaces are sketched in Fig. 2.19b. The longitudinal and transverse effective masses, m_L and m_T respectively, have the following values

$$\frac{m_L}{m_0} = 0.92 \quad \frac{m_T}{m_0} = 0.19 \quad \text{for Si} \tag{2.13a}$$

$$\frac{m_L}{m_0} = 7.25 \quad \frac{m_T}{m_0} = 0.21 \quad \text{for GaP} \tag{2.13b}$$

for these two indirect-gap semiconductors. Germanium has its band minimum at points L of the Brillouin zone sketched in Fig. 2.14, and its Fermi surface is the set of ellipsoids centered at the L points with their axis along the Λ or (111) directions, as shown in Fig. 2.19a. The longitudinal and transverse effective masses for these ellipsoids in Ge are $m_L/m_0 = 1.58$ and $m_T/m_0 = 0.081$, respectively. Cyclotron resonance techniques together with the application of stress to the samples can be used to determine these effective masses for the indirect-bandgap semiconductors.

2.3. LOCALIZED PARTICLES

2.3.1. Donors, Acceptors, and Deep Traps

When a type V atom such as P, As, or Sb, which has five electrons in its outer or valence electron shell, is a substitutional impurity in Si it uses four of these electrons to satisfy the valence requirements of the four nearest-neighbor silicons, and the one remaining electron remains weakly bound. The atom easily donates or passes on this electron to the conduction band, so it is called a *donor*, and the electron is called a *donor electron*. This occurs because the donor energy levels lie in the forbidden region close to the conduction band edge by the amount ΔE_D relative to the thermal energy value $k_B T$, as indicated in Fig. 2.12. A Si atom substituting for Ga plays the role of a donor in GaAs, Al substituting for Zn in ZnSe serves as a donor, and so on.

A type III atom such as Al or Ga, called an *acceptor* atom, which has three electrons in its valence shell, can serve as a substitutional defect in Si, and in this role it requires four valence electrons to bond with the tetrahedron of nearest-neighbor Si atoms. To accomplish this, it draws or accepts an electron from the valence band, leaving behind a hole at the top of this band. This occurs easily because the energy levels of the acceptor atoms are in the forbidden gap slightly above the valence band edge by the amount ΔE_A relative to $k_B T$, as indicated in Fig. 2.12. In other words, the excitation energies needed to ionize the donors and to add electrons to the acceptors are much less than the thermal energy at room temperature $T = 300$ K, that is, ΔE_D, $\Delta E_A \ll k_B T$, so virtually all donors are positively ionized and virtually all acceptors are negatively ionized at room temperature.

The donor and acceptor atoms that we have been discussing are known as *shallow centers*, that is, shallow traps of electrons or holes, because their excitation energies are much less than that of the bandgap (ΔE_D, $\Delta E_A \ll E_g$). There are other centers with energy levels that lie deep within the forbidden gap, often closer to its center than to the top or bottom, in contrast to the case with shallow donors and acceptors. Since generally $E_g \gg k_B T$, these traps are not extensively ionized, and the energies involved in exciting or ionizing them are not small. Examples of deep centers are defects associated with broken bonds, or strain involving displacements of atoms. In Chapter 8 we discuss how deep centers can produce characteristic optical spectroscopic effects.

2.3.2. Mobility

Another important parameter of a semiconductor is the mobility μ or charge carrier drift velocity v per unit electric field E, given by the expression $\mu = |v|/E$. This parameter is defined as positive for both electrons and holes. Table B.9 lists the mobilities μ_e and μ_h for electrons and holes, respectively, in the semiconductors under consideration. The electrical conductivity σ is the sum of contributions from the concentrations of electrons n and of holes p in accordance with the expression

$$\sigma = (ne\mu_e + pe\mu_h) \tag{2.14}$$

where e is the electronic charge. The mobilities have a weak power-law temperature dependence T^n, and the pronounced T dependence of the conductivity is due principally to the dependence of the electron and hole concentrations on the temperature. In doped semiconductors this generally arises mainly from the Boltzmann factor $\exp(-E_i/k_B T)$ associated with the ionization energies E_i of the donors or acceptors. Typical ionization energies for donors and acceptors in Si and Ge listed in Table B.10 are in the range from 0.0096 to 0.16 eV, which is much less than the bandgap energies 1.11 eV and 0.66 eV of Si and Ge, respectively. Figure 2.12 shows the locations of donor and acceptor levels on an energy band plot, and makes clear that their respective ionization energies are much less that E_g. The thermal energy $k_B T = 0.026$ eV at room temperature (300 K) is often comparable to the ionization energies. In intrinsic or undoped materials the main contribution is from the exponential factor $\exp(-E_g/2k_B T)$ in the following expression from the law of mass action

$$n_i = p_i = 2\left(\frac{k_B T}{2\pi\hbar^2}\right)^{3/2} (m_e m_h)^{3/4} \exp\frac{-E_g}{2k_B T} \tag{2.15}$$

where the intrinsic concentrations of electrons n_i and holes p_i are equal to each other because the thermal excitation of n_i electrons to the conduction band leaves behind the same number p_i of holes in the valence band, that is $n_i = p_i$. We see that the expression (2.15) contains the product $m_e m_h$ of the effective masses m_e and m_h of the electrons and holes, respectively, and the ratios m_e/m_0 and m_h/m_0 of these effective masses to the free-electron mass m_0 are presented in Table B.8. These effective masses strongly influence the properties of excitons to be discussed next.

2.3.3. Excitons

An ordinary negative electron and a positive electron, called a *positron*, situated a distance r apart in free space experience an attractive force called the *Coulomb force*, which has the value $-e^2/4\pi\varepsilon_0 r^2$, where e is their charge and ε_0 is the dielectric constant of free space. A quantum-mechanical calculation shows that the electron and positron interact to form an atom called *positronium* which has bound-state energies given by the Rydberg formula introduced by Niels Bohr in 1913 to explain the hydrogen atom

$$E = -\frac{e^2}{8\pi\varepsilon_0 a_0 n^2} = -\frac{6.8}{n^2}\,\mathrm{eV} \qquad (2.16)$$

where a_0 is the Bohr radius given by $a_0 = 4\pi\varepsilon_0\hbar^2/m_0 e^2 = 0.0529\,\mathrm{nm}$, m_0 is the free-electron (and positron) mass, and the quantum number n takes on the values $n = 1, 2, 3, \ldots, \infty$. For the lowest energy or ground state, which has $n = 1$, the energy is 6.8 eV, which is exactly half the ground-state energy of a hydrogen atom, since the effective mass of the bound electron–positron pair is half of that of the bound electron–proton pair in the hydrogen atom. Figure 2.20 shows the energy levels of positronium as a function of the quantum number n. This set of energy levels is often referred to as a *Rydberg series*. The continuum at the top of the figure is the region of positive energies where the electron and hole are so far away from each other that the Coulomb interaction no longer has an appreciable effect, and the energy is all of the kinetic type, $\frac{1}{2}mv^2 = p^2/2m$, or energy of motion, where v is the velocity and $p = mv$ is the momentum.

The analog of positronium in a solid such as a semiconductor is the bound state of an electron–hole pair, called an *exciton*. For a semiconductor the electron is in the conduction band, and the hole is in the valence band. The electron and hole both have effective masses m_e and m_h, respectively, which are less than that (m_0) of a free electron, so the effective mass m^* is given by $m^* = m_e m_h/(m_e + m_h)$. When the electron effective mass is appreciably less than the hole effective mass, $m_e \ll m_h$, the relationship between them is conveniently written in the form

$$m^* = \frac{m_e}{1 + (m_e/m_h)} \qquad (2.17)$$

which shows that for this case m^* becomes comparable with the electron mass. For example, if $m_e/m_h = 0.2$, then $m^* = 0.83 m_e$. A comparison of the data in Table B.8 shows that this situation is typical for GaAs-type semiconductors. We also see from

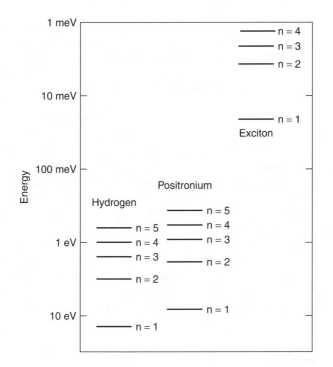

Figure 2.20. First few energy levels in the Rydberg series of a hydrogen atom (left), positronium (center), and a typical exciton (right).

Table B.11 that the relative dielectric constant $\varepsilon/\varepsilon_0$ has the range of values $7.2 < \varepsilon/\varepsilon_0 < 17.7$ for these materials, where ε_0 is the dielectric constant of free space. Both of these factors have the effect of decreasing the exciton energy E_{ex} from that of positronium, and as a result this energy is given by

$$E_{ex} = \frac{m^*/m_0}{(\varepsilon/\varepsilon_0)^2} \frac{e^2}{4\pi\varepsilon_0 a_0 n^2} = \frac{13.6 m^*/m_0}{(\varepsilon/\varepsilon_0)^2 n^2} \text{ eV} \tag{2.18}$$

as shown plotted in Fig. 2.20. These same two factors also increase the effective Bohr radius of the electron orbit, and it becomes

$$a_{eff} = \frac{\varepsilon/\varepsilon_0}{m^*/m_0} a_0 = \frac{0.0529\varepsilon/\varepsilon_0}{m^*/m_0} \text{ nm} \tag{2.19}$$

Using the GaAs electron effective mass and the heavy-hole effective mass values from Table B.8, Eq. (2.17) gives $m^*/m_0 = 0.059$. Utilizing the dielectric constant value from Table B.11, we obtain, with the aid of Eqs. (2.18) and (2.19), for GaAs

$$E_0 = 4.6 \, \text{meV} \qquad a_{\text{eff}} = 11.8 \, \text{nm} \qquad (2.20)$$

where E_0 is the ground-state ($n = 1$) energy. This demonstrates that an exciton extends over quite a few atoms of the lattice, and its radius in GaAs is comparable with the dimensions of a typical nanostructure. An exciton has the properties of a particle; it is mobile and able to move around the lattice. It also exhibits characteristic optical spectra. Figure 2.20 plots the energy levels for an exciton with the ground-state energy $E_0 = 18 \, \text{meV}$.

Technically speaking, the exciton that we have just discussed is a weakly bound electron–hole pair called a *Mott–Wannier exciton*. A strongly or tightly bound exciton, called a *Frenkel exciton*, is similar to a long-lived excited state of an atom or a molecule. It is also mobile, and can move around the lattice by the transfer of the excitation or excited-state charge between adjacent atoms or molecules. Almost all the excitons encountered in semiconductors and in nanostructures are of the Mott–Wannier type, so they are the only ones discussed in this book.

FURTHER READING

K. Boer, ed., *Semiconductor Physics*, Vols. 1 and 2, Wiley, New York, 2001.

G. Burns, *Solid State Physics*, Academic Press, San Diego, 1985.

C. Kittel, *Introduction to Solid State Physics*, 7th ed., Wiley, New York, 1996.

T. P. Martin, T. Bergmann, H. Gohlich, and T. Lange, *Chem. Phys. Lett.* **172**, 209 (1990).

C. P. Poole, Jr. and H. A. Farach, "Chemical Bonding," in *Semiconductor Physics*, Vol. 1, K. Boer, ed., Wiley, New York, 2001, Chapter 2.

S. Sugano and H. Koizumi, *Microcluster Physics*, Springer, Berlin, 1998.

P. Y. Yu and M. Cardona, *Fundamentals of Semiconductors*, 3rd ed., Springer-Verlag, Berlin, 2001.

3

METHODS OF MEASURING PROPERTIES

3.1. INTRODUCTION

The current revolution in nanoscience was brought about by the concomitant development of several advances in technology. One of them has been the progressive ability to fabricate smaller and smaller structures, and another has been the continual improvement in the precision with which such structures are made. One widely used method for the fabrication of nanostructures is *lithography*, which makes use of a radiation-sensitive layer to form well-defined patterns on a surface. *Molecular-beam epitaxy*, the growth of one crystalline material on the surface of another, is a second technique that has become perfected. There are also chemical methods, and the utilization of self-assembly, the spontaneous aggregation of molecular groups.

Since the early 1970s the continual advancement in technology has followed Moore's law (Moore 1975) whereby the number of transistors incorporated on a memory chip doubles every year and a half. This has resulted from continual improvements in design factors such as interconnectivity efficiency, as well as from continual decreases in size. Economic considerations driving this revolution are the need for more and greater information storage capacity, and the need for faster and broader information dispersal through communication networks. Another major

Introduction to Nanotechnology, by Charles P. Poole Jr. and Frank J. Owens.
ISBN 0-471-07935-9. Copyright © 2003 John Wiley & Sons, Inc.

factor responsible for the nanotechnology revolution has been the improvement of old and the introduction of new instrumentation systems for evaluating and characterizing nanostructures. Many of these systems are very large and expensive, in the million-dollar price range, often requiring specialists to operate them. The aim of the present chapter is to explain the principles behind the operation of some of these systems, and to delineate their capabilities.

In the following sections we describe instruments for determining the positions of atoms in materials, instruments for observing and characterizing the surfaces of structures, and various spectroscopic devices for obtaining information of the properties of nanostructures [see e.g., Whan (1986)].

3.2. STRUCTURE

3.2.1. Atomic Structures

To understand a nanomaterial we must, first, learn about its structure, meaning that we must determine the types of atoms that constitute its building blocks and how these atoms are arranged relative to each other. Most nanostructures are crystalline, meaning that their thousands of atoms have a regular arrangement in space on what is called a *crystal lattice*, as explained in Section 2.1.2 (of Chapter 2). This lattice can be described by assigning the positions of the atoms in a unit cell, so the overall lattice arises from the continual replication of this unit cell throughout space. Figure 2.1 presents sketches of the unit cells of the four crystal systems in two dimensions, and the characteristics of the parameters a, b, and θ for these systems are listed in the four top rows of Table 3.1. There are 17 possible types of crystal structures called *space groups*, meaning 17 possible arrangements of atoms in unit cells in two dimensions, and these are divided between the four crystal systems in the manner indicated in column 4 of the table. Of particular interest is the most efficient way to arrange identical atoms on a surface, and this corresponds to the hexagonal system shown in Fig. 2.4a.

In three dimensions the situation is much more complicated, and some particular cases were described in Chapter 2. There are now three lattice constants a, b, and c, for the three dimensions x, y, z, with the respective angles α, β, and γ between them (α is between b and c, etc.). There are seven crystal systems in three dimensions with a total of 230 space groups divided among the systems in the manner indicated in column 4 of Table 3.1. The objective of a crystal structure analysis is to distinguish the symmetry and space group, to determine the values of the lattice constants and angles, and to identify the positions of the atoms in the unit cell.

Certain special cases of crystal structures are important for nanocrystals, such as those involving simple cubic (SC), body-centered cubic (BCC), and face-centered cubic (FCC) unit cells, as shown in Fig. 2.3. Another important structural arrangement is formed by stacking planar hexagonal layers in the manner sketched in Fig. 2.4b, which for a monatomic (single-atom) crystal provides the highest density or closest-packed arrangement of identical spheres. If the third layer is placed directly above the first layer, the fourth directly above the second, and so on, in an A–B–A–B–··· type sequence, the hexagonal close-packed (HCP) structure results.

Table 3.1. Crystal systems, and associated number of space groups, in two and three dimensions[a]

Dimension	System	Conditions	Space Groups
2	Oblique	$a \neq b,\ \gamma \neq 90°$ (or $a = b,\ \gamma \neq 90°,\ 120$)	2
2	Rectangular	$a \neq b,\ \gamma = 90°$	7
2	Square	$a = b,\ \gamma = 90°$	3
2	Hexagonal	$a = b,\ \gamma = 120°$	5
3	Triclinic	$a \neq b \neq c$ $\alpha \neq \beta \neq \gamma$	2
3	Monoclinic	$a \neq b \neq c$ $\alpha = \gamma = 90° \neq \beta$	13
3	Orthorhombic	$a \neq b \neq c$ $\alpha = \beta = \gamma = 90°$	59
3	Tetragonal	$a = b \neq c$ $\alpha = \beta = \gamma = 90°$	68
3	Trigonal	$a = b = c$ $\alpha = \beta = \gamma < 120°,\ \neq 90°$	25
3	Hexagonal	$a = b \neq c$ $\alpha = \beta = 90°,\ \gamma = 120°$	27
3	Cubic	$a = b = c$ $\alpha = \beta = \gamma = 90°$	36

[a]There are 17 two-dimensional space groups and 230 three-dimensional space groups.

If, on the other hand, this stacking is carried out by placing the third layer in a third position and the fourth layer above the first, and so forth, the result is an A–B–C–A–B–C–A–··· sequence, and the structure is FCC, as explained in Chapter 2. The latter arrangement is more commonly found in nanocrystals.

Some properties of nanostructures depend on their crystal structure, while other properties such as catalytic reactivity and adsorption energies depend on the type of exposed surface. Epitaxial films prepared from FCC or HCP crystals generally grow with the planar close-packed atomic arrangement just discussed. Face-centered cubic crystals tend to expose surfaces with this same hexagonal two-dimensional atomic array.

3.2.2. Crystallography

To determine the structure of a crystal, and thereby ascertain the positions of its atoms in the lattice, a collimated beam of X rays, electrons, or neutrons is directed at the crystal, and the angles at which the beam is diffracted are measured. We will explain the method in terms of X rays, but much of what we say carries over to the other two radiation sources. The wavelength λ of the X rays expressed in nanometers is related to the X-ray energy E expressed in the units kiloelectronvolts (keV) through the expression

$$\lambda = \frac{1.240}{E} \text{ nm} \tag{3.1}$$

Ordinarily the beam is fixed in direction and the crystal is rotated through a broad range of angles to record the X-ray spectrum, which is also called a *diffractometer recording* or *X-ray-diffraction scan*. Each detected X-ray signal corresponds to a coherent reflection, called a *Bragg reflection*, from successive planes of the crystal for which Bragg's law is satisfied

$$2d \sin \theta = n\lambda \tag{3.2}$$

as shown in Fig. 3.1, where d is the spacing between the planes, θ is the angle that the X-ray beam makes with respect to the plane, λ is the wavelength of the X rays, and $n = 1,2,3,\ldots$ is an integer that usually has the value $n = 1$.

Each crystallographic plane has three indices h,k,l, and for a cubic crystal they are ratios of the points at which the planes intercept the Cartesian coordinate axes x, y, z. The distance d between parallel crystallographic planes with indices hkl for a simple cubic lattice of lattice constant a has the particularly simple form

$$d = \frac{a}{(h^2 + k^2 + l^2)^{1/2}} \tag{3.3}$$

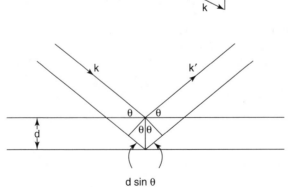

d sin θ

Figure 3.1. Reflection of X-ray beam incident at the angle θ off two parallel planes separated by the distance d. The difference in pathlength $2d \sin \theta$ for the two planes is indicated. (From C. P. Poole Jr., *The Physics Handbook*, Wiley, New York, 1998, p. 333.)

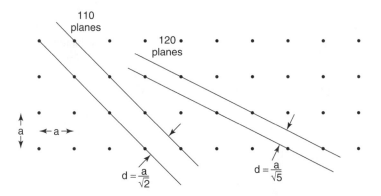

Figure 3.2. Two-dimensional cubic lattice showing projections of pairs of (110) and (120) planes (perpendicular to the surface) with the distances d between them indicated.

so higher index planes have larger Bragg diffraction angles θ. Figure 3.2 shows the spacing d for 110 and 120 planes, where the index $l = 0$ corresponds to planes that are parallel to the z direction. It is clear from this figure that planes with higher indices are closer together, in accordance with Eq. (3.3), so they have larger Bragg angles θ from Eq. (3.2). The amplitudes of the X-ray lines from different crystallographic planes also depend on the indices hkl, with some planes having zero amplitude, and these relative amplitudes help in identifying the structure type. For example, for a body-centered monatomic lattice the only planes that produce observed diffraction peaks are those for which $h + k + l = n$, an even integer, and for a face-centered cubic lattice the only observed diffraction lines either have all odd integers or all even integers.

To obtain a complete crystal structure, X-ray spectra are recorded for rotations around three mutually perpendicular planes of the crystal. This provides comprehensive information on the various crystallographic planes of the lattice. The next step in the analysis is to convert these data on the planes to a knowledge of the positions of the atoms in the unit cell. This can be done by a mathematical procedure called *Fourier transformation*. Carrying out this procedure permits us to identify which one of the 230 crystallographic space groups corresponds to the structure, together with providing the lengths of the lattice constants a,b,c of the unit cell, and the values of the angles α,β,γ between them. In addition, the coordinates of the positions of each atom in the unit cell can be deduced.

As an example of an X-ray diffraction structure determination, consider the case of nanocrystalline titanium nitride prepared by chemical vapor deposition with the grain size distribution shown in Fig. 3.3. The X-ray diffraction scan, with the various lines labeled according to their crystallographic planes, is shown in Fig. 3.4. The fact that all the planes have either all odd or all even indices identifies the structure as face-centered cubic. The data show that TiN has the FCC NaCl structure sketched in Fig. 2.3c, with the lattice constant $a = 0.42417$ nm.

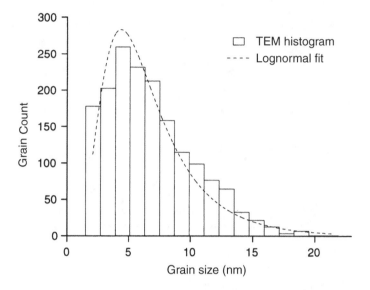

Figure 3.3. Histogram of grain size distribution in nanocrystalline TiN determined from a TEM micrograph. The fit parameters for the dashed curve are $D_0 = 5.8$ nm and $\sigma = 1.71$ nm. [From C. E. Krill et al., in Nalwa (2000), Vol. 2, Chapter 5, p. 207.]

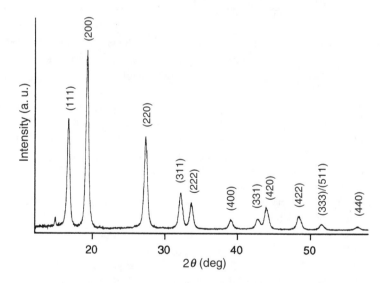

Figure 3.4. X-ray diffraction scan of nanocrystalline TiN with the grain size distribution shown in Fig. 3.3. Molybdenum K_α radiation was used with the wavelength $\lambda = 0.07093$ nm calculated from Eq. (3.1). The X-ray lines are labeled with their respective crystallographic plane indices (*hkl*). Note that these indices are either all even or all odd, as expected for a FCC structure. The nonindexed weak line near $2\theta = 15°$ is due to an unidentified impurity. [From C. E. Krill et al., in Nalwa (2000), Vol. 2, Chapter 3, p. 200.]

The widths of the Bragg peaks of the X-ray scan of Fig. 3.4 can be analyzed to provide information on the average grain size of the TiN sample. Since the widths arise from combinations or convolutions of grain size, microcrystalline strain, and instrumental broadening effects, it is necessary to correct for the instrumental broadening and to sort out the strain components to determine the average grain size. The assumption was made of spherical grains with the diameter D related to the volume V by the expression

$$D = \left(\frac{6V}{\pi}\right)^{1/3} \tag{3.4}$$

and several ways of making the linewidth corrections provided average grain size values between 10 to 12 nm, somewhat larger than expected from the TEM histogram of Fig. 3.3. Thus X-ray diffraction can estimate average grain sizes, but a transmission electron microscope is needed to determine the actual distribution of grain sizes shown in Fig. 3.3.

An alternate approach for obtaining the angles θ that satisfy the Bragg condition (3.2) in a powder sample is the Debye–Scherrer method sketched in Fig. 3.5. It employs a monochromatic X-ray beam incident on a powder sample generally contained in a very fine-walled glass tube. The tube can be rotated to smooth out the recorded diffraction pattern. The conical pattern of X rays emerging for each angle 2θ, with θ satisfying the Bragg condition (3.2), is incident on the film strip in arcs, as shown. It is clear from the figure that the Bragg angle has the value $\theta = S/4R$, where S is the distance between the two corresponding reflections on the film and R is the radius of the film cylinder. Thus a single exposure of the powder to the X-ray beam provides all the Bragg angles at the same time. The Debye–Scherrer powder technique is often used for sample identification. To facilitate the identification,

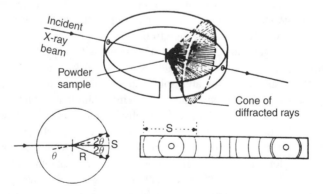

Figure 3.5. Debye–Scherrer powder diffraction technique, showing a sketch of the apparatus (top), an X-ray beam trajectory for the Bragg angle θ (lower left), and images of arcs of the diffraction beam cone on the film plate (lower right). (From G. Burns, *Solid State Physics*, Academic Press, Boston, 1985, p. 81.)

the Debye–Scherrer diffraction results for over 20,000 compounds are available to researchers in a JCPDS powder diffraction card file. This method has been widely used to obtain the structures of powders of nanoparticles.

X-ray crystallography is helpful for studying a series of isomorphic crystals, that is, crystals with the same crystal structure but different lattice constants, such as the solid solution series $Ga_{1-x}In_xAs$ or $GaAs_{1-x}Sb_x$, where x can take on the range of values $0 \leq x \leq 1$. For these cubic crystals the lattice constant a will depend on x since indium (In) is larger than gallium (Ga), and antimony (Sb) is larger than arsenic (As), as the data in Table B.1 indicate. For this case Vegard's law, Eq. (2.8) of Section 2.1.4, is a good approximation for estimating the value of a if x is known, or the value of x if a is known.

3.2.3. Particle Size Determination

In the previous section we discussed determining the sizes of grains in polycrystalline materials via X-ray diffraction. These grains can range from nanoparticles with size distributions such as that sketched in Fig. 3.3 to much larger micrometer-sized particles, held together tightly to form the polycrystalline material. This is the bulk or clustered grain limit. The opposite limit is that of grains or nanoparticles dispersed in a matrix so that the distances between them are greater than their average diameters or dimensions. It is of interest to know how to measure the sizes, or ranges of sizes, of these dispersed particles.

The most straightforward way to determine the size of a micrometer-sized grain is to look at it in a microscope, and for nanosized particles a transmission electron microscope (TEM), to be discussed in Section 3.3.1, serves this purpose. Figure 3.6 shows a TEM micrograph of polyaniline particles with diameters close to 100 nm dispersed in a polymer matrix.

Another method for determining the sizes of particles is by measuring how they scatter light. The extent of the scattering depends on the relationship between the particle size d and the wavelength λ of the light, and it also depends on the polarization of the incident light beam. For example, the scattering of white light, which contains wavelengths in the range from 400 nm for blue to 750 nm for red, off the nitrogen and oxygen molecules in the atmosphere with respective sizes $d = 0.11$ and 0.12 nm, explains why the light reflected from the sky during the day appears blue, and that transmitted by the atmosphere at sunrise and sunset appears red.

Particle size determinations are made using a monochromatic (single-wavelength) laser beam scattered at a particular angle (usually $90°$) for parallel and perpendicular polarizations. The detected intensities can provide the particle size, the particle concentration, and the index of refraction. The Rayleigh–Gans theory is used to interpret the data for particles with sizes d less than 0.1λ, which corresponds to the case for nanoparticles measured by optical wavelengths. The example of a laser beam nanoparticle determination shown in Fig. 3.7 shows an organic solvent dispersion with sizes ranging from 9 to 30 nm, peaking at 12 nm. This method is applicable for use with nanoparticles that have diameters above 2 nm, and for smaller nanoparticles other methods must be used.

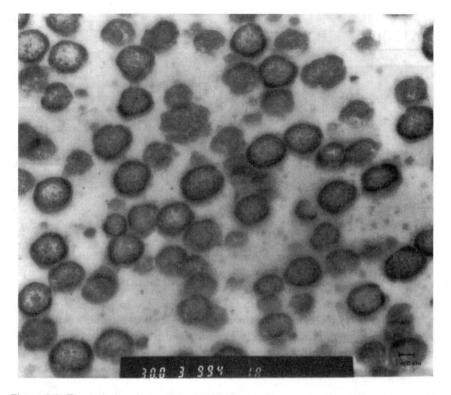

Figure 3.6. Transmission electron micrograph of polyaniline nanoparticles in a polymer matrix. (From B. Wessling, in *Handbook of Organic Conductive Molecules and Polymers*, H. S. Nalwa, ed., Wiley, New York, 1997, Vol. 3, pp. 497.)

The sizes of particles $< 2\,\mathrm{nm}$ can be conveniently determined by the method of mass spectrometry. The typical gas mass spectrometer sketched in Fig. 3.8 ionizes the nanoparticles to form positive ions by impact from electrons emitted by the heated filament (f) in the ionization chamber (I) of the ion source. The newly formed ions are accelerated through the potential drop in voltage V between the repeller (R) and accelerator (A) plates, then focused by lenses L, and collimated by slits S during their transit to the mass analyzer. The magnetic field B of the mass analyzer, oriented normal to the page, exerts the force $F = qvB$, which bends the ion beam through an angle of $90°$ at the radius r, after which they are detected at the ion collector. The mass m : charge q ratio is given by the expression

$$\frac{m}{q} = \frac{B^2 r^2}{2V} \tag{3.5}$$

The bending radius r is ordinarily fixed in a particular instrument, so either the magnetic field B or the accelerating voltage V can be scanned to focus the ions of

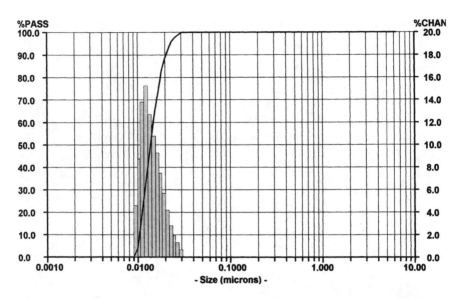

Figure 3.7. Laser light Doppler measurement of size distribution of conductive polymer particles dispersed in an organic solvent. The sizes range from 9 to 30 nm, with a peak at 12 nm. [From B. Wessling, in N. S. Nalwa (2000), Vol. 5, Chapter 10, p. 508.]

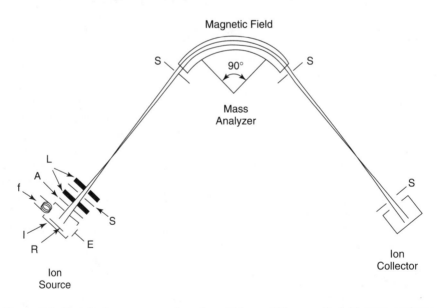

Figure 3.8. Sketch of a mass spectrometer utilizing a 90° magnetic field mass analyzer, showing details of the ion source: A—accelerator or extractor plate, E—electron trap, f—filament, I—ionization chamber, L—focusing lenses, R—repeller, S—slits. The magnetic field of the mass analyzer is perpendicular to the plane of the page.

various masses at the detector. The charge q of the nanosized ion is ordinarily known, so in practice it is the mass m that is determined. Generally the material forming the nanoparticle is known, so its density $\rho = m/V$ is also a known quantity, and its size or linear dimension d can be estimated or evaluated as the cube root of the volume: $d = (V)^{1/3} = (m/\rho)^{1/3}$.

The mass spectrometer that has been described made use of the typical magnetic field mass analyzer. Modern mass spectrometers generally employ other types of mass analyzers, such as the quadrupole model, or the time-of-flight type in which each ion acquires the same kinetic energy $\frac{1}{2}mv^2$ during its acceleration out of the ionization chamber, so the lighter mass ions move faster and arrive at the detector before the heavier ions, thereby providing a separation by mass.

Figure 3.9 gives an example of a time-of-flight mass spectrum obtained from soot produced by laser vaporization of a lanthanum–carbon target. The upper mass spectrum (a) of the figure, taken from the initial crude extract of the soot, shows lines from several fullerene molecules: C_{60}, C_{70}, C_{76}, C_{78}, C_{82}, C_{84}, and LaC_{82}. The latter corresponds to an endohedral fullerene, namely, C_{84} with a lanthanum atom inside the fullerene cage. The second (b) and third (c) mass spectra were obtained by successively separating LaC_{82} from the other fullerenes using a technique called *high-performance liquid chromatography*.

3.2.4. Surface Structure

To obtain crystallographic information about the surface layers of a material a technique called low-energy electron diffraction (LEED) can be employed because at low energies (10–100 eV) the electrons penetrate only very short distances into the surface, so their diffraction pattern reflects the atomic spacings in the surface layer. If the diffraction pattern arises from more than one surface layer, the contribution of lower-lying crystallographic planes will be weaker in intensity. The electron beam behaves like a wave and reflects from crystallographic planes in analogy with an X-ray beam, and its wavelength λ, called the *de Broglie wavelength*, depends on the energy E expressed in the units of electron volts through the expression

$$\lambda = \frac{1.226}{\sqrt{E}} \text{ nm} \tag{3.6}$$

which differs from Eq. (3.1) for X rays. Thus an electron energy of 25.2 eV gives a de Broglie wavelength λ equal to the Ga—As bond distance in gallium arsenide ($3^{1/2}a/4 = 0.2442$ nm), where the lattice constant $a = 0.565$ nm, so we see that low energies are adequate for crystallographic electron diffraction measurements. Another technique for determining surface layer lattice constants is reflection high-energy electron diffraction (RHEED) carried out at grazing incidence angles where the surface penetration is minimal. When θ is small in the Bragg expression (3.2), then λ must be small, so the energy E of Eq. (3.6) must be large, hence the need for higher energies for applying RHEED diffraction at grazing incidence.

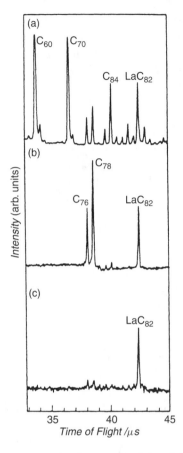

Figure 3.9. Time-of-flight mass spectra from soot produced by laser vaporization of a lanthanum–carbon target, showing the presence of the fullerene molecules C_{60}, C_{70}, C_{76}, C_{78}, C_{82}, C_{84}, and LaC_{84}. The spectra correspond, successively, to (a) the initial crude soot extract, (b) a fraction after passage through a chromatographic column, and (c) a second fraction after passage of the first fraction through another column to isolate and concentrate the endohedral compound LaC_{82}. [From K. Kikuchi, S. Suzuki, Y. Nakao, N. Yakahara, T. Wakabayashi, I. Ikemoto, and Y Achiba, *Chem. Phys. Lett.* **216**, 67 (1993).]

3.3. MICROSCOPY

3.3.1. Transmission Electron Microscopy

Electron beams not only are capable of providing crystallographic information about nanoparticle surfaces but also can be used to produce images of the surface, and they play this role in electron microscopes. We will discuss several ways to use electron beams for the purpose of imaging, using several types of electron microscopes.

In a transmission electron microscope (TEM) the electrons from a source such as an electron gun enter the sample, are scattered as they pass through it, are focused by

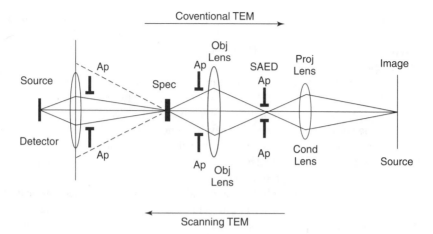

Figure 3.10. Ray diagram of a conventional transmission electron microscope (top path) and of a scanning transmission electron microscope (bottom path). The selected area electron diffraction (SAED) aperture (Ap) and the sample or specimen (Spec) are indicated, as well as the objective (Obj) and projector (Proj) or condenser (Cond) lenses. (Adapted from P. R. Buseck, J. M. Cowley, and L. Eyring, *High-Resolution Transmission Electron Microscopy*, Oxford Univ. Press, New York, 1988, p. 6.)

an objective lens, are amplified by a magnifying (projector) lens, and finally produce the desired image, in the manner reading from left to right (CTEM direction) in Fig. 3.10. The wavelength of the electrons in the incident beam is given by a modified form of Eq. (3.6)

$$\lambda = \frac{0.0388}{\sqrt{V}} \text{ nm} \qquad (3.7)$$

where the energy acquired by the electrons is $E = eV$ and V is the accelerating voltage expressed in kilovolts. If widely separated heavy atoms are present, they can dominate the scattering, with average scattering angles θ given by the expression $\theta \sim \lambda/d$, where d is the average atomic diameter. For an accelerating voltage of 100 kV and an average atomic diameter of 0.15 nm, we obtain $\theta \sim 0.026$ radians or 1.5°. Images are formed because different atoms interact with and absorb electrons to a different extent. When individual atoms of heavy elements are farther apart than several lattice parameters, they can sometimes be resolved by the TEM technique.

Electrons interact much more strongly with matter than do X rays or neutrons with comparable energies or wavelengths. For ordinary elastic scattering of 100-keV electrons the average distance traversed by electrons between scattering events, called the *mean free path*, varies from a few dozen nanometers for light elements to tens or perhaps hundreds of nanometers for heavy elements. The best results are obtained in electron microscopy by using film thicknesses that are comparable with the mean free path. Much thinner films exhibit too little scattering to provide useful images, and in thick films multiple scattering events dominate, making the image blurred and difficult to interpret. Thick specimens can be studied by detecting back-scattered electrons.

A transmission electron microscope can form images by the use of the selected-area electron diffraction (SAED) aperture located between the objective and projector lenses shown in Fig. 3.10. The main part of the electron beam transmitted by the sample consists of electrons that have not undergone any scattering. The beam also contains electrons that have lost energy through inelastic scattering with no deviation of their paths, and electrons that have been reflected by various *hkl* crystallographic planes. To produce what is called a *bright-field image*, the aperture is inserted so that it allows only the main undeviated transmitted electron beam to pass, as shown in Fig. 3.11. The bright-field image is observed at the detector or viewing screen. If the aperture is positioned to select only one of the beams reflected from a particular *hkl* plane, the result is the generation of a dark-field image at the viewing screen. The details of the dark-field image that is formed can depend on the particular diffracted beam (particular *hkl* plane) that is selected for the imaging. Figure 3.11 shows the locations of the bright-field (BF) and dark-field (DF) aperture positions. To illustrate this imaging technique we present in Fig. 3.12 images of an iron base superalloy with a FCC austenite structure containing 2–3-nm γ' preci-

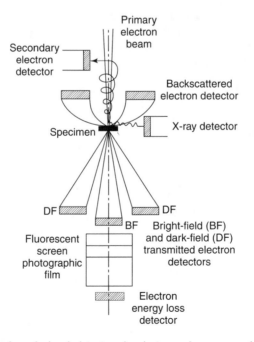

Figure 3.11. Positioning of signal detectors in electron microscope column. (From D. B. Williams, *Practical Analytical Electron Microscopy in Materials Science*, Phillips Electronic Instruments, Mahwah NJ, 1984.)

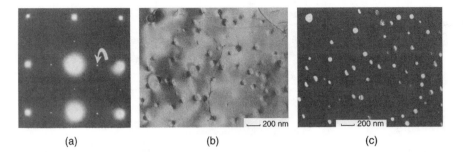

(a)   (b)   (c)

Figure 3.12. Transmission electron microscope figures from 2–3-nm-diameter γ' precipitates of $Ni_2(Ti, Al)$ in iron-base superalloy, showing (a) [100] FCC zone diffraction pattern displaying large bright spots from the superalloy and weak spots from the γ' precipitates, (b) bright-field image showing 25-nm elastic strain fields around marginally visible γ' phase particles, and (c) dark-field image obtained using an SAED aperture to pass the γ' diffraction spot marked with the arrow in (a). Image (c) displays the γ' precipitate particles. [From T. J. Headley, cited in A. D. Romig, Jr., chapter in R. E. Whan (1986), p. 442.]

pitates of $Ni_3(Ti, Al)$ with a FCC structure. The diffraction pattern in Fig. 3.12a obtained without the aid of the filter exhibits large, bright spots from the superalloy, and very small, dim spots from the γ' nanoparticles. In the bright-field image displayed in Fig. 3.12b the γ' particles are barely visible, but the 25-nm-diameter elastic strain fields generated by them are clearly seen. If the γ' diffraction spot beam indicated by the arrow in Fig. 3.12a is selected for passage by the SAED aperture, the resulting dark-field image presented in Fig. 3.12c shows very clearly the positions of the γ' precipitates.

A technique called image processing can be used to increase the information obtainable from a TEM image, and enhance some features that are close to the noise level. If the image is Fourier-transformed by a highly efficient technique called a *fast Fourier transform*, then it provides information similar to that in the direct diffraction pattern. An example of the advantages of image processing is given by the sequence of images presented in Fig. 3.13 for a Ni nanoparticle supported on a SiO_2 substrate. Figure 3.13a shows the original image, and Fig. 3.13b presents the fast Fourier transform, which has the appearance of a diffraction pattern. Figures 3.13c–3.13e illustrate successive steps in the image processing, and Fig. 3.13f is an image of the SiO_2 substrate obtained by subtracting the particle image. Finally Fig. 3.13g presents the nanoparticle reconstruction from the processed data.

In addition to the directly transmitted and the diffracted electrons, there are other electrons in the beam that undergo inelastic scattering and lose energy by creating excitations in the specimen. This can occur by inducing vibrational motions in the atoms near their path, and thereby creating phonons or quantized lattice vibrations that will propagate through the crystal. If the sample is a metal, then the incoming electron can scatter inelastically by producing a plasmon, which is a collective excitation of the free-electron gas in the conduction band. A third very important source of inelastic scattering occurs when the incoming electron induces a single-

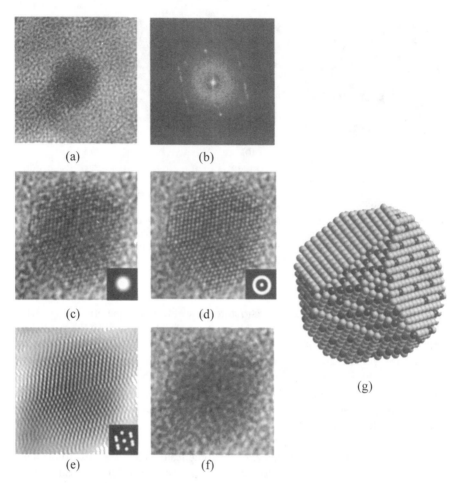

Figure 3.13. Transmission electron microscope image processing for a Ni particle on a SiO_2, substrate, showing (a) original bright-field image, (b) fast Fourier transform diffraction-pattern-type image, (c) processed image with aperture filter shown in inset, (d) image after further processing with aperture filter in the inset, (e) final processed image, (f) image of SiO_2 substrate obtained by subtracting out the particle image, and (g) model of nanoparticle constructed from the processed data. [From Benaissa and Diaz, cited by M. José-Yacamán and J. A. Ascencio, in Nalwa (2000), Vol. 2, Chapter 8, p. 405.]

electron excitation in an atom. This might involve inner core atomic levels of atoms, such as inducing a transition from a K level ($n = 1$) or L level ($n = 2$) of the atom to a higher energy that might be a discrete atomic level with a larger quantum number n, an electron band, or result in total removal (ionization) from the solid. Lower levels of energy loss occur when the excited electron is in the valence band of a semiconductor. This excitation can decay via the return of excited electrons to their ground states, thereby producing secondary radiation, and the nature of the secondary radiation can often give useful information about the sample. These types of transitions are utilized in various branches of electron spectroscopy. They can be surface-sensitive because of the short penetration distance of the electrons into the material.

3.3.2. Field Ion Microscopy

Another instrumental technique in which the resolution approaches interatomic dimensions is field ion microscopy. In a field ion microscope a wire with a fine tip located in a high-vacuum chamber is given a positive charge. The electric field and electric field gradient in the neighborhood of the tip are both quite high, and residual gas molecules that come close to the tip become ionized by them, transferring electrons to the tip, thereby acquiring a positive charge. These gaseous cations are repelled by the tip and move directly outward toward a photographic plate where, on impact, they create spots. Each spot on the plate corresponds to an atom on the tip, so the distribution of dots on the photographic plate represents a highly enlarged image of the distribution of atoms on the tip. Figure 3.14 presents a field ion micrograph from a tungsten tip, and Fig. 3.15 provides a stereographic projection of a cubic crystal with the orientation corresponding to the micrograph of Fig. 3.14. The *International Tables for Crystallography*, edited by T. Hahn (Hahn 1996), provide stereographic projections for various point groups and crystal classes.

3.3.3. Scanning Microscopy

An efficient way to obtain images of the surface of a specimen is to scan the surface with an electron beam in a raster pattern, similar to the way an electron gun scans the

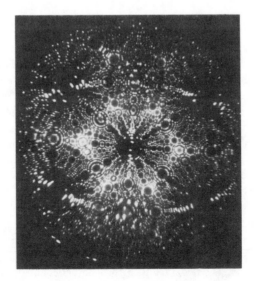

Figure 3.14. Field ion micrograph of a tungsten tip (T. J. Godfrey), explained by the stereographic projection of Fig. 3.15. [From G. D. W. Smith, chapter in Whan (1986), p. 585.]

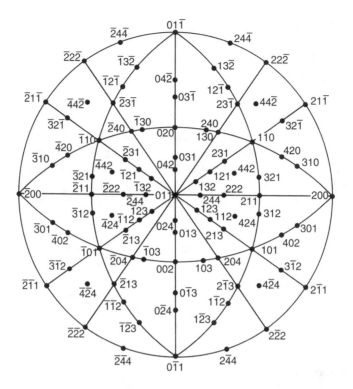

Figure 3.15. Stereographic [011] projection of a cubic crystal corresponding to the field ion tungsten micrograph in Fig. 3.14. [From G. D. W. Smith, chapter in Whan (1986), p. 583.]

screen in a television set. This is a systematic type scan. Surface information can also be obtained by a scanning probe in which the trajectory of the electron beam is directed across the regions of particular interest on the surface. The scanning can also be carried out by a probe that monitors electrons that tunnel between the surface and the probe tip, or by a probe that monitors the force exerted between the surface and the tip. We shall describe in turn the instrumentation systems that carry out these three respective functions: the scanning transmission electron microscope (SEM), the scanning tunneling microscope (STM), and the atomic force microscope (AFM).

It was mentioned above that the electron optics sketched in Fig. 3.10 for a conventional transmission electron microscope is similar to that of a scanning electron microscope, except that in the former TEM case the electrons travel from left to right, and in the latter SEM they move in the opposite direction, from right to left, in the figure. Since quite a bit has been said about the workings of an electron microscope, we describe only the electron deflection system of a scanning electron microscope, which is sketched in Fig. 3.16. The deflection is done magnetically through magnetic fields generated by electric currents flowing through coils, as

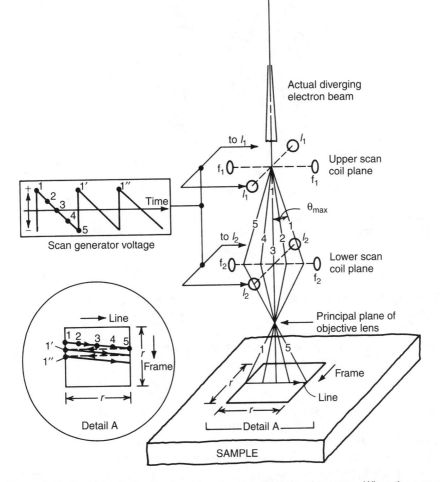

Figure 3.16. Double-deflection system of a scanning electron microscope. When the upper scan coils l_1–l_1 deflect the beam through an angle θ, the lower scan coils l_2–l_2 deflect it back through the angle 2θ, and the electrons sequentially strike the sample along the indicated line. The upper inset on the left shows the sawtooth voltage that controls the current through the l_i scanning coils. The lower inset on the left shows the sequence of points on the sample for various electron trajectories 1, 2, 3, 4, 5 downward along the microscope axis. Scan coils f_1–f_1 and f_2–f_2 provide the beam deflection for the sequence of points 1, 1′, 1″ shown in detail A. [From J. D. Verhoeven, chapter in Whan (1986), p, 493.]

occurs in many television sets. The magnetic field produced by a coil is proportional to the voltage V applied to the coil. We see from the upper inset on the left side of Fig. 3.16 that a sawtooth voltage is applied to the pairs of coils I_1,I_1 and I_2,I_2. The magnetic field produced by the coils exerts a force that deflects the electron beam from left to right along the direction of the line drawn at the bottom on the sample.

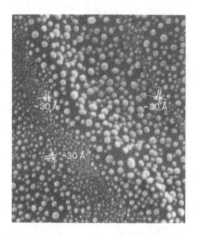

Figure 3.17. Micrograph of 3-nm-diameter (30-Å) gold particles on a carbon substrate taken with a scanning electron microscope. [From J. D. Verhoeven, chapter in Whan (1986), p. 497.]

The varying magnetic fields in the coil pairs f_1, f_1 and f_2, f_2 produce the smaller deflections from points 1 to 1′ to 1″ shown in the detail A inset. Thus the electron beam scans repeatedly from left to right across the sample in a raster pattern that eventually covers the entire $r \times r$ frame area on the sample. Figure 3.17 shows 3-nm gold particles on a carbon substrate resolved by an SEM.

A scanning tunneling microscope utilizes a wire with a very fine point. This fine point is positively charged and acts as a probe when it is lowered to a distance of about 1 nm above the surface under study. Electrons at individual surface atoms are attracted to the positive charge of the probe wire and jump (tunnel) up to it, thereby generating a weak electric current. The probe wire is scanned back and forth across the surface in a raster pattern, in either a constant-height mode, or a constant-current mode, in the arrangements sketched in Fig. 3.18. In the constant current mode a feedback loop maintains a constant probe height above the sample surface profile, and the up/down probe variations are recorded. This mode of operation assumes a constant tunneling barrier across the surface. In the constant-probe-height mode the tip is constantly changing its distance from the surface, and this is reflected in variations of the recorded tunneling current as the probe scans. The feedback loop establishes the initial probe height, and is then turned off during the scan. The scanning probe provides a mapping of the distribution of the atoms on the surface.

The STM often employs a piezoelectric tripod scanner, and an early design of this scanner, built by Binning and Rohrer, is depicted in Fig. 3.19. A piezoelectric is a material in which an applied voltage elicits a mechanical response, and the reverse. Applied voltages induce piezo transducers to move the scanning probe (or the sample) in nanometer increments along the x, y or z directions indicated on the arms (3) of the scanner in the figure. The initial setting is accomplished with the aid of a stepper motor and micrometer screws. The tunneling current, which varies expo-

STM - Constant Height Mode (Feedback Off)

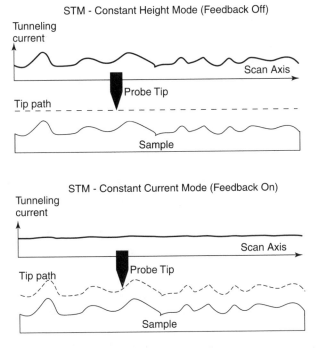

STM - Constant Current Mode (Feedback On)

Figure 3.18. Constant-height (top sketch) and constant-current (bottom sketch) imaging mode of a scanning tunneling microscope. [From T. Bayburt, J. Carlson, B. Godfrey, M. Shank-Retzlaff, and S. G. Sligar, in Nalwa (2000), Vol. 5, Chapter 12, p. 641.]

nentially with the probe–surface atom separation, depends on the nature of the probe tip and the composition of the sample surface. From a quantum-mechanical point of view, the current depends on the dangling bond state of the tip apex atom and on the orbital states of the surface atoms.

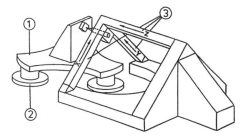

Figure 3.19. Scanning mechanism for scanning tunneling microscope, showing (1) the piezo-electric baseplate, (2) the three feet of the baseplate and (3) the piezoelectric tripod scanner holding the probe tip that points toward the sample. (From R. Wiesendanger, *Scanning Probe Microscopy and Spectroscopy*, Cambridge Univ. Press, Cambridge, UK, 1994, p. 81.)

The third technique in wide use for nanostructure surface studies is atomic force microscopy, and Fig. 3.20 presents a diagram of a typical atomic force microscope (AFM). The fundamental difference between the STM and the AFM is that the former monitors the electric tunneling current between the surface and the probe tip

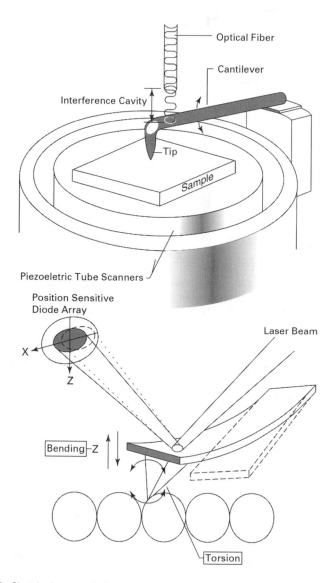

Figure 3.20. Sketch of an atomic force microscope (AFM) showing the cantilever arm provided with a probe tip that traverses the sample surface through the action of the piezoelectric scanner. The upper figure shows an interference deflection sensor, and the lower enlarged view of the cantilever and tip is provided with a laser beam deflection sensor. The sensors monitor the probe tip elevations upward from the surface during the scan.

and the latter monitors the force exerted between the surface and the probe tip. The AFM, like the STM, has two modes of operation. The AFM can operate in a close contact mode in which the core-to-core repulsive forces with the surface dominate, or in a greater separation "noncontact" mode in which the relevant force is the gradient of the van der Waals potential. As in the STM case, a piezoelectric scanner is used. The vertical motions of the tip during the scanning may be monitored by the interference pattern of a light beam from an optical fiber, as shown in the upper diagram of the figure, or by the reflection of a laser beam, as shown in the enlarged view of the probe tip in the lower diagram of the figure. The atomic force microscope is sensitive to the vertical component of the surface forces. A related but more versatile device called a *friction force microscope*, also sometimes referred to as a *lateral force microscope*, simultaneously measures both normal and lateral forces of the surface on the tip.

All three of these scanning microscopes can provide information on the topography and defect structure of a surface over distances close to the atomic scale. Figure 3.21 shows a three-dimensional rendering of an AFM image of chromium deposited on a surface of SiO_2. The surface was prepared by the laser-focused deposition of atomic chromium in the presence of a Gaussian standing wave that reproduced the observed regular array of peaks and valleys on the surface. When the laser-focused chromium deposition was carried out in the presence of two plane waves displaced by $90°$ relative to each other, the two-dimensional arrangement AFM image shown in Fig. 3.22 was obtained. Note that the separation between the peaks, 212.78 nm, is the same in both images. The peak heights are higher (13 nm) in the two-dimensional array (8 nm) than in the linear one.

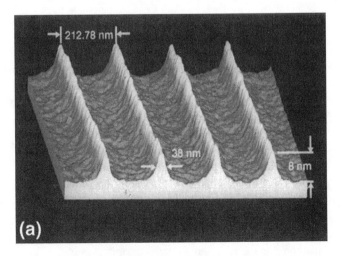

Figure 3.21. Three-dimensional rendering of an AFM image of nanostructure formed by laser focused atomic Cr deposition in a Gaussian standing wave on an SiO_2 surface. [From J. J. McClelland, R. Gupta, Z. J. Jabbour, and R. L. Celotta, *Aust J. Phys.* **49**, 555 (1996).]

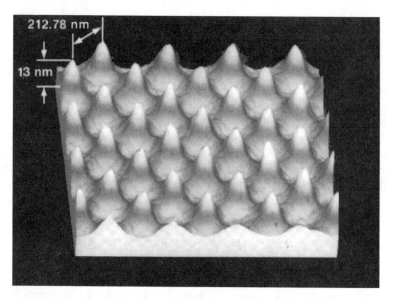

Figure 3.22. AFM image of nanostructure array formed when the laser focused Cr deposition is carried out in two standing waves oriented at 90° relative to each other. [From R. Gupta, J. J. McClelland, Z. J. Jabbour, and R. L. Celotta, *Appl. Phys. Lett.* **67**, 1378 (1995).]

3.4. SPECTROSCOPY

3.4.1. Infrared and Raman Spectroscopy

Vibrational spectroscopy involves photons that induce transitions between vibrational states in molecules and solids, typically in the infrared (IR) frequency range from 2 to 12×10^{13} Hz. Section 2.1.5 discusses the normal modes of vibration of molecules and solids. The energy gaps of many semiconductors are in this same frequency region, and can be studied by infrared techniques.

In infrared spectroscopy an IR photon $h\nu$ is absorbed directly to induce a transition between two vibrational levels E_n and $E_{n'}$, where

$$E_n = \left(n + \tfrac{1}{2}\right)h\nu_0 \qquad (3.8)$$

The vibrational quantum number $n = 0, 1, 2, \ldots$ is a positive integer, and ν_0 is the characteristic frequency for a particular normal mode. In accordance with the selection rule $\Delta n = \pm 1$, infrared transitions are observed only between adjacent vibrational energy levels, and hence have the frequency ν_0. In Raman spectroscopy a vibrational transition is induced when an incident optical photon of frequency $h\nu_{\text{inc}}$ is absorbed and another optical photon $h\nu_{\text{emit}}$ is emitted:

$$E_n = |h\nu_{\text{inc}} - h\nu_{\text{emit}}| \qquad (3.9)$$

From Eq. (3.8) the frequency difference is given by $|v_{inc} - v_{emit}| = |n' - n''|v_0 = v_0$ since the same infrared selection rule $\Delta n = \pm 1$ is obeyed. Two cases are observed: (1) $v_{inc} > v_{emit}$ corresponding to a *Stokes line*, and (2) $v_{inc} < v_{emit}$ for an *anti-Stokes line*. Infrared active vibrational modes arise from a change in the electric dipole moment μ of the molecule, while Raman-active vibrational modes involve a change in the polarizability $P = \mu_{ind}/E$, where the electric vector E of the incident light induces the dipole moment μ_{ind} in the sample. Thus some vibrational modes are IR-active, that is, measurable by infrared spectroscopy, and some are Raman-active.

Infrared and optical spectroscopy is often carried out by reflection, and the measurements of nanostructures provide the reflectance (or reflectivity) R, which is the fraction of reflected light. For normal incidence we have

$$R = \frac{I_r}{I_0} = \frac{|\sqrt{\varepsilon} - 1|}{|\sqrt{\varepsilon} + 1|} \tag{3.10}$$

where ε is the dimensionless dielectric constant of the material. The dielectric constant $\varepsilon(v)$ has real and imaginary parts, $\varepsilon = \varepsilon'(v) + \varepsilon''(v)$, where the real or dispersion part ε' provides the frequencies of the IR bands, and the imaginary part ε'' measures the energy absorption or loss. A technique called *Kramers–Kronig analysis* is used to extract the frequency dependences of $\varepsilon'(v)$ and $\varepsilon''(v)$ from measured IR reflection spectra.

The classical way to carry out infrared spectroscopy is to scan the frequency of the incoming light to enable the detector to record changes in the light intensity for those frequencies at which the sample absorbs energy. A major disadvantage of this method is that the detector records meaningful information only while the scan is passing through absorption lines, while most of the time is spent scanning between lines when the detector has nothing to record. To overcome this deficiency, modern infrared spectrometers irradiate the sample with a broad band of frequencies simultaneously, and then carry out a mathematical analysis of the resulting signal called a *Fourier transformation* to convert the detected signal back into the classical form of the spectrum. The resulting signal is called a Fourier transform infrared (FTIR) spectrum. The Fourier transform technique is also widely used in nuclear magnetic resonance, discussed below, and in other branches of spectroscopy.

Figure 3.23 presents an FTIR spectrum of silicon nitride (Si_3N_4) nanopowder showing the vibrational absorption lines corresponding to the presence of hydroxyl Si—OH, amino Si—NH_2, and imido Si—NH—Si groups on the surface. Figure 3.24 shows a similar FTIR spectrum of silicon carbonitride nanopowder, (SiCN), revealing the presence of several chemical species on the surface after activation at 873 K, and their removal by heating for an hour under dry oxygen at 773 K.

Figure 3.25 shows how the width of the Raman spectral lines of germanium nanocrystals embedded in SiO_2 thin films exhibit a pronounced broadening when the particle size decreases below about 20 nm. Figure 3.26 shows a Raman spectrum recorded for germanium nanocrystals that arises from a distribution of particle sizes around an average value of 65 nm. These nanocrystals were prepared by chemical reduction, precipitation, and subsequent annealing of the phase $Si_xGe_yO_z$, and the

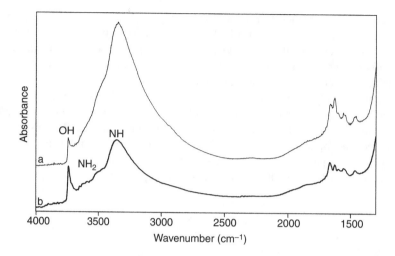

Figure 3.23. Fourier transform infrared spectrum of silicon nitride nanopowder recorded at room temperature under vacuum (tracing a) and after activation at 773 K (tracing b). (From G. Ramis, G. Busca, V. Lorenzelli, M.-I. Baraton, T. Merle-Méjean, and P. Quintard, in *Surfaces and Interfaces of Ceramic Materials*. L. C. Dufour et al., eds., Kluwer Academic, Dordrecht, 1989, p. 173.)

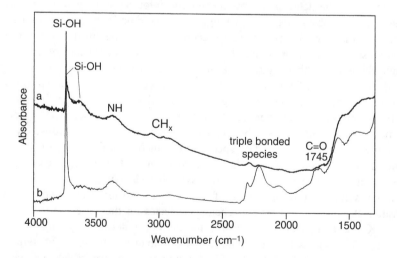

Figure 3.24. Fourier transform infrared spectrum of silicon carbonitride nanopowder recorded after activation at 873 K (tracing a) and after subsequent heating in dry oxygen at 773 K for one hour (tracing b). [From M.-I. Baraton, in Nalwa (2000), Vol. 2, Chapter 2, p. 131; see also M.-I. Baraton, W. Chang, and B. H. Kear, *J. Phys. Chem.* **100**, 16647 (1996).]

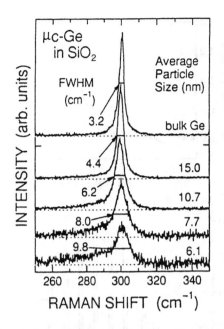

Figure 3.25. Dependence of the Raman spectra of germnanium microcrystals (μ_C-Ge) embedded in SiO_2 thin films on the crystallite size. The average particle size and the full width at half-maximum height (FWHM) of the Raman line of each sample are indicated. [From M. Fujii, S. Hayashi, and K. Yamamoto, *Jpn. J. Appl. Phys.* **30**, 657 (1991).]

histogram shown on the figure was determined from a scanning electron micrograph. Further Raman spectroscopy studies demonstrated that larger average particle sizes are obtained when the annealing is carried out for longer times and at higher temperatures.

We have been discussing what is traditionally known as *Raman scattering of light* or *Raman spectroscopy*. This is spectroscopy in which the phonon vibration of Eq. (3.8), corresponding to the energy difference of Eq. (3.9), is an optical phonon of the type discussed in the previous section, namely, a phonon with a frequency of vibration in the infrared region of the spectrum, corresponding to about $\cong 400\,\mathrm{cm}^{-1}$, or a frequency of $\cong 1.2 \times 10^{13}\,\mathrm{Hz}$. When a low-frequency acoustic phonon is involved in the scattering expressed in Eq. (3.9), then the process is referred to as *Brillouin scattering*. Acoustic phonons can have frequencies of vibration or energies that are a factor of 1000 less than those of optical phonons. Typical values are $\cong 1.5 \times 10^{10}\,\mathrm{Hz}$ or $\cong 0.5\,\mathrm{cm}^{-1}$. Brillouin spectroscopy, which involves both Stokes and anti-Stokes lines, as does Raman spectroscopy, is discussed in Chapter 8.

Chapter 8 is devoted to the infrared and optical spectroscopy of nanomaterials, so in this chapter we will restrict ourselves to commenting on the representative optical spectra presented in Fig. 3.27, which were obtained from colloidal cadmium selenide semiconductor nanocrystals that are transparent for photon energies below the bandgap, and that absorb light above the gap. Colloidal nanocrystal synthesis

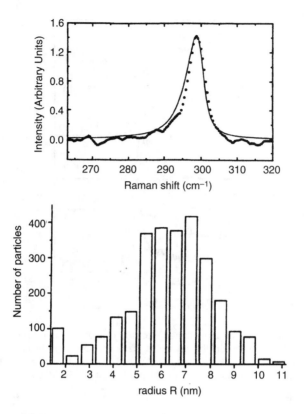

Figure 3.26. Histogram of the distribution of Ge nanocrystal particle sizes (lower figure) around the mean radius 6.5 nm, used to simulate (solid line) the Raman spectrum (dotted line) depicted at the top. The size distribution was obtained by transmission electron microscopy. [From C. E. Bottani, C. Mantini, P. Milani, M. Manfredini, A. Stella, P. Tognini, P. Cheyssac, and R. Kofman, *Appl. Phys. Lett.* **69**, 2409 (1996).]

provides a narrow range of size distributions, so optical spectra of the type shown in Fig. 3.27 can accurately determine the dependence of the gap on the average particle radius and this dependence is shown in Fig. 3.28 for CdSe. We see that the energy gap data for the colloidal samples plotted on this figure agree with the gaps evaluated from optical spectra of CdSe nanoparticle glass samples. The solid line in the figure is a theoretical plot made using a parabolic band model for the vibrational potential, and the measured energy is asymptotic to the bulk value for large particle sizes. The increase in the gap for small nanoparticle radii is due to quantum confinement effects.

3.4.2. Photoemission and X-Ray Spectroscopy

Photoemission spectroscopy (PES) measures the energy distribution of electrons emitted by atoms and molecules in various charge and energy states. A material

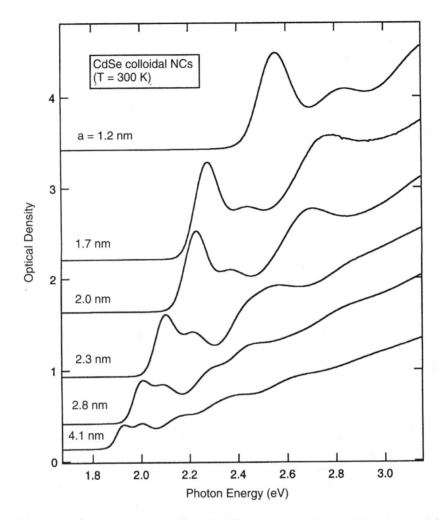

Figure 3.27. Room-temperature optical absorption spectra for CdSe colloidal nanocrystals (NCs) with mean radii in the range from 1.2 to 4.1 nm. [From V. I. Klimov, in Nalwa (2000), Vol. 4, Chapter 7, p. 464.]

irradiated with ultraviolet light (UPS) or X rays (XPS) can emit electrons called *photoelectrons* from atomic energy levels with a kinetic energy equal to the difference between the photon energy $h\nu_{ph}$ and the ionization energy E_{ion}, which is the energy required to completely remove an electron from an atomic energy level

$$KE = h\nu_{ph} - E_{ion} \qquad (3.11)$$

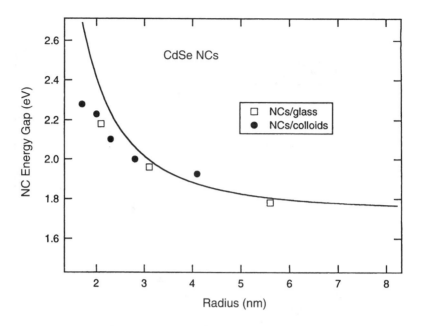

Figure 3.28. Dependence of the semiconductor energy gap of CdSe nanocrystals (Ncs), in the form of a glass (□) and a colloidal suspension (), on the mean particle radius. The colloidal nanocrystal spectral data are presented in Fig. 3.27. The solid curve gives the fit to the data using a parabolic band model for the vibrational potential. [From V. I. Klimov in Nalwa (2000), Vol. 4, Chapter 7, p. 465.]

The photoelectron spectrometer sketched in Fig. 3.29 shows, at the lower left side, X-ray photons incident on a specimen. The emitted photoelectrons then pass through a velocity analyzer that allows electrons within only a very narrow range of velocities to move along trajectories that take them from the entrance slit on the left, and allows them to pass through the exit slit on the right, and impinge on the detector. The detector measures the number of emitted electrons having a given kinetic energy, and this number is appreciable for kinetic energies for which Eq. (3.11) is satisfied.

The energy states of atoms or molecular ions in the valence band region have characteristic ionization energies that reflect perturbations by the surrounding lattice environment, so this environment is probed by the measurement. Related spectroscopic techniques such as inverse photoelectron spectroscopy (IPS), Bremsstrahlung isochromat spectroscopy (BIS), electron energy-loss spectroscopy (EELS), and Auger electron spectroscopy provide similar information.

As an example of the usefulness of X-ray photoemission spectroscopy, the ratio of Ga to N in a GaN sample was determined by measuring the gallium $3d$ XPS peak at 1.1185 keV and the nitrogen $1s$ peak at 0.3975 keV, and the result gave the average composition $Ga_{0.95}N$. An XPS study of 10-nm InP provided the indium $3d_{5/2}$ asymmetric line shown in Fig. 3.30a, and this was analyzed to reveal the

X-RAY PHOTOELECTRON SPECTROMETER

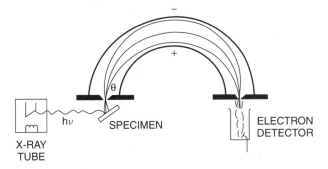

Figure 3.29. X-ray photoelectron spectrometer showing photons $h\nu$ generated by an X-ray tube incident on the specimen where they produce photoelectrons e⁻ characteristic of the specimen material, which then traverse a velocity analyzer, and are brought to focus at an electron detector that measures their kinetic energy.

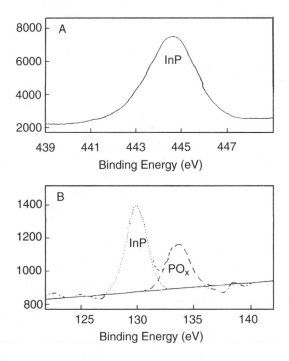

Figure 3.30. X-ray photoelectron spectra (XPS) of nanocrystalline InP showing (a) $3d_{5/2}$ indium signal and (b) $2p$ phosphorus peaks. [From Q. Yitai, in Nalwa (2000), Vol. 1, Chapter 9, p. 427.]

presence of two superposed lines; a main one at 444.6 eV arising from In in InP, and a weaker one at 442.7 attributed to In in the oxide In_2O_3. The phosphorus $2p$ XPS spectrum of Fig. 3.30b exhibits two well-resolved phosphorus peaks, one from InP and one from a phosphorus-oxygen species.

Inner electron or core-level transitions from levels n_1 to n_2 have frequencies v approximated by the well-known Rydberg formula, Eq. (2.16)

$$hv = \frac{me^4 Z^2}{32\pi^2 \varepsilon_0^2 \hbar^2} \left(\frac{1}{n_1^2} - \frac{1}{n_2^2} \right) \tag{3.12}$$

where Z is the atomic number and the other symbols have their usual meanings. The atomic number dependence of the frequency v of the K_α line, the innermost X-ray transition from $n_1 = 1$ to $n_2 = 2$

$$\sqrt{v} = a_K(Z - 1) \tag{3.13}$$

is called *Moseley's law.* The factor $(Z - 1)$ appears in Eq. (3.13) instead of Z to take into account the shielding of the nucleus by the remaining $n_1 = 1$ electron, which lowers its apparent charge to $(Z - 1)$. A similar expression applies to the next-highest-frequency L_α line, which has $n_1 = 2$ and $n_2 = 3$. Figure 3.31 presents a plot of $\sqrt{v}$ versus the atomic number Z for the experimentally measured K_α and L_α lines of the elements in the periodic table from $Z \cong 15$ to $Z = 60$. Measurements based on Moseley's law can provide information on the atom content of nano-material for all except the lightest elements. The ionization energies of the outer electrons of atoms are more dependent on the number of electrons outside of closed shells than they are on the atomic number, as shown by the data in Fig. 3.32. These ionization energies are in the visible or near-ultraviolet region.

An energetic photon is capable of removing electrons from all occupied atomic energy levels that have ionization energies less than the incoming energy. When the photon energy drops below the largest ionization energy corresponding to the K level, then the $n = 1$ electron can no longer be removed, and the X-ray absorption coefficient abruptly drops. It does not, however, drop to zero because the incoming energy is still sufficient to raise the $n = 1$ electron to a higher unoccupied level, such as a $3d$ or a $4p$ level, or to knock out electrons in the L ($n = 2$), M ($n = 3$), and other levels. The abrupt drop in absorption coefficient is referred to as an absorption edge; in this case it is a K-absorption edge. It is clear from the relative spacings of the energy levels of Fig. 3.33 that transitions of this type are close in energy to the ionization energy, and they provide what is called "fine structure on the absorption edge." They give information on the bonding states of the atom in question. The resolution of individual fine-structure transitions can be improved with the use of polarized X-ray beams. Several related X-ray absorption spectroscopy techniques are available for resolving fine structure.

Another way to obtain information on absorption edges is via electron energy-loss spectroscopy (EELS). This involves irradiating a thin film with a beam of

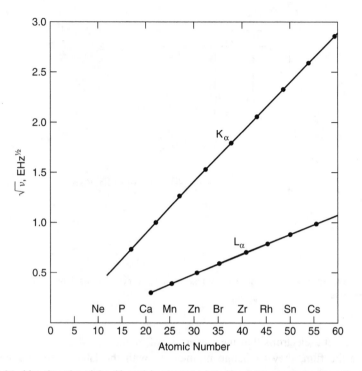

Figure 3.31. Moseley plot of the K_α and L_α characteristic X-ray frequencies versus the atomic number Z. (From C. P. Poole, Jr, H. A. Farach, and R. J. Creswick, *Superconductivity*, Academic Press, Boston, 1995, p. 515.)

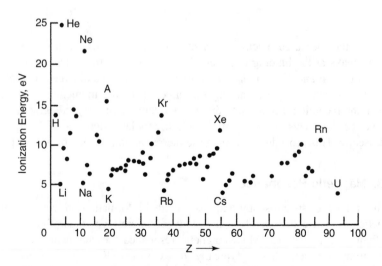

Figure 3.32. Experimentally determined ionization energy of the outer electron in various elements. (From R. Eisberg and R. Resnick, *Quantum Physics*, Wiley, New York, 1994, p. 364.)

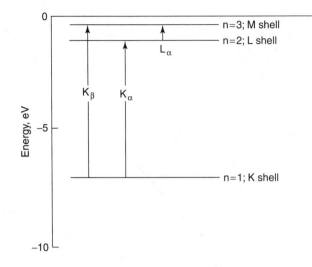

Figure 3.33. Energy-level diagram of molybdenum showing the transitions of the K and L series of X-ray lines.

monoenergetic electrons that have energies of, perhaps, 170 keV. As the electrons traverse the film, they exchange momentum with the lattice and lose energy by exciting or ionizing atoms, and an electron energy analyzer is employed to measure the amount of energy E_{abs} that is absorbed. This energy corresponds to a transition of the type indicated in Fig. 3.33, and is equal to the difference between the kinetic energy KE_0 of the incident electrons and that KE_{SC} of the scattered electrons

$$E_{abs} = KE_0 - KE_{SC} \qquad (3.14)$$

A plot of the measured electron intensity as a function of the absorbed energy contains peaks at the binding energies of the various electrons in the sample. The analog of optical and X-ray polarization experiments can be obtained with electron energy-loss spectroscopy by varying the direction of the momentum transfer $\Delta \mathbf{p}$ between the incoming electron and the lattice relative to the crystallographic c axis. This vector $\Delta \mathbf{p}$ plays the role of the electric polarization vector $\mathbf{E}$ in photon spectroscopy. This procedure can increase the resolution of the absorption peaks.

3.4.3. Magnetic Resonance

Another branch of spectroscopy that has provided information on nanostructures is magnetic resonance that involves the study of microwave (radar frequency) and radiofrequency transitions. Most magnetic resonance measurements are made in fairly strong magnetic fields, typically $B \approx 0.33\,\mathrm{T}$ (3300 Gs) for electron spin resonance (ESR), and $B \approx 10\,\mathrm{T}$ for nuclear magnetic resonance (NMR). Several types of magnetic resonance are mentioned below.

NMR involves the interaction of a nucleus possessing a nonzero nuclear spin I with an applied magnetic field B_{app} to give the energy-level splitting into $2I + 1$ lines with the energies

$$E_m = \hbar \gamma B_{app} m \tag{3.15}$$

where γ is the gyromagnetic ratio, sometimes called the *magnetogyric ratio*, characteristic of the nucleus, and m assumes integer or half-integer values in the range $-I < m < +I$ depending on whether I is an integer or a half-integer. The value of γ is sensitive to the local chemical environment of the nucleus, and it is customary to report the chemical shift δ of γ relative to a reference value γ_R, that is, $\delta = (\gamma - \gamma_R)/\gamma_R$. Chemical shifts are very small, and are usually reported in parts per million (ppm). The most favorable nuclei for study are those with $I = \frac{1}{2}$, such as H, ^{19}F, ^{31}P, and ^{13}C; the latter isotope is only 1.1% abundant.

Fullerene molecules such as C_{60} and C_{70} are discussed in Chapter 5. The well-known buckyball C_{60} has the shape of a soccer ball with 12 regular pentagons and 20 hexagons. The fact that all of its carbon atoms are equivalent was determined unequivocally by the ^{13}C NMR spectrum that contains only a single narrow line. In contrast to this, the rugby-ball-shaped C_{70} fullerene molecule, which contains 12 pentagons (2 regular) and 25 hexagons, has five types of carbons, and this is confirmed by the ^{13}C NMR spectrum presented at the top of Fig. 3.34. The five NMR lines from the a, b, c, d, and e carbons, indicated in the upper figure, have the intensity ratios $10 : 10 : 20 : 20 : 10$ corresponding to the number of each carbon type in the molecule. Thus NMR provided a confirmation of the structures of these two fullerene molecules.

Electron paramagnetic resonance (EPR), sometimes called *electron spin resonance* (ESR), detects unpaired electrons in transition ions, especially those with odd numbers of electrons such as Cu^{2+} ($3d^9$) and Gd^{3+} ($4f^7$). Free radicals such as those associated with defects or radiation damage can also be detected. The energies or resonant frequencies are three orders of magnitude higher than NMR for the same magnetic field. A different notation is employed for the energy $E_m = g\mu_B B_{app} m$, where μ_B is the Bohr magneton and g is the dimensionless g factor, which has the value 2.0023 for a free electron. For the unpaired electron with spin $S = \frac{1}{2}$ on a free radical EPR measures the energy difference $\Delta E = E_{1/2} - E_{-1/2}$ between the levels $m = \pm\frac{1}{2}$, to give a single-line spectrum at the energy level

$$\Delta E = g\mu_B B_{app} \tag{3.16}$$

Equations (3.15) and (3.16) are related through the expression $g\mu_B = \hbar\gamma$. If the unpaired electron interacts with a nuclear spin of magnitude I, then $2I + 1$ hyperfine structure lines appear at the energies

$$\Delta E(m_I) = g\mu_B B_{app} + A m_I \tag{3.17}$$

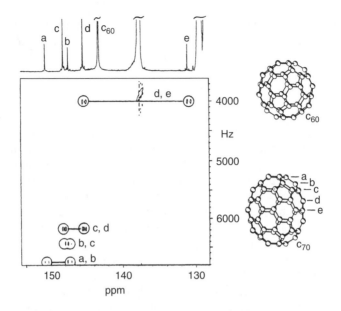

Figure 3.34. NMR spectrum from ^{13}C-isotope-enriched C_{70} fullerene molecules. The five NMR lines from the $a, b, c, d,$ and e carbons, indicated in the molecule on the lower right, have the intensity ratios $10:10:20:20:10$ in the upper spectrum corresponding to the number of each carbon type in the molecule. The so-called two-dimensional NMR spectrum at the lower left, which is a plot of the spin–spin coupling constant frequencies versus the chemical shift in parts per million (ppm), shows doublets corresponding to the indicated carbon bond pairs. The ppm chemical shift scale at the bottom applies to both the lower two-dimensional spectrum, and the upper conventional NMR spectrum. One of the NMR lines arises from the fullerene species C_{60}, which is present as an impurity. [From K. Kikuchi, S. Suzuki, Y. Nakao, N. Nakahara, T. Wakabayashi, H. Shiromaru, I. Ikemoto, and Y. Achiba, *Chem. Phys. Lett.* **216**, 67 (1993).]

where A is the hyperfine coupling constant, and m_I takes on the $2I + 1$ values in the range $-I \leq m_I \leq I$. Figure 3.35 shows an example of such a spectrum arising from the endohedral fullerene compound LaC_{82}, which was detected in the mass spectrum shown in Fig. 3.9. The lanthanum atom inside the C_{82} cage has the nuclear spin line $I = \frac{7}{2}$, and the unpaired electron delocalized throughout the C_{82} fullerene cage interacts with this nuclear spin to produce the eight-line hyperfine multiplet shown in the figure.

EPR has been utilized to study conduction electrons in metal nanoparticles, and to detect the presence of conduction electrons in nanotubes to determine whether the tubes are metals or very narrowband semiconductors. This technique has been employed to identify trapped oxygen holes in colloidal TiO_2 semiconductor nanoclusters. It has also been helpful in clarifying spin–flip resonance transitions and Landau bands in quantum dots.

The construction of new nanostructured biomaterials is being investigated by seeking to understand the structure and organization of supermolecular assemblies by studying the interaction of proteins with phospholipid bilayers. This can be

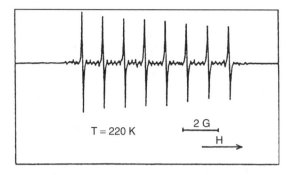

Figure 3.35. Electron paramagnetic resonance spectrum of the endohedral fullerene molecule LaC_{82} dissolved in toluene at 220 K. The unpaired electron delocalized on the carbon cage interacts with the $I = \frac{7}{2}$ nuclear spin of the lanthanum atom inside the cage to produce the observed eight-line hyperfine multiplet. [From R. D. Johnson, D. S. Bethune, C. S. Yannoni, *Acc. Chem. Res.* **25**, 169 (1992).]

conveniently studied by attaching spin labels (e.g., paramagnetic nitroxides) to the lipids and using EPR to probe the restriction of the spin-label motions arising from phospholipids associated with membrane-inserted domains of proteins.

Microwaves or radar waves can also provide useful information about materials when employed under nonresonant conditions in the absence of an applied magnetic field. For example, energy gaps that occur in the microwave region can be estimated by the frequency dependence of the microwave absorption signal. Microwaves have been used to study photon-assisted single-electron tunneling and Coulomb blockades in quantum dots.

FURTHER READING

J. M. Cowley and J. C. H. Spence, "Nanodiffraction," in Nalwa (2000), Vol. 2, Chapter 1, 2000.

T. Hahn, ed., *International Tables for Crystallography*, 4th ed., Kluwer, Dordrecht, 1996.

M. José-Yacamán and J. A. Ascencio, "Electron Microscopy Study of Nanostructured and Ancient Materials," in Nalwa (2000), Vol. 2, Chapter 8, 2000.

C. E. Krill, R. Haberkorn, and R. Birringer, "Specification of Microstructure and Characterization by Scattering Techniques," Nalwa (2000), Vol. 2, Chapter 3, 2000.

G. E. Moore, *IEEE IEDM Tech. Digest.* 11–13 (1975).

H. S. Nalwa, ed., *Handbook of Nanostructured Materials and Nanotechnology*, Vols. 1–5, Academic Press, Boston, 2000.

R. E. Whan, *Materials Characterization*, Vol. 10 of *Metals Handbook*, American Society for Metals, Metals Park OH, 1986.

4

PROPERTIES OF INDIVIDUAL NANOPARTICLES

4.1. INTRODUCTION

The purpose of this chapter is to describe the unique properties of individual nanoparticles. Because nanoparticles have 10^6 atoms or less, their properties differ from those of the same atoms bonded together to form bulk materials. First, it is necessary to define what we mean by a nanoparticle. The words *nanoparticle* and *nanotechnology* are relatively new. However, nanoparticles themselves had been around and studied long before the words were coined. For example, many of the beautiful colors of stained-glass windows are a result of the presence of small metal oxide clusters in the glass, having a size comparable to the wavelength of light. Particles of different sizes scatter different wavelengths of light, imparting different colors to the glass. Small colloidal particles of silver are a part of the process of image formation in photography. Water at ambient temperature consists of clusters of hydrogen-bonded water molecules. Nanoparticles are generally considered to be a number of atoms or molecules bonded together with a radius of <100 nm. A nanometer is 10^{-9} m or 10 Å, so particles having a radius of about ≤ 1000 Å can be considered to be nanoparticles. Figure 4.1 gives a somewhat arbitrary

Introduction to Nanotechnology, by Charles P. Poole Jr. and Frank J. Owens.
ISBN 0-471-07935-9. Copyright © 2003 John Wiley & Sons, Inc.

NUMBER OF ATOMS   RADIUS (nm)

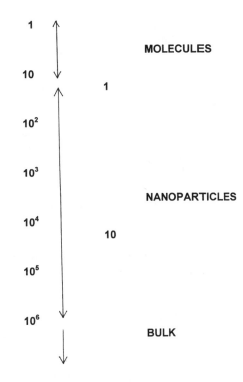

Figure 4.1. Distinction between molecules, nanoparticles, and bulk according to the number of atoms in the cluster.

classification of atomic clusters according to their size showing the relationship between the number of atoms in the cluster and its radius. For example, a cluster of one nanometer radius has approximately 25 atoms, but most of the atoms are on the surface of the cluster. This definition based on size is not totally satisfactory because it does not really distinguish between molecules and nanoparticles. Many molecules contain more than 25 atoms, particularly biological molecules. For example, the heme molecule, $FeC_{34}H_{32}O_4N_4$, which is incorporated in the hemoglobin molecule in human blood and transports oxygen to the cells, contains 75 atoms. In truth, there is no clear distinction. They can be built by assembling individual atoms or subdividing bulk materials. What makes nanoparticles very interesting and endows them with their unique properties is that their size is smaller than critical lengths that characterize many physical phenomena. Generally, physical properties of materials can be characterized by some critical length, a thermal diffusion length, or a scattering length, for example. The electrical conductivity of a metal is strongly determined by the distance that the electrons travel between collisions with the vibrating atoms or impurities of the solid. This distance is called the *mean free path* or the *scattering length*. If the sizes of particles are less than these characteristic lengths, it is possible that new physics or chemistry may occur.

Perhaps a working definition of a nanoparticle is an aggregate of atoms between 1 and 100 nm viewed as a subdivision of a bulk material, and of dimension less than the characteristic length of some phenomena.

4.2. METAL NANOCLUSTERS

4.2.1. Magic Numbers

Figure 4.2 is an illustration of a device used to form clusters of metal atoms. A high-intensity laser beam is incident on a metal rod, causing evaporation of atoms from the surface of the metal. The atoms are then swept away by a burst of helium and passed through an orifice into a vacuum where the expansion of the gas causes cooling and formation of clusters of the metal atoms. These clusters are then ionized by UV radiation and passed into a mass spectrometer that measures their mass : charge ratio. Figure 4.3 shows the mass spectrum data of lead clusters formed in such an experiment where the number of ions (counts) is plotted as a function of the number of atoms in the cluster. (Usually mass spectra data are plotted as counts versus mass over charge.) The data show that clusters of 7 and 10 atoms are more likely than other clusters, which means that these clusters are more stable than clusters of other sizes. Figure 4.4a is a plot of the ionization potential of atoms as a function of their atomic number Z, which is the number of electrons in the atom. The ionization potential is the energy necessary to remove the outer electron from the

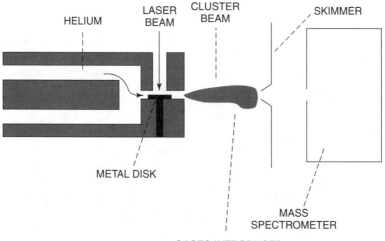

Figure 4.2. Apparatus to make metal nanoparticles by laser induced evaporation of atoms from the surface of a metal. Various gases such as oxygen can be introduced to study the chemical interaction of the nanoparticles and the gases. (With permission from F. J. Owens and C. P. Poole, Jr., *New Superconductors*, Plenum Press, 1999.)

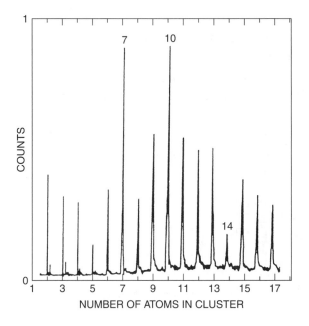

Figure 4.3. Mass spectrum of Pb clusters. [Adapted from M. A. Duncan and D. H. Rouvray, *Sci. Am.* 110 (Dec. 1989).]

atom. The maximum ionization potentials occur for the rare-gas atoms ^{2}He, ^{10}Ne, and ^{18}Ar because their outermost s and p orbitals are filled. More energy is required to remove electrons from filled orbitals than from unfilled orbitals. Figure 4.4b shows the ionization potential of sodium clusters as a function of the number of atoms in a cluster. Peaks are observed at clusters having two and eight atoms. These numbers are referred to as *electronic magic numbers*. Their existence suggests that clusters can be viewed as superatoms, and this result motivated the development of the jellium model of clusters. In the case of larger clusters stability, as discussed in Chapter 2, is determined by structure and the magic numbers are referred to as *structural magic numbers*.

4.2.2. Theoretical Modeling of Nanoparticles

The jellium model envisions a cluster of atoms as a large atom. The positive nuclear charge of each atom of the cluster is assumed to be uniformly distributed over a sphere the size of the cluster. A spherically symmetric potential well is used to represent the potential describing the interaction of the electron with the positive spherical charge distribution. Thus the energy levels can be obtained by solving the Schrödinger equation for this system in a fashion analogous to that for the hydrogen atom. Figure 4.5 compares the energy level scheme for the hydrogen atom and the energy-level scheme for a spherical positive-charge distribution. The superscripts refer to the number of electrons that fill a particular energy level. The electronic magic number corresponds to the total number of electrons on the superatom when

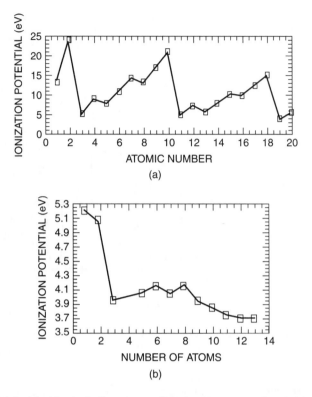

Figure 4.4. (a) A plot of the ionization energy of single atoms versus the atomic number. The ionization energy of the sodium atom at atomic number 11 is 5.14 eV (b) plot of the ionization energy of sodium nanoparticles versus the number of atoms in the cluster. [Adapted from A. Herman et al., *J. Chem. Phys.* **80**, 1780 (1984).]

the top level is filled. Notice that the order of the levels in the jellium model is different from that of the hydrogen atom. In this model the magic numbers correspond to those clusters having a size in which all the energy levels are filled.

An alternative model that has been used to calculate the properties of clusters is to treat them as molecules and use existing molecular orbital theories such as density functional theory to calculate their properties. This approach can be used to calculate the actual geometric and electronic structure of small metal clusters. In the quantum theory of the hydrogen atom, the electron circulating about the nucleus is described by a wave. The mathematical function for this wave, called the *wavefunction* ψ, is obtained by solving the Schrödinger equation, which includes the electrostatic potential between the electron and the positively charged nucleus. The square of the amplitude of the wavefunction represents the probability of finding the electron at some position relative to the nucleus. The wavefunction of the lowest level of the hydrogen atom designated the $1s$ level has the form

$$\psi(1s) = A \exp\left(-\frac{r}{\rho}\right) \tag{4.1}$$

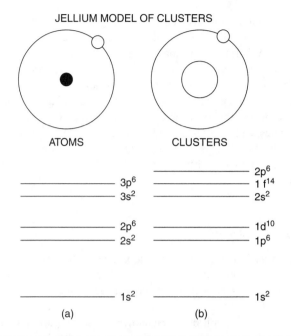

Figure 4.5. A comparison of the energy levels of the hydrogen atom and those of the jellium model of a cluster. The electronic magic numbers of the atoms are 2, 10, 18, and 36 for He, Ne, Ar, and Kr, respectively (the Kr energy levels are not shown on the figure) and 2, 18, and 40 for the clusters. [Adapted from B. K. Rao et al., *J. Cluster Sci.* **10**, 477 (1999).]

where r is the distance of the electron from the nucleus and ρ is the radius of the first Bohr orbit. This comes from solving the Schrödinger equation for an electron having an electrostatic interaction with a positive nucleus given by e/r. The equation of the hydrogen atom is one of the few exactly solvable problems in physics, and is one of the best understood systems in the Universe. In the case of a molecule such as the H_2^+ ion, molecular orbital theory assumes that the wavefunction of the electron around the two H nuclei can be described as a linear combination of the wavefunction of the isolated H atoms. Thus the wavefunction of the electrons in the ground state will have the form,

$$\psi = a\psi(1)_{1s} + a\psi(2)_{1s} \tag{4.2}$$

The Schrödinger equation for the molecular ion is

$$\left[\left(\frac{-\hbar^2}{2m}\right)\nabla^2 - \frac{e^2}{r_a} - \frac{e^2}{r_b}\right]\psi = E\psi \tag{4.3}$$

The symbol ∇^2 denotes a double differentiation operation. The last two terms in the brackets are the electrostatic attraction of the electron to the two positive nuclei, which are at distances r_a and r_b from the electron. For the hydrogen molecule, which has two electrons, a term for the electrostatic repulsion of the two electrons would be added. The Schrödinger equation is solved with this linear combination [Eq. (4.2)] of wavefunctions. When there are many atoms in the molecule and many electrons, the problem becomes complex, and many approximations are used to obtain the solution. Density functional theory represents one approximation. With the development of large fast computer capability and new theoretical approaches, it is possible using molecular orbital theory to determine the geometric and electronic structures of large molecules with a high degree of accuracy. The calculations can find the structure with the lowest energy, which will be the equilibrium geometry. These molecular orbital methods with some modification have been applied to metal nanoparticles.

4.2.3. Geometric Structure

Generally the crystal structure of large nanoparticles is the same as the bulk structure with somewhat different lattice parameters. X-ray diffraction studies of 80-nm aluminum particles have shown that it has the face-centered cubic (FCC) unit cell shown in Fig. 4.6a, which is the structure of the unit cell of bulk aluminum. However, in some instances it has been shown that small particles having diameters of <5 nm may have different structures. For example, it has been shown that 3–5-nm gold particles have an icosahedral structure rather than the bulk FCC structure. It is of interest to consider an aluminum cluster of 13 atoms because this is a magic number. Figure 4.6b shows three possible arrangements of atoms for the cluster. On the basis of criteria of maximizing the number of bonds and minimizing the number of atoms on the surface, as well as the fact that the structure of bulk aluminum is FCC, one might expect the structure of the particle to be FCC. However, molecular

(a)

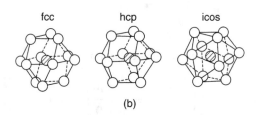

(b)

Figure 4.6. (a) The unit cell of bulk aluminum; (b) three possible structures of Al_{13}: a face-centered cubic structure (FCC), an hexagonal close-packed structure (HCP), and an icosahedral (ICOS) structure.

orbital calculations based on the density functional method predict that the icosahedral form has a lower energy than the other forms, suggesting the possibility of a structural change. There are no experimental measurements of the Al_{13} structure to verify this prediction. The experimental determination of the structure of small metal nanoparticles is difficult, and there are not many structural data available. In the late 1970s and early 1980s, G. D. Stien was able to determine the structure of Bi_N, Pb_N, In_N, and Ag_N nanoparticles. The particles were made using an oven to vaporize the metal and a supersonic expansion of an inert gas to promote cluster formation. Deviations from the face-centered cubic structure were observed for clusters smaller than 8 nm in diameter. Indium clusters undergo a change of structure when the size is smaller than 5.5 nm. Above 6.5 nm, a diameter corresponding to about 6000 atoms, the clusters have a face-centered tetragonal structure with a c/a ratio of 1.075. In a tetragonal unit cell the edges of the cell are perpendicular, the long axis is denoted by c, and the two short axes by a. Below ~6.5 nm the c/a ratio begins to decrease, and at 5 nm $c/a = 1$, meaning that the structure is face-centered cubic. Figure 4.7 is a plot of c/a versus the diameter of indium nanoparticles. It needs to be pointed out that the structure of isolated nanoparticles may differ from that of ligand-stabilized structures. *Ligand stabilization* refers to associating nonmetal ion groups with metal atoms or ions. The structure of these kinds of nanostructured materials is discussed in Chapter 10. A different structure can result in a

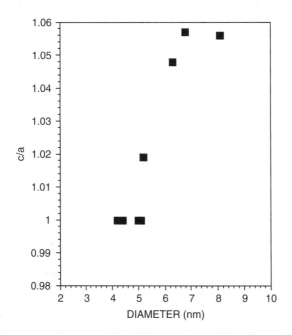

Figure 4.7. Plot of the ratio of the length of the c axis to the a axis of the tetragonal unit cell of indium nanoparticles versus the diameter of nanoparticles. [Plotted from data in A. Yokozeki and G. D. Stein, *J. Appl. Phys.* **49**, 224 (1978).]

Table 4.1. Calculated binding energy per atom and atomic separation in some aluminum nanoparticles compared with bulk aluminum

Cluster	Binding Energy (eV)	Al Separation (Å)
Al_{13}	2.77	2.814
Al_{13}^{-}	3.10	2.75
Bulk Al	3.39	2.86

change in many properties. One obvious property that will be different is the electronic structure. Table 4.1 gives the result of density functional calculations of some of the electronic properties of Al_{13}. Notice that binding energy per atom in Al_{13} is less than in the bulk aluminum. The Al_{13} cluster has an unpaired electron in the outer shell. The addition of an electron to form $Al_{13}(-)$ closes the shell with a significant increase in the binding energy. The molecular orbital approach is also able to account for the dependence of the binding energy and ionization energy on the number of atoms in the cluster. Figure 4.8 shows some examples of the structure of boron nanoparticles of different sizes calculated by density functional theory. Figures 4.6 and 4.8 illustrate another important property of metal nanoparticles. For these small particles all the atoms that make up the particle are on the surface. This has important implications for many of the properties of the nanoparticles such as their vibrational structure, stability, and reactivity. Although in this chapter we are discussing metal nanoparticles as though they can exist as isolated entities, this is not always the case. Some nanoparticles such as aluminum are highly reactive. If one were to have an isolated aluminum nanoparticle exposed to air, it would immediately react with oxygen, resulting in an oxide coating of Al_2O_3 on the surface. X-ray photoelectron spectroscopy of oxygen-passivated, 80-nm, aluminum nanoparticles indicates that they have a 3–5-nm layer of Al_2O_3 on the surface. As we will see later, nanoparticles can be made in solution without exposure to air. For example, aluminum nanoparticles can be made by decomposing aluminum hydride in certain heated solutions. In this case the molecules of the solvent may be bonded to the surface of the nanoparticle, or a surfactant (surface-active agent) such as oleic acid can be added to the solution. The surfactant will coat the particles and prevent them from aggregating. Such metal nanoparticles are said to be passivated, that is, coated with some other chemical to which they are exposed. The chemical nature of this layer will have a significant influence on the properties of the nanoparticle. Self-assembled monolayers (SAMs) can also be used to coat metal nanoparticles. The concept of self-assembly will be discussed in more detail in later chapters. Gold nanoparticles have been passivated by self-assembly using octadecylthiol, which produces a SAM, $C_{18}H_{37}S-Au$. Here the long hydrocarbon chain molecule is tethered at its end to the gold particle Au by the thio head group SH, which forms a strong S—Au bond. Attractive interactions between the molecules produce a symmetric ordered arrangement of them about the particle. This symmetric arrangement of the molecules around the particle is a key characteristic of the SAMs.

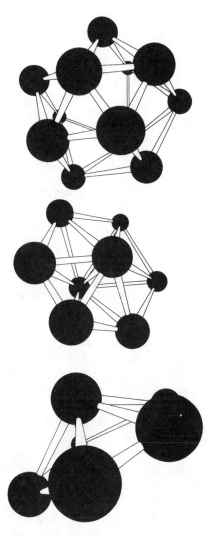

Figure 4.8. Illustration of some calculated structures of small boron nanoparticles. (F. J. Owens, unpublished.)

4.2.4. Electronic Structure

When atoms form a lattice, the discrete energy levels of the atoms are smudged out into energy bands. The term *density of states* refers to the number of energy levels in a given interval of energy. For a metal, the top band is not totally filled. In the case of a semiconductor the top occupied band, called the *valence band*, is filled, and there is a small energy separation referred to as the *band gap* between it and the next higher unfilled band. When a metal particle having bulk properties is reduced in size to a few hundred atoms, the density of states in the conduction band, the top band

CHANGE IN VALENCE ENERGY BAND LEVELS
WITH SIZE

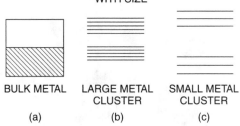

BULK METAL	LARGE METAL CLUSTER	SMALL METAL CLUSTER
(a)	(b)	(c)

Figure 4.9. Illustration of how energy levels of a metal change when the number of atoms of the material is reduced: (a) valence band of *bulk metal*; (b) *large metal cluster* of 100 atoms showing opening of a band gap; (c) *small metal cluster* containing three atoms.

containing electrons, changes dramatically. The continuous density of states in the band is replaced by a set of discrete energy levels, which may have energy level spacings larger than the thermal energy $k_B T$, and a gap opens up. The changes in the electronic structure during the transition of a bulk metal to a large cluster, and then down to a small cluster of less than 15 atoms, are illustrated in Fig. 4.9. The small cluster is analogous to a molecule having discrete energy levels with bonding and antibonding orbitals. Eventually a size is reached where the surfaces of the particles are separated by distances which are in the order of the wavelengths of the electrons. In this situation the energy levels can be modeled by the quantum-mechanical treatment of a particle in a box. This is referred to as the *quantum size effect*. The emergence of new electronic properties can be understood in terms of the Heisenberg uncertainty principle, which states that the more an electron is spatially confined the broader will be its range of momentum. The average energy will not be determined so much by the chemical nature of the atoms, but mainly by the dimension of the particle. It is interesting to note that the quantum size effect occurs in semiconductors at larger sizes because of the longer wavelength of conduction electrons and holes in semiconductors due the larger effective mass. In a semiconductor the wavelength can approach one micrometer, whereas in a metal it is in the order of 0.5 nm.

The color of a material is determined by the wavelength of light that is absorbed by it. The absorption occurs because electrons are induced by the photons of the incident light to make transitions between the lower-lying occupied levels and higher unoccupied energy levels of the materials. Clusters of different sizes will have different electronic structures, and different energy-level separations. Figure 4.10 compares the calculated energy levels of some excited states of boron clusters B_6, B_8, and B_{12} showing the difference in the energy-level separations. Light-induced transitions between these levels determines the color of the materials. This means that clusters of different sizes can have different colors, and the size of the cluster can be used to engineer the color of a material. We will come back to this when we discuss semiconducting clusters.

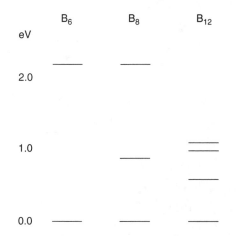

Figure 4.10. Density functional calculation of excited state energy levels of B_6, B_8, and B_{12} nanoparticles. Photon-induced transitions between the lowest level and the upper levels determine the color of the particles. (F. J. Owens, unpublished.)

One method of studying the electronic structure of nanoparticles is UV photo-electron spectroscopy, which is described in more detail in Chapter 3. An incident UV photon removes electrons from the outer valence levels of the atom. The electrons are counted and their energy measured. The data obtained from the measurement are the number of electrons (counts) versus energy. Because the clusters have discrete energy levels, the data will be a series of peaks with separations corresponding to the separations of the energy levels of the cluster. Figure 4.11 shows the UV photoelectron spectrum of the outer levels of copper clusters having 20 and 40 atoms. It is clear that the electronic structure in the valence region varies with the size of the cluster. The energy of the lowest peak is a measure of the electron affinity of the cluster. The *electron affinity* is defined as the increase in electronic energy of the cluster when an electron is added to it. Figure 4.12 is a plot of the measured electron affinities versus the size of Cu clusters, again showing peaks at certain cluster sizes.

4.2.5. Reactivity

Since the electronic structure of nanoparticles depends on the size of the particle, the ability of the cluster to react with other species should depend on cluster size. This has important implications for the design of catalytic agents.

There is experimental evidence for the effect of size on the reactivity of nanoparticles. Their reactivity with various gases can be studied by the apparatus sketched in Fig. 4.2, in which gases such as oxygen are introduced into the region of the cluster beam. A laser beam aimed at a metal disk dislodges metallic particles that are carried along to a mass spectrometer by a flow of helium gas. Down stream

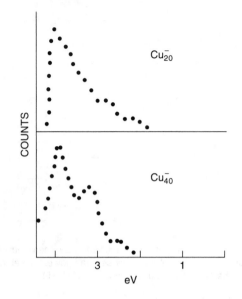

Figure 4.11. UV photoelectron spectrum in the valence band region of copper nanoparticles having 20 and 40 atoms. [Adapted from C. L. Pettiete et al., *J. Chem. Phys.* **88**, 5377 (1988).]

before the particles in the cluster beam enter the mass spectrometer, various gases are introduced, as shown in the figure. Figure 4.13 shows the results of studies of the reaction of oxygen with aluminum particles. The top figure is the mass spectrum of the aluminum particles before the oxygen is introduced. The bottom figure shows the spectrum after oxygen enters the chamber. The results show that two peaks have

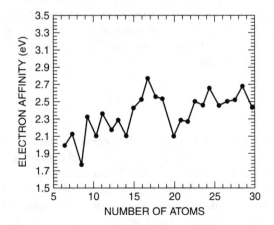

Figure 4.12. Plot of measured electron affinity of copper versus size of nanoparticle. [Adapted from C. L. Pettiete et al., *J. Chem. Phys.* **88**, 5377 (1988).]

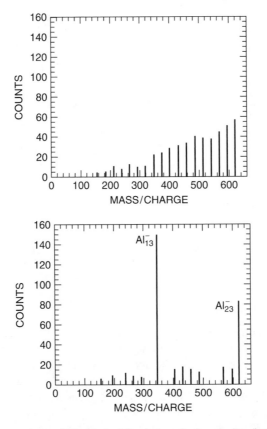

Figure 4.13. Mass spectrum of Al nanoparticles before (top) and after (bottom) exposure to oxygen gas. [Adapted from R. E. Leuchtner et al., *J. Chem. Phys.*, **91**, 2753 (1989).]

increased substantially and certain peaks (12, 14, 19, and 20) have disappeared. The Al_{13} and Al_{23} peaks have increased substantially, and peaks from Al_{15} to Al_{22} have decreased.

These data provide clear evidence for the dependence of the reactivity of aluminum clusters on the number of atoms in the cluster. Similar size dependences have been observed for the reactivity of other metals. Figure 4.14 plots the reaction rate of iron with hydrogen as a function of the size of the iron nanoparticles. The data show that particles of certain sizes such as the one with 10 atoms and sizes greater than 18 atoms are more reactive with hydrogen than others.

A group at Osaka National research Institute in Japan discovered that high catalytic activity is observed to switch on for gold nanoparticles smaller than 3–5 nm, where the structure is icosahedral instead of the bulk FCC arrangement. This work has led to the development of odor eaters for bathrooms based on gold nanoparticles on a Fe_2O_3 substrate.

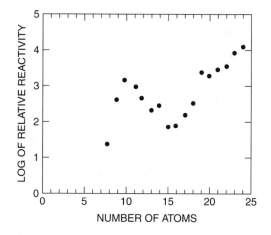

Figure 4.14. Reaction rate of hydrogen gas with iron nanoparticles versus the particle size. [Adapted from R. L. Whetten et al., *Phys. Rev. Lett.* **54**, 1494 (1985).]

4.2.6. Fluctuations

Very small nanoparticles, as is clear from the sketches in Figs. 4.6 and 4.8, have all or almost all of their atoms on the surface. Surface atoms are less restricted in their ability to vibrate than those in the interior, and they are able to make larger excursions from their equilibrium positions. This can lead to changes in the structure of the particle. Observations of the changes in the geometry with time of gold clusters have been made using an electron microscope. The gold clusters of 10–100-Å radii are prepared in vacuum and deposited on a silicon substrate, which is then covered with an SiO_2 film. The electron microscope pictures of gold nanoparticles presented in Fig. 4.15, which were taken at different times, show a number of fluctuation-induced changes in the structure brought about by the particles transforming between different structural arrangements. At higher temperatures these fluctuations can cause a breakdown in the symmetry of the nanoparticle, resulting in the formation of a liquid-like droplet of atoms.

4.2.7. Magnetic Clusters

Although it is not rigorously correct and leads to incorrect predictions of some properties, an electron in an atom can be viewed as a point charge orbiting the nucleus. Its motion around the nucleus gives it orbital angular momentum and produces a magnetic field (except for s states). The magnetic field pattern arising from this movement resembles that of a bar magnet. The electron is said to have an orbital magnetic moment. There is also another contribution to the magnetic moment arising from the fact that the electron has a spin. Classically one can think of the electron as a spherical charge rotating about some axis. Thus there is both a spin and an orbital magnetic moment, which can be added to give the total magnetic

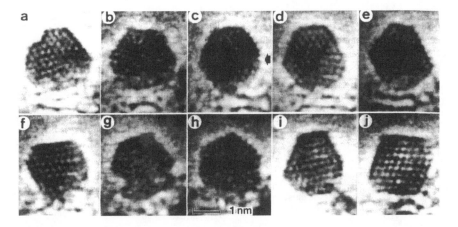

Figure 4.15. A series of electron microscope pictures of gold nanoparticles containing approximately 460 atoms taken at various times showing fluctuation-induced changes in the structure. (With permission from S. Sugano and H. Koizumi, in *Microcluster Physics*, Springer, Berlin, 1998, p. 18.)

moment of the electron. The total moment of the atom will be the vector sum of the moments of each electron of the atom. In energy levels filled with an even number of electrons the electron magnetic moments are paired oppositely, and the net magnetic moment is zero. Thus most atoms in solids do not have a net magnetic moment. There are, however, some transition ion atoms such as iron, manganese, and cobalt, which have partially filled inner d-orbital levels, and hence they posess a net magnetic moment. Crystals of these atoms can become ferromagnetic when the magnetic moments of all the atoms are aligned in the same direction. In this section we discuss the magnetic properties of nanoclusters of metal atoms that have a net magnet moment. In a cluster the magnetic moment of each atom will interact with the moments of the other atoms, and can force all the moments to align in one direction with respect to some symmetry axis of the cluster. The cluster will have a net moment, and is said to be magnetized. The magnetic moment of such clusters can be measured by means of the Stern–Gerlach experiment illustrated in Fig. 4.16. The cluster particles are sent into a region where there is an inhomogeneous magnetic field, which separates the particles according to whether their magnetic moment is up or down. From the beam separation, and a knowledge of the strength and gradient of the magnetic field, the magnetic moment can be determined. However, for magnetic nanoparticles the measured magnetic moment is found to be less than the value for a perfect parallel alignment of the moments in the cluster. The atoms of the cluster vibrate, and this vibrational energy increases with temperature. These vibrations cause some misalignment of the magnetic moments of the individual atoms of the cluster so that the net magnetic moment of the cluster is less than what it would be if all the atoms had their moments aligned in the same direction. The component of the magnetic moment of an individual cluster will

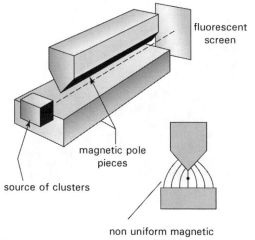

Figure 4.16. Illustration of the Stern–Gerlach experiment used to measure the magnetic moments of nanoparticles. A beam of metal clusters from a source is sent between the poles of permanent magnets shaped to produce a uniform gradient DC magnetic field, which produces a net force on the magnetic dipole moments of the clusters, thereby deflecting the beam. The magnetic moment can be determined by the extent of the deflection, which is measured on a photographic plate or fluorescent screen.

interact with an applied DC magnetic field, and is more likely to align parallel than antiparallel to the field. The overall net moment will be lower at higher temperatures; more precisely, it is inversely proportional to the temperature, an effect called "superparamagnetism." When the interaction energy between the magnetic moment of a cluster and the applied magnetic field is greater than the vibrational energy, there is no vibrational averaging, but because the clusters rotate, there is some averaging. This is called "locked moment magnetism."

One of the most interesting observed properties of nanoparticles is that clusters made up of nonmagnetic atoms can have a net magnetic moment. For example, clusters of rhenium show a pronounced increase in their magnetic moment when they contain less than 20 atoms. Figure 4.17 is a plot of the magnetic moment versus the size of the rhenium cluster. The magnetic moment is large when n is less than 15.

4.2.8. Bulk to Nanotransition

At what number of atoms does a cluster assume the properties of the bulk material? In a cluster with less than 100 atoms, the amount of energy needed to ionize it, that is, to remove an electron from the cluster, differs from the work function. The work function is the amount of energy needed to remove an electron from the bulk solid.

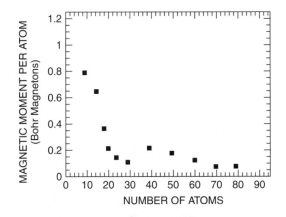

Figure 4.17. Plot of the magnetic moment per atom of rhenium nanoparticles versus the number of atoms in the particle. [Adapted from A. J. Cox et al., *Phys. Rev.* **B49**, 12295 (1994).]

Clusters of gold have been found to have the same melting point of bulk gold only when they contain 1000 atoms or more. Figure 4.18 is a plot of the melting temperature of gold nanoparticles versus the diameter of the particle. The average separation of copper atoms in a copper cluster approaches the value of the bulk material when the clusters have 100 atoms or more. In general, it appears that different physical properties of clusters reach the characteristic values of the solid at different cluster sizes. The size of the cluster where the transition to bulk behavior occurs appears to depend on the property being measured.

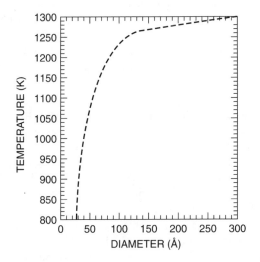

Figure 4.18. Melting temperature of gold nanoparticles versus particle diameter ($10\,\text{Å} = 1\,\text{nm}$). [Adapted from J. P. Borel et al., *Surface Sci.* **106**, I (1981).]

4.3. SEMICONDUCTING NANOPARTICLES

4.3.1. Optical Properties

Because of their role in quantum dots, nanoparticles made of the elements, which are normal constituents of semiconductors, have been the subject of much study, with particular emphasis on their electronic properties. The title of this section, "semiconducting nanoparticles," is somewhat misleading. Nanoparticles made of cadmium, germanium, or silicon are not themselves semiconductors. A nanoparticle of Si_n can be made by laser evaporation of a Si substrate in the region of a helium gas pulse. The beam of neutral clusters is photolyzed by a UV laser producing ionized clusters whose mass to charge ratio is then measured in a mass spectrometer. The most striking property of nanoparticles made of semiconducting elements is the pronounced changes in their optical properties compared to those of the bulk material. There is a significant shift in the optical absorption spectra toward the blue (shorter wavelength) as the particle size is reduced.

In a bulk semiconductor a bound electron–hole pair, called an *exciton*, can be produced by a photon having an energy greater than that of the band gap of the material. The band gap is the energy separation between the top filled energy level of the valence band and the nearest unfilled level in the conduction band above it. The photon excites an electron from the filled band to the unfilled band above. The result is a hole in the otherwise filled valence band, which corresponds to an electron with an effective positive charge. Because of the Coulomb attraction between the positive hole and the negative electron, a bound pair, called an *exciton*, is formed that can move through the lattice. The separation between the hole and the electron is many lattice parameters. The existence of the exciton has a strong influence on the electronic properties of the semiconductor and its optical absorption. The exciton can be modeled as a hydrogen-like atom and has energy levels with relative spacings analogous to the energy levels of the hydrogen atom but with lower actual energies, as explained in Section 2.3.3. Light-induced transitions between these hydrogen-like energy levels produce a series of optical absorptions that can be labeled by the principal quantum numbers of the hydrogen energy levels. Figure 4.19 presents the optical absorption spectra of cuprous oxide (Cu_2O), showing the absorption spectra due to the exciton. We are particularly interested in what happens when the size of the nanoparticle becomes smaller than or comparable to the radius of the orbit of the electron–hole pair. There are two situations, called the *weak-confinement* and the *strong-confinement regimes*. In the weak regime the particle radius is larger than the radius of the electron-hole pair, but the range of motion of the exciton is limited, which causes a blue shift of the absorption spectrum. When the radius of the particle is smaller than the orbital radius of the electron–hole pair, the motion of the electron and the hole become independent, and the exciton does not exist. The hole and the electron have their own set of energy levels. Here there is also a blue shift, and the emergence of a new set of absorption lines. Figure 4.20 shows the optical absorption spectra of a CdSe nanoparticle at two different sizes measured at 10 K. One can see that the lowest energy absorption region, referred to as the *absorption edge*, is

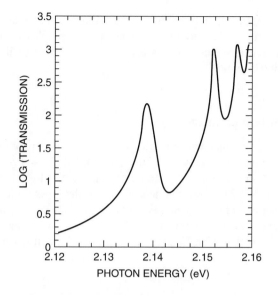

Figure 4.19. Optical absorption spectrum of hydrogen-like transitions of excitons in Cu_2O. [Adapted from P. W. Baumeister, *Phys. Rev.* **121**, 359 (1961).]

shifted to higher energy as the particle size decreases. Since the absorption edge is due to the band gap, this means that the band gap increases as particle size decreases. Notice also that the intensity of the absorption increases as the particle size is reduced. The higher energy peaks are associated with the exciton, and they shift to higher energies with the decrease in particle size. These effects are a result of

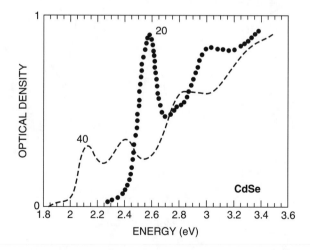

Figure 4.20. Optical absorption spectrum of CdSe for two nanoparticles having sizes 20 Å and 40 Å, respectively. [Adapted from D. M. Mittleman, *Phys. Rev.* **B49**, 14435 (1994).]

the confinement of the exciton that was discussed above. Essentially, as the particle size is reduced, the hole and the electron are forced closer together, and the separation between the energy levels changes. This subject will be discussed in greater detail in Chapter 9.

4.3.2. Photofragmentation

It has been observed that nanoparticles of silicon and germanium can undergo fragmentation when subjected to laser light from a Q-switched Nd:YAG laser. The products depend on the size of the cluster, the intensity of the laser light, and the wavelength. Figure 4.21 shows the dependence of the cross section for photofragmentation (a measure of the probability for breakup of the cluster) with 532 nm laser light, versus the size of a Si fragment. One can see that certain sized fragments are more likely to dissociate than others. Some of the fissions that have been observed are

$$Si_{12} + hv \rightarrow Si_6 + Si_6 \tag{4.4}$$

$$Si_{20} + hv \rightarrow Si_{10} + Si_{10} \tag{4.5}$$

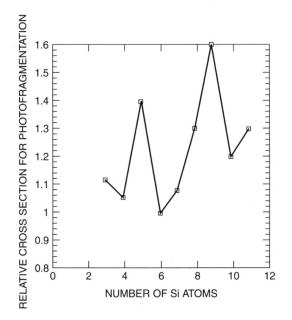

Figure 4.21. Photodissociation cross section of silicon nanoparticles versus number of atoms in particle. [Adapted from L. Bloomfield et al., *Phys. Rev. Lett.* **54**, 2266 (1985).]

where hv is a photon of light energy. Similar results have been obtained for germanium nanoparticles. When the cluster size is greater than 30 atoms, the fragmentation has been observed to occur explosively.

4.3.3. Coulombic Explosion

Multiple ionization of clusters causes them to become unstable, resulting in very rapid high-energy dissociation or explosion. The fragment velocities from this process are very high. The phenomena is called *Coulombic explosion*. Multiple ionizations of a cluster cause a rapid redistribution of the charges on the atoms of the cluster, making each atom more positive. If the strength of the electrostatic repulsion between the atoms is greater than the binding energy between the atoms, the atoms will rapidly fly apart from each other with high velocities. The minimum number of atoms N required for a cluster of charge Q to be stable depends on the kinds of atoms, and the nature of the bonding between the atoms of the cluster. Table 4.2 gives the smallest size that is stable for doubly charged clusters of different types of atoms and molecules. The table also shows that larger clusters are more readily stabilized at higher degrees of ionizations. Clusters of inert gases tend to be larger because their atoms have closed shells that are held together by much weaker forces called *van der Waals forces*.

The attractive forces between the atoms of the cluster can be overcome by the electrostatic repulsion between the atoms when they become positively charged as a result of photoionization. One of the most dramatic manifestations of Coulombic explosion reported in the journal *Nature* is the observation of nuclear fusion in deuterium clusters subjected to femtosecond laser pulses. A femtosecond is 10^{-15} seconds. The clusters were made in the usual way described above, and then subjected to a high-intensity femtosecond laser pulse. The fragments of the dissociation have energies up to one million electron volts (MeV). When the

Table 4.2. Some examples of the smallest obtainable multiply charged clusters of different kinds (smaller clusters will explode)

Atom	Charge		
	$+2$	$+3$	$+4$
Kr	Kr_{73}		
Xe	Xe_{52}	Xe_{114}	Xe_{206}
CO_2	$(CO_2)_{44}$	$(CO_2)_{106}$	$(CO_2)_{216}$
Si	Si_3		
Au	Au_3		
Pb	Pb_7		

deuterium fragments collide, they have sufficient energy to undergo nuclear fusion by the following reaction:

$$D + D \Rightarrow {}^3He + \text{neutron} \tag{4.6}$$

This reaction releases a neutron of 2.54 MeV energy. Evidence for the occurrence of fusion is the detection of the neutrons using neutron scintillation detectors coupled to photomultiplier tubes.

4.4. RARE GAS AND MOLECULAR CLUSTERS

4.4.1. Inert-Gas Clusters

Table 4.2 lists a number of different kinds of nanoparticles. Besides metal atoms and semiconducting atoms, nanoparticles can be assembled from rare gases such as krypton and xenon, and molecules such as water. Xenon clusters are formed by adiabatic expansion of a supersonic jet of the gas through a small capillary into a vacuum. The gas is then collected by a mass spectrometer, where it is ionized by an electron beam, and its mass : charge ratio measured. As in the case of metals, there are magic numbers, meaning that clusters having a certain number of atoms are more stable than others. For the case of xenon, the most stable clusters occur at particles having 13, 19, 25, 55, 71, 87, and 147 atoms. Argon clusters have similar structural magic numbers. Since the inert-gas atoms have filled electronic shells, their magic numbers are structural magic numbers as discussed in Chapter 2. The forces that bond inert-gas atoms into clusters are weaker than those that bond metals and semiconducting atoms. Even though inert-gas atoms have filled electron shells, because of the movement of the electrons about the atoms, they can have an instantaneous electric dipole moment, P_1. An electric dipole moment occurs when a positive charge and a negative charge are separated by some distance. This dipole produces an electric field $2P_1/R^3$ at another atom a distance R away. This, in turn, induces a dipole moment, P_2, on the second atom, $2\alpha P_1/R^3$, where α is called the *electronic polarizability*. Thus two inert-gas atoms will have an attractive potential

$$U(R) = \frac{2P_1 P_2}{R^3} = \frac{-4\alpha P_1^2}{R^6} \tag{4.7}$$

This is known as the *van der Waals potential*, and it is effective at relatively large separations of the atoms. As the two atoms get much closer together, there will be repulsion between the electronic cores of each atom. Experimentally this has been shown to have the form B/R^{12}. Thus the overall interaction potential between two inert-gas atoms has the form

$$U(R) = \frac{B}{R^{12}} - \frac{C}{R^6} \tag{4.8}$$

This is known as Lennard-Jones potential, and it is used to calculate the structure of inert gas clusters. The force between the atoms arising from this potential is a minimum for the equilibrium distance $R_{min} = (2B/C)^{1/6}$, which is attractive for larger separations and repulsive for smaller separations of the atoms. More generally, it is weaker than the forces that bind metal and semiconducting atoms into clusters.

4.4.2. Superfluid Clusters

Clusters of ^{4}He and ^{3}He atoms formed by supersonic free-jet expansion of helium gas have been studied by mass spectrometry, and magic numbers are found at cluster sizes of $N = 7,10,14,23,30$ for ^{4}He, and $N = 7,10,14,21,30$ for ^{3}He. One of the more unusual properties displayed by clusters is the observation of superfluidity in He clusters having 64 and 128 atoms. Superfluidity is the result of the difference in the behavior of atomic particles having half-integer spin, called *fermions* and particles having integer spin called *bosons*. The difference between them lies in the rules that determine how they occupy the energy levels of a system. Fermions such as electrons are only allowed to have two particles in each energy level with their spins oppositely aligned. Bosons on the other hand do not have this restriction. This means as the temperature is lowered and more and more of the lower levels become occupied bosons can all occupy the lowest level, whereas fermions will be distributed in pairs at the lowest sequence of levels. Figure 4.22 illustrates the difference. The case where all the bosons are in the lowest level is referred to as *Bose–Einstein condensation*. When this occurs the wavelength of each boson is the same as every other, and all of the waves are in phase.

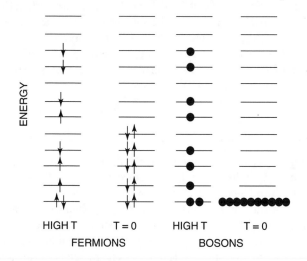

Figure 4.22. Illustration of how fermions and bosons distribute over the energy levels of a system at high and low temperature.

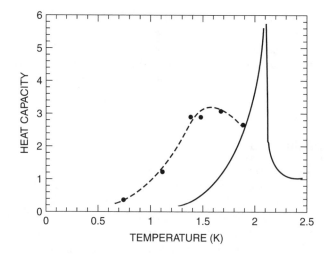

Figure 4.23. Specific heat versus temperature for liquid helium (solid line) and a liquid consisting of clusters of 64 helium atoms (dark circles). The peak corresponds to the transition to the superfluid state. [Adapted from P. Sindzingre, *Phys. Rev. Lett.* **63**, 1601 (1989).]

When boson condensation occurs in liquid He^4 at the temperature 2.2 K, called the *lambda point* (λ point), the liquid helium becomes a superfluid, and its viscosity drops to zero. Normally when a liquid is forced though a small thin tube, it moves slowly because of friction with the walls, and increasing the pressure at one end increases the velocity. In the superfluid state the liquid moves quickly through the tube, and increasing the pressure at one end does not change the velocity. The transition to the superfluid state at 2.2 K is marked by a discontinuity in the specific heat known as the *lambda transition*. The specific heat is the amount of heat energy necessary to raise the temperature of one gram of the material by 1 K. Figure 4.23 shows a plot of the specific heat versus temperature for bulk liquid helium, and for a helium cluster of 64 atoms, showing that clusters become superfluid at a lower temperature than the bulk liquid of He atoms.

4.4.3. Molecular Clusters

Individual molecules can form clusters. One of the most common examples of this is the water molecule. It has been known since the early 1970s, long before the invention of the word *nanoparticle*, that water does not consist of isolated H_2O molecules. The broad Raman spectra of the O—H stretch of the water molecule in the liquid phase at 3200–3600 cm^{-1} has been shown to be due to a number of overlapping peaks arising from both isolated water molecules and water molecules hydrogen-bonded into clusters. The H atom of one molecule forms a bond with the oxygen atom of another. Figure 4.24 shows the structure of one such water cluster. At ambient conditions 80% of water molecules are bonded into clusters, and as the

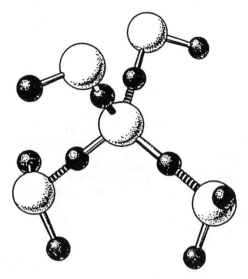

Figure 4.24. A hydrogen-bonded cluster of five water molecules. The large spheres are oxygen, and the small spheres are hydrogen atoms.

temperature is raised, the clusters dissociate into isolated H_2O molecules. In the complex shown in Fig. 4.24 the H atom is not equidistant between the two oxygens. Interestingly, it has been predicted that under 9 GPa of shock loading pressure a new form of water might exist called *symmetrically hydrogen-bonded water*, where the H atom is equally shared between both oxygens. It is possible that such water could have properties different from those of normal water. There are other examples of molecular clusters such as $(NH_3)_n{}^+$, $(CO_2)_{44}$ and $(C_4H_8)_{30}$.

4.5. METHODS OF SYNTHESIS

Earlier in the chapter we described one method of making nanoparticles using laser evaporation in which a high intensity laser beam is incident on a metal rod, causing atoms to be evaporated from the surface of the metal. These metal atoms are then cooled into nanoparticles. There are, however, other ways to make nanoparticles, and we will describe several of them.

4.5.1. RF Plasma

Figure 4.25 illustrates a method of nanoparticle synthesis that utilizes a plasma generated by RF heating coils. The starting metal is contained in a pestle in an evacuated chamber. The metal is heated above its evaporation point using high voltage RF coils wrapped around the evacuated system in the vicinity of the pestle. Helium gas is then allowed to enter the system, forming a high temperature plasma

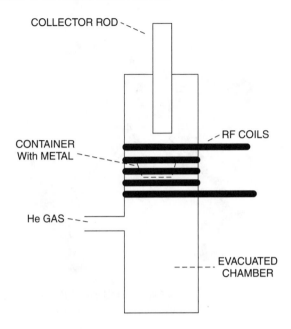

Figure 4.25. Illustration of apparatus for the synthesis of nanoparticles using an RF-produced plasma.

in the region of the coils. The metal vapor nucleates on the He gas atoms and diffuses up to a colder collector rod where nanoparticles are formed. The particles are generally passivated by the introduction of some gas such as oxygen. In the case of aluminum nanoparticles the oxygen forms a layer of aluminum oxide about the particle.

4.5.2. Chemical Methods

Probably the most useful methods of synthesis in terms of their potential to be scaled up are chemical methods. There are a number of different chemical methods that can be used to make nanoparticles of metals, and we will give some examples. Several types of reducing agents can be used to produce nanoparticles such as $NaBEt_3H$, $LiBEt_3H$, and $NaBH_4$ where Et denotes the ethyl ($\cdot C_2H_5$) radical. For example, nanoparticles of molybdenum (Mo) can be reduced in toluene solution with $NaBEt_3H$ at room temperature, providing a high yield of Mo nanoparticles having dimensions of 1–5 nm. The equation for the reaction is

$$MoCl_3 + 3NaBEt_3H \Rightarrow Mo + 3NaCl + 3BEt_3 + (3/2)H_2 \qquad (4.9)$$

Nanoparticles of aluminum have been made by decomposing $Me_2EtNAlH_3$ in toluene and heating the solution to 105°C for 2 h (Me is methyl, $\cdot CH_3$). Titanium isopropoxide is added to the solution. The titanium acts as a catalyst for the reaction. The choice of catalyst determines the size of the particles produced. For instance, 80-nm particles have been made using titanium. A surfactant such as oleic acid can be added to the solution to coat the particles and prevent aggregation.

4.5.3. Thermolysis

Nanoparticles can be made by decomposing solids at high temperature having metal cations, and molecular anions or metal organic compounds. The process is called *thermolysis*. For example, small lithium particles can be made by decomposing lithium azide, LiN_3. The material is placed in an evacuated quartz tube and heated to 400°C in the apparatus shown in Fig. 4.26. At about 370°C the LiN_3 decomposes, releasing N_2 gas, which is observed by an increase in the pressure on the vacuum gauge. In a few minutes the pressure drops back to its original low value, indicating that all the N_2 has been removed. The remaining lithium atoms coalesce to form small colloidal metal particles. Particles less than 5 nm can be made by this method. Passivation can be achieved by introducing an appropriate gas.

The presence of these nanoparticles can be detected by electron paramagnetic resonance (EPR) of the conduction electrons of the metal particles. Electron paramagnetic resonance, which is described in more detail in Chapter 3, measures the energy absorbed when electromagnetic radiation such as microwaves induces a transition between the spin states m_s split by a DC magnetic field. Generally the experiment measures the derivative of the absorption as a function of an increasing DC magnetic field. Normally because of the low penetration depth of the microwaves into a metal, it is not possible to observe the EPR of the conduction electrons.

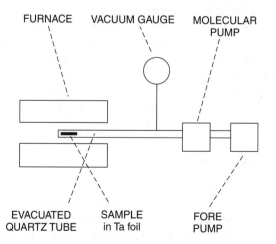

Figure 4.26. Apparatus used to make metal nanoparticles by thermally decomposing solids consisting of metal cations and molecular anions, or metal organic solids. (F. J. Owens, unpublished.)

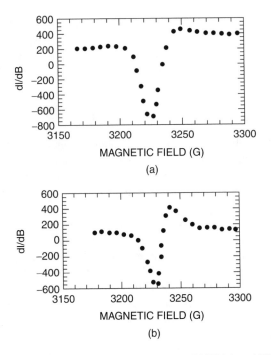

Figure 4.27. Electron paramagnetic resonance spectra at 300 K (a) and 77 K (b) arising from conduction electrons in lithium nanoparticles formed from the thermal decomposition of LiN₃. (F. J. Owens, unpublished.)

However, in a collection of nanoparticles there is a large increase in surface area, and the size is of the order of the penetration depth, so it is possible to detect the EPR of the conduction electrons. Generally EPR derivative signals are quite symmetric, but for the case of conduction electrons, relaxation effects make the lines very asymmetric, and the extent of the asymmetry is related to the small dimensions of the particles. This asymmetry is quite temperature dependent as indicated in Fig. 4.27, which shows the EPR spectra of lithium particles at 300 and 77 K that had been made by the process described above. It is possible to estimate the size of the particles from the g-factor shift and line width of the spectra.

4.5.4. Pulsed Laser Methods

Pulsed lasers have been used in the synthesis of nanoparticles of silver. Silver nitrate solution and a reducing agent are flowed through a blenderlike device. In the blender there is a solid disk, which rotates in the solution. The solid disk is subjected to pulses from a laser beam creating hot spots on the surface of the disk. The apparatus is illustrated in Fig. 4.28. Silver nitrate and the reducing agent react at these hot spots, resulting in the formation of small silver particles, which can be separated from the solution using a centrifuge. The size of the particles is controlled by the

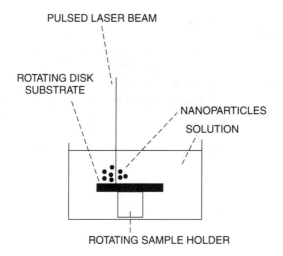

PULSED LASER BEAM

ROTATING DISK
SUBSTRATE

NANOPARTICLES
SOLUTION

ROTATING SAMPLE HOLDER

Figure 4.28. Apparatus to make silver nanoparticles using a pulsed laser beam that creates hot spots on the surface of a rotating disk. [Adapted from J. Singh, *Mater. Today* **2**, 10 (2001).]

energy of the laser and the rotation speed of the disk. This method is capable of a high rate of production of 2–3 g/min.

4.6. CONCLUSION

In this chapter a number of examples have been presented showing that the physical, chemical, and electronic properties of nanoparticles depend strongly on the number and kind of atoms that make up the particle. We have seen that color, reactivity, stability, and magnetic behavior all depend on particle size. In some instances entirely new behavior not seen in the bulk has been observed such as magnetism in clusters that are constituted from nonmagnetic atoms. Besides providing new research challenges for scientists to understand the new behavior, the results have enormous potential for applications, allowing the design of properties by control of particle size. It is clear that nanoscale materials can form the basis of a new class of atomically engineered materials.

FURTHER READING

R. P. Anders et al., "Research Opportunities in Clusters and Cluster Assembled Materials," *J. Mater. Res.* **4**, 704 (1989).

W. A. De Heer, "Physics of Simple Metal Clusters," *Rev. Mod. Phys.* **65**, 611 (1993).

M. A. Duncan and D. H. Rouvray, "Microclusters," *Sci. Am.* 110 (Dec. 1989).

S. N. Khanna, *Handbook of Nano Phase Materials*, in A N. Goldstein, ed., Marcel Decker, New York, 1997, Chapter 1.

J. Lue, "A Review of Characterization and Physical Property Studies of Metallic Nano-particles," *J. Phys. Chem. Solids* **62**, 1599 (2001).

M. Morse, "Clusters of Transition Atoms," *Chem. Rev.* **86**, 1049 (1986).

S. Sugano and H. Koizumi, *Microcluster Physics*, Springer-Verlag, Heidelberg, 1998.

5

CARBON NANOSTRUCTURES

5.1. INTRODUCTION

This chapter is concerned with various nanostructures of carbon. A separate chapter is devoted to carbon because of the important role of carbon bonding in the organic molecules of life (see Chapter 12), and the unique nature of the carbon bond itself. It is the diverse nature of this bond that allows carbon to form some of the more interesting nanostructures, particularly carbon nanotubes. Possibly more than any of the other nanostructures, these carbon nanotubes have enormous applications potential which we will discuss in this chapter.

5.2. CARBON MOLECULES

5.2.1. Nature of the Carbon Bond

In order to understand the nature of the carbon bond it is necessary to examine the electronic structure of the carbon atom. Carbon contains six electrons, which are distributed over the lowest energy levels of the carbon atom. The structure is designated as follows $(1s)^2$, $(2s)$, $(2p_x)$, $(2p_y)$, $(2p_z)$ when bonded to atoms in molecules. The lowest energy level $1s$ with the quantum number $N = 1$ contains two

Introduction to Nanotechnology, by Charles P. Poole Jr. and Frank J. Owens.
ISBN 0-471-07935-9. Copyright © 2003 John Wiley & Sons, Inc.

electrons with oppositely paired electron spins. The electron charge distribution in an s state is spherically symmetric about the nucleus. These $1s$ electrons do not participate in the chemical bonding. The next four electrons are in the $N=2$ energy state, one in a spherically symmetric s orbital, and three in p_x, p_y, and p_z orbitals, which have the very directed charge distributions shown in Fig. 5.1a, oriented perpendicular to each other. This outer s orbital together with the three p orbitals form the chemical bonds of carbon with other atoms. The charge distribution associated with these orbitals mixes (or overlaps) with the charge distribution of each other atom being bonded to carbon. In effect, one can view the electron charge between the two atoms of a bond as the glue that holds the atoms together. On the basis of this simple picture the methane molecule, CH_4, might have the structure shown in Fig. 5.1b, where the H−C bonds are at right angles to each other. However,

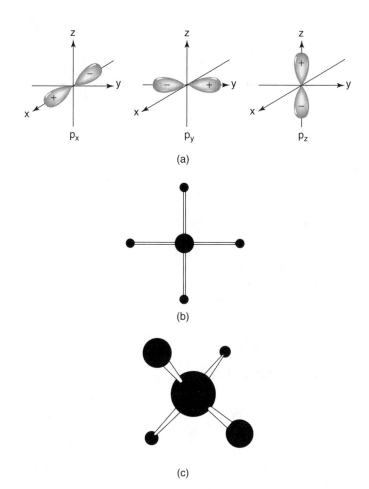

Figure 5.1. (a) Illustration of p_x, p_y, and p_z orbitals of the carbon atom; (b) structure of methane CH_4 assuming that the valence orbitals of carbon are pure p_x, p_y, and p_z orbitals; (c) actual structure of CH_4, which is explained by sp^3 hybridization.

methane does not have this configuration; rather, it has the tetrahedral structure shown in Fig. 5.1c, where the carbon bonds make angles of $109°28'$ with each other. The explanation lies in the concept of hybridization. In the carbon atom the energy separation between the $2s$ level and the $2p$ levels is very small, and this allows an admixture of the $2s$ wavefunction with one or more of the $2p_i$ wavefunctions. The un-normalized wavefunction Ψ in a valence state can be designated by the expression

$$\Psi = s + \lambda p \tag{5.1}$$

where p indicates an admixture of p_i orbitals. With this hybridization the directions of the p lobes and the angles between them changes. The angles will depend on the relative admixture λ of the p states with the s state. Three kinds of hybridization are identified in Table 5.1, which shows the bond angles for the various possibilities, which are $180°$, $120°$, and $109°28'$ for the linear compound acetylene ($H-C \equiv C-H$), the planar compound ethylene ($H_2C=CH_2$), and tetrahedral methane (CH_4), respectively. In general, most of the bond angles for carbon in organic molecules have these values. For example, the carbon bond angle in diamond is $109°$, and in graphite and benzene it is $120°$.

Solid carbon has two main structures called *allotropic forms* that are stable at room temperature: diamond and graphite. Diamond consists of carbon atoms that are tetrahedrally bonded to each other through sp^3 hybrid bonds that form a three-dimensional network. Each carbon atom has four nearest-neighbor carbons. Graphite has a layered structure with each layer, called a *graphitic sheet*, formed from hexagons of carbon atoms bound together by sp^2 hybrid bonds that make $120°$ angles with each other. Each carbon atom has three nearest-neighbor carbons in the planar layer. The hexagonal sheets are held together by weaker van der Waals forces, discussed in the previous chapter.

5.2.2. New Carbon Structures

Until 1964 it was generally believed that no other carbon bond angles were possible in hydrocarbons, that is, compounds containing only carbon and hydrogen atoms. In that year Phil Eaton of the University of Chicago synthesized a square carbon molecule, C_8H_8, called *cubane*, shown in Fig. 5.2a. In 1983 L. Paquette of Ohio State University synthesized a $C_{20}H_{20}$ molecule having a dodecahedron shape,

Table 5.1. Types of sp^n hybridization, the resulting bond angles, and examples of molecules

Type of Hybridization	Digonal sp	Trigonal sp^2	Tetrahedral sp^3
Orbitals used for bond	s, p_x	s, p_x, p_y	s, p_x, p_y, p_z
Example	Acetylene C_2H_2	Ethylene C_2H_4	Methane CH_4
Value of λ	1	$2^{1/2}$	$3^{1/2}$
Bond angle	$180°$	$120°$	$109°28'$

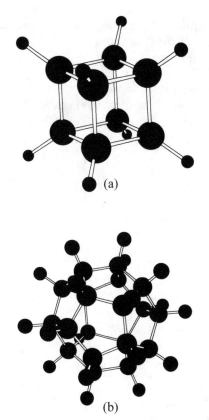

Figure 5.2. (a) Carbon cube structure of cubane C_8H_8 and (b) carbon dodecahedron structure of $C_{20}H_{20}$.

shown in Fig. 5.2b, formed by joining carbon pentagons, and having C–C bond angles ranging from 108° to 110°. The synthesis of these hydrocarbon molecules with carbon bond angles different from the standard hybridization values of Table 5.1 has important implications for the formation of carbon nanostructures, which would also require different bonding angles.

5.3. CARBON CLUSTERS

5.3.1. Small Carbon Clusters

Laser evaporation of a carbon substrate using the apparatus shown in Fig. 4.2 in a pulse of He gas can be used to make carbon clusters. The neutral cluster beam is photoionized by a UV laser and analyzed by a mass spectrometer. Figure 5.3 shows

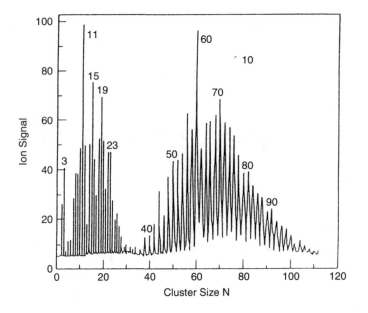

Figure 5.3. Mass spectrum of carbon clusters. The C_{60} and C_{70} fullerene peaks are evident. (With permission from S. Sugano and H. Koizuni, in *Microcluster Physics*, Springer-Verlag, Heidelberg, 1998.)

the typical mass spectrum data obtained from such an experiment. For the number of atoms N less than 30, there are clusters of every N, although some are more prominent than others. Calculations of the structure of small clusters by molecular orbital theory show the clusters have linear or closed nonplanar monocyclic geometries, as shown in Fig. 5.4. The linear structures that have *sp* hybridization occur when N is odd, and the closed structures form when N is even. The open structures with 3, 11, 15, 19, and 23 carbons have the usual bond angles and are more prominent and more stable. The closed structures have angles between the carbon bonds that differ from those predicted by the conventional hybridization concept. Notice in Fig. 5.3 that there is a very large mass peak at 60. The explanation of this peak and its structure earned a Nobel Prize.

5.3.2. Discovery of C_{60}

The discovery of the existence of a soccer ball-like molecule containing 60 carbon atoms was a somewhat fortuitous result of research on the nature of matter in outer space involving studies of light transmission through interstellar dust, the small particles of matter that fill the regions of outer space between stars and galaxies. When light from a distant star passes through the cosmos and arrives on Earth, the intensity of the light is reduced. This is referred to as *optical extinction*. It occurs because of the absorption and scattering of the light from the interstellar dust lying in

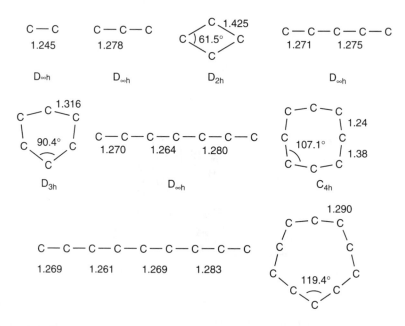

Figure 5.4. Some examples of the structures of small carbon clusters. [With permission from K. Raghavacari et al., *J. Chem. Phys.* **87**, 2191 (I987).]

the pathway of the light on its way to Earth. Scientists study this extinction by measuring the intensity of light coming from the stars at different wavelengths, that is, with different colors. When these studies were made, it was noted that there was an increased extinction or absorption in the ultraviolet region at a wavelength of 220 nm (5.6 eV), which was attributed to light scattered from small particles of graphite that were believed to be present in the regions between the stars. Figure 5.5 shows a plot of this extinction versus the photon energy. This explanation for the optical extinction in the 220-nm region has been widely accepted by astronomers.

Donald Huffman of the University of Arizona and Wolfgang Kratschmer of the Max Planck Institute of Nuclear Physics in Heidelberg were not convinced of this explanation, and decided to study the question further. Their approach was to simulate the graphite dust in the laboratory and investigate light transmission through it. They made smokelike particles by striking an arc between two graphite electrodes in a helium gas environment, and then condensing the smoke on quartz glass plates. Various spectroscopic methods such as infrared and Raman spectroscopy, which can measure the vibrational frequencies of molecules, were used to investigate the condensed graphite. They did indeed obtain the spectral lines known to arise from graphite, but they also observed four additional IR absorption bands that did not originate from graphite, and they found this very puzzling.

Although a soccer ball-like molecule consisting of 60 carbon atoms with the chemical formula C_{60} had been envisioned by theoretical chemists for a number of

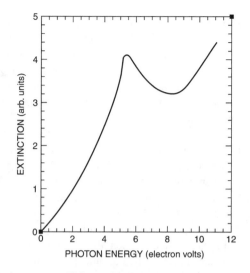

Figure 5.5. Optical spectrum of light coming from stars in outer space. The peak at 5.6 eV is due to absorption from C_{60} present in interstellar dust. (With permission from F. J. Owens and C. P. Poole, in *New Superconductors*, Plenum Press, 1998.)

years, no evidence had ever been found for its existence. Many detailed properties of the molecule had, however, been calculated by the theorists, including a prediction of what the IR absorption spectrum of the molecule would look like. To the amazement of Huffman and Kratschmer the four bands observed in the condensed "graphite" material corresponded closely to those predicted for a C_{60} molecule. Could the extinction of UV light coming from stars be due to the existence of C_{60} molecules? To further verify this, the scientists studied the IR absorption spectrum using carbon arcs made of the 1% abundant ^{13}C isotope, and compared it to their original spectrum which arose from the usual ^{12}C isotope. It was well known that this change in isotope would shift the IR spectrum by the square root of the ratio of the masses, which in this case is

$$\left(\frac{13}{12}\right)^{1/2} = 1.041 \tag{5.2}$$

corresponding to a shift of 4.1%. This is exactly what was observed when the experiment was performed. The two scientists now had firm evidence for the existence of an intriguing new molecule consisting of 60 carbon atoms bonded in the shape of a sphere. Other experimental methods such as mass spectroscopy were used to verify this conclusion, and the results were published in *Nature* in 1990.

Other research groups were also approaching the existence of the C_{60} molecule by different methods, although ironically cosmological issues were also driving their research. Harlod Kroto, a chemist from the University of Sussex in England, was part of a team that had found evidence for the presence of long linear carbon chain molecules, like those shown in Fig. 5.4, in outer space. He was interested in how these chains came to be, and had speculated that such molecules might be created in the outer atmosphere of a type of star called a "red giant." In order to test his hypothesis, he wanted to re-create the conditions of the outer atmosphere of the star in a laboratory setting to determine whether the linear carbon chains might be formed. He knew that high-powered pulsed lasers could simulate the conditions of hot carbon vapor that might exist in the outer surface of red giants. He contacted Professor Richard Smalley of Rice University in Houston, who had built the apparatus depicted in Fig. 4.2, to make small clusters of atoms using high-powered pulsed lasers. In this experiment a graphite disk is heated by a high-intensity laser beam that produces a hot vapor of carbon. A burst of helium gas then sweeps the vapor out through an opening where the beam expands. The expansion cools the atoms and they condense into clusters. This cooled cluster beam is then narrowed by a skimmer and fed into a mass spectrometer, which is a device designed to measure the mass of molecules in the clusters. When the experiment was done using a graphite disk, the mass spectrometer yielded an unexpected result. A mass number of 720 that would consist of 60 carbon atoms, each of mass 12, was observed. Evidence for a C_{60} molecule had been found! Although the data from this experiment did not give information about the structure of the carbon cluster, the scientists suggested that the molecule might be spherical, and they built a geodesic dome model of it.

5.3.3. Structure of C_{60} and Its Crystal

The C_{60} molecule has been named fullerene after the architect and inventor R. Buckminister Fuller, who designed the geodesic dome that resembles the structure of C_{60}. Originally the molecule was called *buckminsterfullerene*, but this name is a bit unwieldy, so it has been shortened to fullerene. A sketch of the molecule is shown in Fig. 5.6. It has 12 pentagonal (5 sided) and 20 hexagonal (6 sided) faces symmetrically arrayed to form a molecular ball. In fact a soccer ball has the same geometric configuration as fullerene. These ball-like molecules bind with each other in the solid state to form a crystal lattice having a face centered cubic structure shown in Fig. 5.7. In the lattice each C_{60} molecule is separated from its nearest neighbor by 1 nm (the distance between their centers is 1 nm), and they are held together by weak forces called van der Waals forces that were discussed in the previous chapter. Because C_{60} is soluble in benzene, single crystals of it can be grown by slow evaporation from benzene solutions.

5.3.4. Alkali-Doped C_{60}

In the face-centered cubic fullerene structure, 26% of the volume of the unit cell is empty, so alkali atoms can easily fit into the empty spaces between the molecular balls of the material. When C_{60} crystals and potassium metal are placed in evacuated tubes and heated to 400°C, potassium vapor diffuses into these empty spaces to form the compound K_3C_{60}. The C_{60} crystal is an insulator, but when doped with an alkali

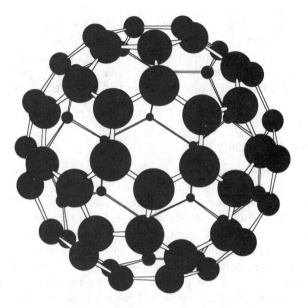

Figure 5.6. Structure of the C_{60} fullerene molecule.

atom it becomes electrically conducting. Figure 5.7 shows the location of the alkali atoms in the lattice where they occupy the two vacant tetrahedral sites and a larger octahedral site per C_{60} molecule. In the tetrahedral site the alkali atom has four surrounding C_{60} balls, and in the octahedral site there are six surrounding C_{60}

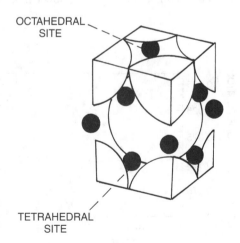

Figure 5.7. Crystal lattice unit cell of C_{60} molecules (large spheres) doped with alkali atoms (dark circles). [From F. J. Owens and C. P. Poole, Jr., *The New Superconductors*, Plenum, 1998.]

molecules. When C_{60} is doped with potassium to form K_3C_{60}, the potassium atoms become ionized to form K^+ and their electrons become associated with the C_{60}, which becomes a $C_{60}{}^{3-}$ triply negative ion. Thus each C_{60} has three extra electrons that are loosely bonded to the C_{60}, and can move through the lattice making C_{60} electrically conducting. In this case the C_{60} is said to be electron-doped.

5.3.5. Superconductivity in C_{60}

Superconductivity is a state of matter in which the resistance of a sample becomes zero, and in which no magnetic field is allowed to penetrate the sample. The latter manifests itself as a reduction of the magnetic susceptibility χ of the sample to $\chi = -1$ (in the MKS system). In 1991, when A. F. Hebard and his co-workers at Bell Telephone Laboratories doped C_{60} crystals with potassium by the methods described above and tested them for superconductivity, to the surprise of all, evidence was found for a superconducting transition at 18 K. Figure 5.8 shows the drop in the magnetization indicative of the presence of superconductivity. A new class of superconducting materials had been found having a simple cubic structure and containing only two elements. Not long after the initial report it was found that many alkali atoms could be doped into the lattice, and the transition temperature increased to as high as 33 K in Cs_2RbC_{60}. As the radius of the dopant alkali atom increases, the cubic C_{60} lattice expands, and the superconducting transition temperature goes up. Figure 5.9 is a plot of the transition temperature versus the lattice parameter.

It was mentioned above that graphite consists of parallel planar graphitic sheets of carbon atoms. It is possible to put other atoms between the planes of these sheets, a procedure called *intercalation*. When intercalated with potassium atoms, crystalline graphite becomes superconducting at the extremely low temperature of a few tenths of a kelvin.

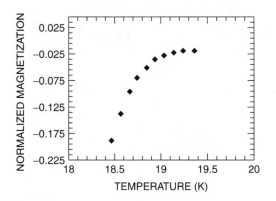

Figure 5.8. Magnetization versus temperature for K_3C_{60} showing the transition to the superconducting state. [Adapted from A. F. Hebard, *Phys. Today* **29** (Nov. 1992).]

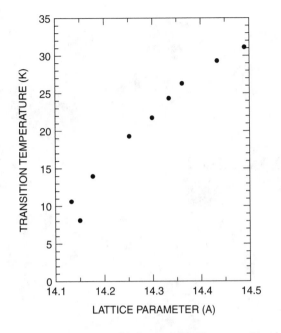

Figure 5.9. Plot of transition temperature of A_3C_{60} versus lattice parameter, where A is an alkali atom. ($10\,\mathring{A} = 1\,nm$). [Adapted from A. F. Hebard, *Phys. Today* **29** (Nov. 1992).]

5.3.6. Larger and Smaller Fullerenes

Larger fullerenes such as C_{70}, C_{76}, C_{80}, and C_{84} have also been found. A C_{20} dodecahedral carbon molecule has been synthesized by gas-phase dissociation of $C_{20}HBr_{13}$. $C_{36}H_4$ has also been made by pulsed laser ablation of graphite. A solid phase of C_{22} has been identified in which the lattice consists of C_{20} molecules bonded together by an intermediate carbon atom. One interesting aspect of the existence of these smaller fullerenes is the prediction that they could be superconductors at high temperatures when appropriately doped.

5.3.7. Other Buckyballs

What about the possibility of buckyballs made of other materials such as silicon or nitrogen? Researchers in Japan have managed to make cage structures of silicon. However, unlike carbon atoms, pure silicon cannot form closed structures. The researchers showed that silicon can form a closed structure around a tungsten atom in the form of an hexagonal cage. Potential applications of such structures are components in quantum computers, chemical catalysts, and new superconducting materials. There are a number of molecular orbital calculations that predict closed stable structures for other atoms. For example, the density functional method has been used to demonstrate that an N_{20} cluster should be stable, with the predicted

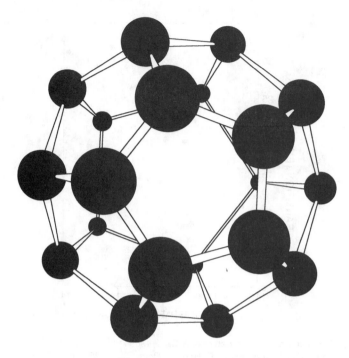

Figure 5.10. Illustration of predicted structure of the N_{20} molecule calculated by density functional theory. (F. J. Owens, unpublished.)

dodecahedral structure shown in Fig. 5.10. The calculation also revealed that the cluster would be a very powerful explosive, about 3 times more powerful than the presently most energetic material. However N_{20} may be very difficult to synthesize.

5.4. CARBON NANOTUBES

Perhaps the more interesting nanostructures with large application potential are carbon nanotubes. One can think of a carbon nanotube as a sheet of graphite rolled into a tube with bonds at the end of the sheet forming the bonds that close the tube. Figure 5.11a shows the structure of a tube formed by rolling the graphite sheet about an axis parallel to C—C bonds. A single-walled nanotube (SWNT) can have a diameter of 2 nm and a length of 100 μm, making it effectively a one dimensional structure called a *nanowire*.

5.4.1. Fabrication

Carbon nanotubes can be made by laser evaporation, carbon arc methods, and chemical vapor deposition. Figure 5.12 illustrates the apparatus for making carbon nanotubes by laser evaporation. A quartz tube containing argon gas and a graphite

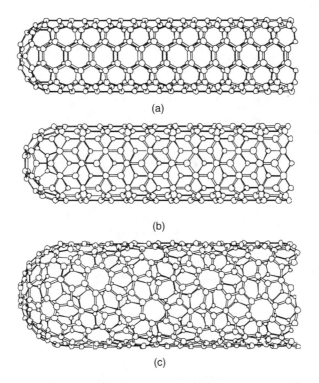

(a)

(b)

(c)

Figure 5.11. Illustration of some possible structures of carbon nanotubes, depending on how graphite sheets are rolled: (a) armchair structure; (b) zigzag structure; (c) chiral structure.

target are heated to 1200°C. Contained in the tube, but somewhat outside the furnace, is a water-cooled copper collector. The graphite target contains small amounts of cobalt and nickel that act as catalytic nucleation sites for the formation of the tubes. An intense pulsed laser beam is incident on the target, evaporating carbon from the graphite. The argon then sweeps the carbon atoms from the high-temperature zone to the colder copper collector on which they condense into nanotubes. Tubes 10–20 nm in diameter and 100 μm long can be made by this method.

Nanotubes can also be synthesized using a carbon arc. A potential of 20–25 V is applied across carbon electrodes of 5–20 μm diameter and separated by 1 mm at 500 torr pressure of flowing helium. Carbon atoms are ejected from the positive electrode and form nanotubes on the negative electrode. As the tubes form, the length of the positive electrode decreases, and a carbon deposit forms on the negative electrode. To produce single-walled nanotubes, a small amount of cobalt, nickel, or iron is incorporated as a catalyst in the central region of the positive electrode. If no catalysts are used, the tubes are nested or multiwalled types (MWNT), which are nanotubes within nanotubes, as illustrated in Fig. 5.13. The carbon arc method can produce single-walled nanotubes of diameters 1–5 nm with a length of 1 μm.

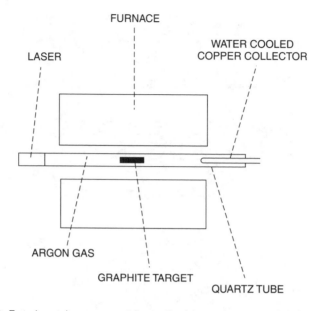

Figure 5.12. Experimental arrangement for synthesizing carbon nanotubes by laser evaporation.

The chemical vapor deposition method involves decomposing a hydrocarbon gas such as methane (CH_4) at $1100°C$. As the gas decomposes, carbon atoms are produced that then condense on a cooler substrate that may contain various catalysts such as iron. This method produced tubes with open ends, which does not occur when other methods are used. This procedure allows continuous fabrication, and may be the most favorable method for scaleup and production.

MULTI WALLED CARBON NANO TUBES

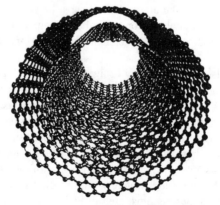

Figure 5.13. Illustration of a nested nanotube in which one tube is inside the another.

The mechanism of nanotube growth is not understood. Since the metal catalyst is necessary for the growth of SWNTs, the mechanism must involve the role of the Co or Ni atoms. One proposal referred to as the "scooter mechanism" suggests that atoms of the metal catalyst attach to the dangling bonds at the open end of the tubes, and that these atoms scoot around the rim of the tube, absorbing carbon atoms as they arrive.

Generally when nanotubes are synthesized, the result is a mix of different kinds, some metallic and some semiconducting. A group at IBM has developed a method to separate the semiconducting from the metallic nanotubes. The separation was accomplished by depositing bundles of nanotubes, some of which are metallic and some semiconducting, on a silicon wafer. Metal electrodes were then deposited over the bundle. Using the silicon wafer as an electrode, a small bias voltage was applied that prevents the semiconducting tubes from conducting, effectively making them insulators. A high voltage is then applied across the metal electrodes, thereby sending a high current through the metallic tubes but not the insulating tubes. This causes the metallic tubes to vaporize, leaving behind only the semiconducting tubes.

5.4.2. Structure

There are a variety of structures of carbon nanotubes, and these various structures have different properties. Although carbon nanotubes are not actually made by rolling graphite sheets, it is possible to explain the different structures by consideration of the way graphite sheets might be rolled into tubes. A nanotube can be formed when a graphite sheet is rolled up about the axis T shown in Fig. 5.14. The C_h vector is called the *circumferential vector*, and it is at right angles to T. Three examples of nanotube structures constructed by rolling the graphite sheet about the T vector

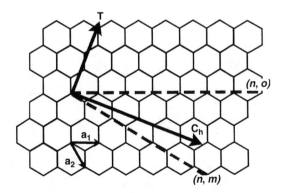

Figure 5.14. Graphitic sheet showing the basis vectors a_1 and a_2 of the two-dimensional unit cell, the axis vector T about which the sheet is rolled to generate the armchair structure nanotube sketched in Fig. 5.11a, and the circumferential vector C_h at right angles to T. Other orientations of T on the sheet generate the zigzag and chiral structures of Figs. 5.11b and 5.11c, respectively.

having different orientations in the graphite sheet are shown in Fig. 5.11. When *T* is parallel to the C—C bonds of the carbon hexagons, the structure shown in Fig. 5.11a is obtained, and it is referred to as the "armchair" structure. The tubes sketched in Figs. 5.11b and 5.11c, referred to respectively as the *zigzag* and the *chiral structures*, are formed by rolling about a *T* vector having different orientations in the graphite plane, but not parallel to C—C bonds. Looking down the tube of the chiral structure, one would see a spiraling row of carbon atoms. Generally nanotubes are closed at both ends, which involves the introduction of a pentagonal topological arrangement on each end of the cylinder. The tubes are essentially cylinders with each end attached to half of a large fullerenelike structure. In the case of SWNTs metal particles are found at the ends of the tubes, which is evidence for the catalytic role of the metal particles in their formation.

5.4.3. Electrical Properties

Carbon nanotubes have the most interesting property that they are metallic or semiconducting, depending on the diameter and chirality of the tube. *Chirality* refers to how the tubes are rolled with respect to the direction of the *T* vector in the graphite plane, as discussed above. Synthesis generally results in a mixture of tubes two-thirds of which are semiconducting and one-third metallic. The metallic tubes have the armchair structure shown in Fig. 5.11a. Figure 5.15 is a plot of the energy gap of semiconducting chiral carbon nanotubes versus the reciprocal of the diameter, showing that as the diameter of the tube increases, the bandgap decreases. Scanning tunneling microscopy (STM), which is described in Chapter 3, has been used to

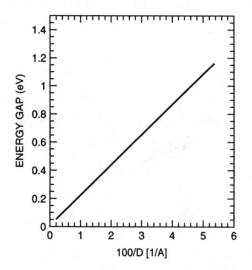

Figure 5.15. Plot of the magnitude of the energy band gap of a semiconducting, chiral carbon nanotube versus the reciprocal of the diameter of the tube ($10\,\text{Å} = 1\,\text{nm}$). [Adapted from M. S. Dresselhaus et al., *Molec. Mater.* **4**, 27 (1994).]

investigate the electronic structure of carbon nanotubes. In this measurement the position of the STM tip is fixed above the nanotube, and the voltage V between the tip and the sample is swept while the tunneling current I is monitored. The measured conductance $G = I/V$ is a direct measure of the local electronic density of states. The density of states, discussed in more detail in Chapter 2, is a measure of how close together the energy levels are to each other. Figure 5.16 gives the STM data plotted as the differential conductance, which is $(dI/dV)/(I/V)$, versus the applied voltage between the tip and carbon nanotube. The data show clearly the energy gap in materials at voltages where very little current is observed. The voltage width of this region measures the gap, which for the semiconducting material shown on the bottom of Fig. 5.16 is 0.7 eV.

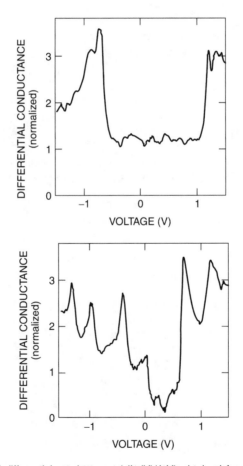

Figure 5.16. Plot of differential conductance $(dI/dV)(I/V)$ obtained from scanning tunneling microscope measurements of the tunneling current of metallic (top figure) and semiconducting (bottom figure) nanotubes. [With permission from C. Dekker, *Phys. Today* 22 (May 1999).]

At higher energies sharp peaks are observed in the density of states, referred to as *van Hove singularities*, and are characteristic of low-dimensional conducting materials. The peaks occur at the bottom and top of a number of subbands. As we have discussed earlier, electrons in the quantum theory can be viewed as waves. If the electron wavelength is not a multiple of the circumference of the tube, it will destructively interfere with itself, and therefore only electron wavelengths that are integer multiples of the circumference of the tubes are allowed. This severely limits the number of energy states available for conduction around the cylinder. The dominant remaining conduction path is along the axis of the tubes, making carbon nanotubes function as one-dimensional quantum wires. A more detailed discussion of quantum wires is presented later, in Chapter 9. The electronic states of the tubes do not form a single wide electronic energy band, but instead split into one-dimensional subbands that are evident in the data in Fig. 5.16. As we will see later, these states can be modeled by a potential well having a depth equal to the length of the nanotube.

Electron transport has been measured on individual single-walled carbon nanotubes. The measurements at a millikelvin ($T = 0.001$ K) on a single metallic nanotube lying across two metal electrodes show steplike features in the current–voltage measurements, as seen in Fig. 5.17. The steps occur at voltages which depend on the voltage applied to a third electrode that is electrostatically coupled to the nanotube. This resembles a field effect transistor made from a carbon nanotube, which is discussed below and illustrated in Fig. 5.21. The step like features in the *I–V* curve are due to single-electron tunneling and resonant tunneling through single molecular orbitals. Single electron tunneling occurs when the capacitance of the nanotube is so small that adding a single electron requires an electrostatic charging

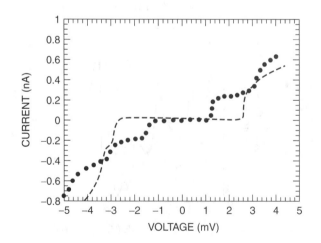

Figure 5.17. Plot of electron transport for two different gate voltages through a single metallic carbon nanotube showing steps in the *I–V* curves. [With permission from C. Dekker, *Phys. Today* 22 (May 1999).]

energy greater than the thermal energy $k_B T$. Electron transport is blocked at low voltages, which is called *Coulomb blockade*, and this is discussed in more detail in Chapter 9 (Section 9.5). By gradually increasing the gate voltage, electrons can be added to the tube one by one. Electron transport in the tube occurs by means of electron tunneling through discrete electron states. The current at each step in Fig. 5.17 is caused by one additional molecular orbital. This means that the electrons in the nanotube are not strongly localized, but rather are spatially extended over a large distance along the tube. Generally in one-dimensional systems the presence of a defect will cause a localization of the electrons. However, a defect in a nanotube will not cause localization because the effect will be averaged over the entire tube circumference because of the doughnut shape of the electron wavefunction.

In the metallic state the conductivity of the nanotubes is very high. It is estimated that they can carry a billion amperes per square centimeter. Copper wire fails at one million amperes per square centimeter because resistive heating melts the wire. One reason for the high conductivity of the carbon tubes is that they have very few defects to scatter electrons, and thus a very low resistance. High currents do not heat the tubes in the same way that they heat copper wires. Nanotubes also have a very high thermal conductivity, almost a factor of 2 more than that of diamond. This means that they are also very good conductors of heat.

Magnetoresistance is a phenomenon whereby the resistance of a material is changed by the application of a DC magnetic field. Carbon nanotubes display magnetoresistive effects at low temperature. Figure 5.18 shows a plot of the magnetic field dependence of the change in resistance ΔR of nanotubes at 2.3 and 0.35 K compared to their resistance R in zero magnetic field. This is a negative

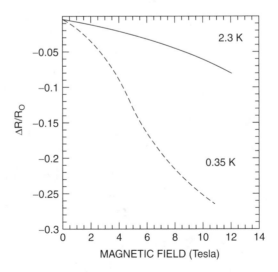

Figure 5.18. Effect of a DC magnetic field on the resistance of nanotubes at the temperatures of 0.35 and 2.3 K. (Adapted from R. Saito, G. Dresselhaus, and M. S. Dresselhaus, *Physical Properties of Nanotubes*, Imperial College Press, 1998.)

magnetoresistance effect because the resistance decreases with increasing DC magnetic field, so its reciprocal, the conductance $G = 1/R$, increases. This occurs because when a DC magnetic field is applied to the nanotubes, the conduction electrons acquire new energy levels associated with their spiraling motion about the field. It turns out that for nanotubes these levels, called *Landau levels*, lie very close to the topmost filled energy levels (the Fermi level). Thus there are more available states for the electrons to increase their energy, and the material is more conducting.

5.4.4. Vibrational Properties

The atoms in a molecule or nanoparticle continually vibrate back and forth. Each molecule has a specific set of vibrational motions, called *normal modes of vibration*, which are determined by the symmetry of the molecule For example carbon dioxide CO_2, which has the structure O=C=O, is a bent molecule with three normal modes. One mode involves a bending of the molecule. Another, called the *symmetric stretch*, consists of an in-phase elongation of the two C=O bonds. The asymmetric

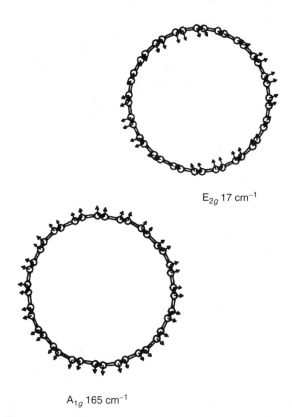

E_{2g} 17 cm^{-1}

A_{1g} 165 cm^{-1}

Figure 5.19. Illustration of two normal modes of vibration of carbon nanotubes.

stretch consists of out-of-phase stretches of the C=O bond length, where one bond length increases while the other decreases. Similarly carbon nanotubes also have normal modes of vibration. Figure 5.19 illustrates two of the normal modes of nanotubes. One mode, labeled A_{1g}, involves an "in and out" oscillation of the diameter of the tube. Another mode, the E_{2g} mode, involves a squashing of the tube where it squeezes down in one direction and expands in the perpendicular direction essentially oscillating between a sphere and an ellipse. The frequencies of these two modes are Raman-active and depend on the radius of the tube. Figure 5.20 is a plot of the frequency of the A_{1g} mode as a function of this radius. The dependence of this frequency on the radius is now routinely used to measure the radius of nanotubes.

5.4.5. Mechanical Properties

Carbon nanotubes are very strong. If a weight W is attached to the end of a thin wire nailed to the roof of a room, the wire will stretch. The stress S on the wire is defined as the load, or the weight per unit cross-sectional area A of the wire:

$$S = \frac{W}{A} \tag{5.3}$$

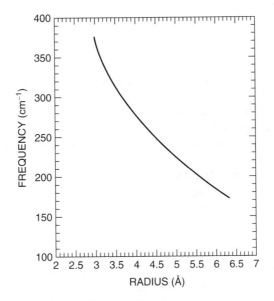

Figure 5.20. Plot of the frequency of the Raman A_{1g} vibrational normal mode versus the radius of the nanotube. ($10\,\text{Å} = 1\,\text{nm}$). (Adapted from R. Saito, G. Dresselhaus, and M. S. Dresselhaus, *Physical Properties of Nanotubes*, Imperial College Press, 1998.)

The strain e is defined as the amount of stretch ΔL of the wire per unit length L

$$e = \frac{\Delta L}{L} \tag{5.4}$$

where L is the length of the wire before the weight is attached. Hooke's law says that the increase in the length of the wire is proportional to the weight at the end of the wire. More generally, we say stress S is proportional to strain e:

$$S = Ee \tag{5.5}$$

The proportionality constant $E = LW/A\,\Delta L$ is Young's modulus, and it is a property of a given material. It characterizes the elastic flexibility of a material. The larger the value of Young's modulus, the less flexible the material. Young's modulus of steel is about 30,000 times that of rubber. Carbon nanotubes have Young's moduli ranging from 1.28 to 1.8 TPa. One terapascal (TPa) is a pressure very close to 10^7 times atmospheric pressure. Young's modulus of steel is 0.21 TPa, which means that Young's modulus of carbon nanotubes is almost 10 times that of steel. This would imply that carbon nanotubes are very stiff and hard to bend. However, this is not quite true because they are so thin. The deflection D of a cylindrical hollow beam of length L with a force F on the end and the inner and outer radii of r_i and r_o, has been shown to be

$$D = \frac{FL^3}{3EI} \tag{5.6}$$

where I is the areal moment of inertia given by $\Pi(r_o^4 - r_i^4)/4$, Since the wall thickness of carbon nanotubes is about 0.34 nm, $r_o^4 - r_i^4$ is very small, somewhat compensating for the large value of E.

When carbon nanotubes are bent, they are very resilient. They buckle like straws but do not break, and can be straightened back without any damage. Most materials fracture on bending because of the presence of defects such as dislocations or grain boundaries. Because carbon nanotubes have so few defects in the structure of their walls, this does not occur. Another reason why they do not fracture is that as they are bent severely, the almost hexagonal carbon rings in the walls change in structure but do not break. This is a unique result of the fact that the carbon–carbon bonds are sp^2 hybrids, and these sp^2 bonds can rehybridize as they are bent. The degree of change and the amount of s–p admixture both depend on the degree of bending of the bonds.

Strength is not the same as stiffness. Young's modulus is a measure of how stiff or flexible a material is. Tensile strength is a measure of the amount of stress needed to pull a material apart. The tensile strength of carbon nanotubes is about 45 billion pascals. High-strength steel alloys break at about 2 billion pascals. Thus carbon nanotubes are about 20 times stronger than steel. Nested nanotubes also have improved mechanical properties, but they are not as good as their single-walled counterparts. For example, multi-walled nanotubes of 200 nm diameter have a tensile strength of 0.007 TPa (i.e., 7 GPa) and a modulus of 0.6 TPa.

5.5. APPLICATIONS OF CARBON NANOTUBES

The unusual properties of carbon nanotubes make possible many applications ranging from battery electrodes, to electronic devices, to reinforcing fibers, which make stronger composites. In this section we describe some of the potential applications that researchers are now working on. However, for the application potential to be realized, methods for large-scale production of single-walled carbon nanotubes will have to be developed. The present synthesis methods provide only small yields, and make the cost of the tubes about $1500 per gram ($680,000 per pound). On the other hand, large-scale production methods based on chemical deposition have been developed for multiwalled tubes, which are presently available for $60 per pound, and as demand increases, this price is expected to drop significantly. The methods used to scale up the multiwalled tubes should provide the basis for scaling up synthesis of single-walled nanotubes. Because of the enormous application potential, it might be reasonable to hope that large-scale synthesis methods will be developed, resulting in a decrease in the cost to the order of $10 per pound.

5.5.1. Field Emission and Shielding

When a small electric field is applied parallel to the axis of a nanotube, electrons are emitted at a very high rate from the ends of the tube. This is called *field emission*. This effect can easily be observed by applying a small voltage between two parallel metal electrodes, and spreading a composite paste of nanotubes on one electrode. A sufficient number of tubes will be perpendicular to the electrode so that electron emission can be observed. One application of this effect is the development of flat panel displays. Television and computer monitors use a controlled electron gun to impinge electrons on the phosphors of the screen, which then emit light of the appropriate colors. Samsung in Korea is developing a flat-panel display using the electron emission of carbon nanotubes. A thin film of nanotubes is placed over control electronics with a phosphor-coated glass plate on top. A Japanese company is using this electron emission effect to make vacuum tube lamps that are as bright as conventional light bulbs, and longer-lived and more efficient. Other researchers are using the effect to develop a way to generate microwaves.

The high electrical conductivity of carbon nanotubes means that they will be poor transmitters of electromagnetic energy. A plastic composite of carbon nanotubes could provide lightweight shielding material for electromagnetic radiation. This is a matter of much concern to the military, which is developing a highly digitized battlefield for command, control, and communication. The computers and electronic devices that are a part of this system need to be protected from weapons that emit electromagnetic pulses.

5.5.2. Computers

The feasibility of designing field-effect transistors (FETs), the switching components of computers, based on semiconducting carbon nanotubes connecting two gold electrodes, has been demonstrated. An illustration of the device is shown in Fig. 5.21. When a small voltage is applied to the gate, the silicon substrate, current flows through the nanotube between the source and the drain. The device is switched on when current is flowing, and off when it is not. It has been found that a small voltage applied to the gate can change the conductivity of the nanotube by a factor of $>1 \times 10^6$, which is comparable to silicon field-effect transistors. It has been estimated that the switching time of these devices will be very fast, allowing clock speeds of a terahertz, which is 10^4 times faster than present processors. The gold sources and drains are deposited by lithographic methods, and the connecting nanotube wire is less than one nanometer in diameter. This small size should allow more switches to be packed on a chip. It should be emphasized that these devices have been built in the laboratory one at a time, and methods to produce them cheaply in large scale on a chip will have to be developed before they can be used in applications such as computers.

A major objective of computer developers is to increase the number of switches on a chip. The approach to this is to use smaller-diameter interconnecting wires and smaller switches, and to pack then more tightly on the chip. However, there are some difficulties in doing this with present metal interconnect wire and available switches. As the cross section of a metal wire, such as copper, decreases, the resistance increases, and the heat generated by current flowing in the wire increases. The heat can reach such a value that it can melt or vaporize the wire. However, carbon nanotubes with diameters of 2 nm have extremely low resistance, and thus can carry large currents without heating, so they could be used as interconnects. Their very high thermal conductivity means that they can also serve as heat sinks, allowing heat to be rapidly transferred away from the chip.

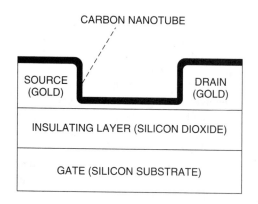

Figure 5.21. A schematic of a field-effect transistor made from a carbon nanotube.

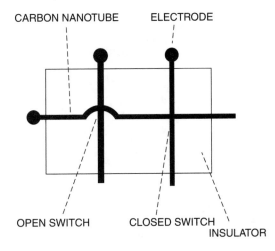

Figure 5.22. Illustration of the concept of a computer switching device made from carbon nanotubes. [Adapted from C. M. Lieber, *Sci. Am.*, 59 (Sept. 2001).]

Another idea that is being pursued is to make a computer out of carbon nanotubes. The computer would be an array of parallel nanotubes on a substrate. Above this, but not touching the lower array and having a small separation from them, are carbon nanotubes tubes oriented perpendicular to the tubes on the substrate. Each tube would be connected to a metal electrode. Figure 5.22 illustrates the concept. The crossing points would represent the switches of the computer. When the tubes are not touching at the crossing points, the switch is off because the resistance is high. In the ON state the tubes are in contact and have a low resistance. The ON and OFF configurations can be controlled by the flow of current in the tubes. The researchers estimate that 10^{12} switches could fit on a square centimeter chip. Present Pentium chips have about 10^8 switches on them. The switching rate of such devices is estimated to be about 100 times faster than that of the present generation of Intel chips. Ideally one would like to have semiconducting nanotubes on the bottom and metallic nanotubes on the top. Then, when contact is made, there would be a metal–semiconductor junction that allows current to flow in only one direction; thus the junction is a rectifier.

5.5.3. Fuel Cells

Carbon nanotubes have applications in battery technology. Lithium, which is a charge carrier in some batteries, can be stored inside nanotubes. It is estimated that one lithium atom can be stored for every six carbons of the tube. Storing hydrogen in nanotubes is another possible application, one that is related to the development of fuel cells as sources of electrical energy for future automobiles. A fuel cell consists of two electrodes separated by a special electrolyte that allows hydrogen ions, but not electrons, to pass through it. Hydrogen is sent to the anode, where it is ionized. The

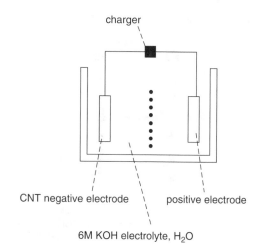

charger

CNT negative electrode   positive electrode

6M KOH electrolyte, H$_2$O

Figure 5.23. An electrochemical cell used to inject hydrogen into carbon nanotubes. The cell consists of an electrolytic solution of KOH with a negative electrode consisting of carbon nanotube (CNT) paper. Application of a voltage between the electrodes causes the H$^+$ ion to be attracted to the negative electrode. (Z. Iqbal, unpublished.)

freed electrons travel through an external circuit wire to the cathode. The hydrogen ions diffuse through the electrolyte to the cathode, where electrons, hydrogen, and oxygen combine to form water. The system needs a source of hydrogen. One possibility is to store the hydrogen inside carbon nanotubes. It is estimated that to be useful in this application, the tubes need to hold 6.5% hydrogen by weight. At present only about 4% hydrogen by weight has been successfully put inside the tubes.

An elegant method to put hydrogen into carbon nanotubes employs the electrochemical cell sketched in Fig. 5.23. Single-walled nanotubes in the form of paper are the negative electrode in a KOH electrolyte solution. A counterelectrode consists of Ni(OH)$_2$. The water of the electrolyte is decomposed into positive hydrogen ions (H$^+$) that are attracted to the negative SWNT electrode. The presence of hydrogen bonded to the tubes is indicated by a decrease in the intensity of a Raman-active vibration, as indicated in Fig. 5.24, which shows the Raman spectrum before and after the material is subjected to the electrochemical process.

5.5.4. Chemical Sensors

A field-effect transistor similar to the one shown in Fig. 5.21 made of the chiral semiconducting carbon nanotubes has been demonstrated to be a sensitive detector of various gases. The transistor was placed in a 500 ml flask having electrical feed throughs and inlet and outlet valves to allow gases to flow over the carbon nanotubes of the FET. Two to 200 parts per million of NO$_2$ flowing at a rate of 700 ml/min for 10 min caused a threefold increase in the conductance of the carbon nanotubes.

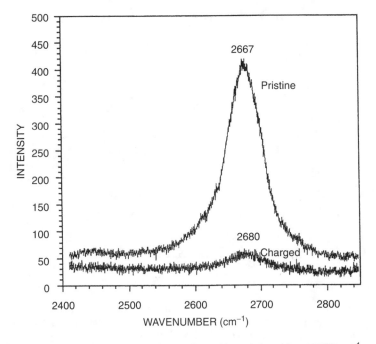

Figure 5.24. Raman spectrum of carbon nanotubes with peak intensity at 2667 cm^{-1} recorded before (pristine) and after (charged) treatment in the electrochemical cell sketched in Fig. 5.23. (From Z. Iqbal, unpublished.)

Figure 5.25 shows the current–voltage relationship before and after exposure to NO_2. These data were taken for a gate voltage of 4 V. The effect occurs because when NO_2 bonds to the carbon nanotube, charge is transferred from the nanotube to the NO_2, increasing the hole concentration in the carbon nanotube and enhancing the conductance.

The frequency of one of the normal-mode vibrations of the nanotubes, which gives a very strong Raman line, is also very sensitive to the presence of other molecules on the surface of the tubes. The direction and the magnitude of the shift depend on the kind of molecule on the surface. This effect could also be the basis of a chemical gas sensor employing nanotubes.

5.5.5. Catalysis

A catalytic agent is a material, typically a metal or alloy, that enhances the rate of a reaction between chemicals. Nanotubes serve as catalysts for some chemical reactions. For example, nested nanotubes with ruthenium metal bonded to the outside have been demonstrated to have a strong catalytic effect in the hydrogenation reaction of cinnamaldehyde ($C_6H_5CH=CHCHO$) in the liquid phase compared with the effect when the same metal Ru is attached to other carbon substrates. Chemical

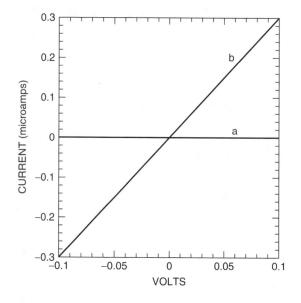

Figure 5.25. Plot of current versus voltage for carbon nanotube field effect transistor before (line a) and after (line b) exposure to NO_2 gas. These data were taken for a 4 V gate voltage. [Adapted from J. Kong et al., *Science* **287**, 622 (2000).]

reactions have also been carried out inside nanotubes, such as the reduction of nickel oxide NiO to the base metal Ni, and the reduction of $AlCl_3$ to its base metal Al. A stream of hydrogen gas H_2 at 475°C partially reduces MoO_3 to MoO_2, with the accompanying formation of steam H_2O, inside multiwalled nanotubes. Cadmium sulfide (CdS) crystals have been formed inside nanotubes by reacting cadmium oxide (CdO) crystals with hydrogen sulfide gas (H_2S) at 400°C.

5.5.6. Mechanical Reinforcement

The use of long carbon fibers such as polyacrylonitrile (PAN) is an established technology to increase the strength of plastic composites. PAN has a tensile strength in the order of 7 Gpa and can have diameters of 1–10 μm. The use of this fiber for reinforcement requires developing methods to have the fibers preferentially oriented and uniformly dispersed in the material. The fiber must be able to survive the processing conditions. Important parameters in determining how effective a fiber is in increasing the strength of a composite are the tensile strength of the fiber and the length : diameter ratio, as well as the ability of the fiber to bind to the matrix. Because of their high tensile strength and large length : diameter ratios, carbon nanotubes should be excellent materials for composite reinforcement. Some preliminary work has been done in this area. Work at General Motors Research and Development Center has shown that adding to polypropylene 11.5% by weight of nested carbon nanotubes having an 0.2 μm diameter approximately doubled the tensile strength of the polypropylene. A study at the University of Tokyo showed that incorporation of 5% by volume of nanotubes in aluminum increased the tensile strength by a factor of 2 compared to pure aluminum subjected to the same processing. The composites were prepared by hot pressing and hot extrusion. Aluminum powder and carbon nanotubes were mixed and heated to over 800 K in a vacuum, and then compressed with steel dies. After this the melt was extruded into rods. This work is very important in that it demonstrates that the carbon nanotubes can be put into aluminum, and are chemically stable through the necessary processing. The researchers believe a substantial increase in the tensile strength can be achieved by producing a more homogenous and an aligned distribution of nanotubes in the material. Theoretical estimates suggest that with optimum fabrication a 10% volume fraction of nanotubes should increase the tensile strength by a factor of 6.

However, the possibility that nanotube walls slide with respect to each other in MWNTs, or that individual SWNTs slip in the bundles of tubes, may mean that the actual strengths that are obtainable will be less than expected. The atomically smooth surfaces of nanotubes may mean that they will not have strong interface interactions with the material being reinforced. On the other hand, carbon nanotubes have been shown to form strong bonds with iron which is the main constituent of steel, suggesting the possibility that nanotubes could be used to increase the tensile strength of steel. Figure 5.26 shows the results of a calculation of the tensile strength of steel versus volume fraction of SWNTs having a 10 nm diameter and a 100 μm length using a formula called the *Kelly–Tyson equation*. The calculation predicts that

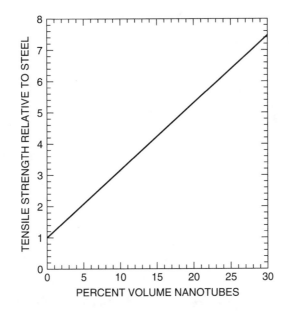

Figure 5.26. Tensile strength of steel versus volume fraction of carbon nanotubes calculated by the Kelly–Tyson formula. The nanotubes were 100 μm long and 10 nm in diameter.

the tensile strength of steel could be increased by 7 times if the steel had 30% by weight of oriented carbon nanotubes in it. While all of these results are promising, there is much work to be done in this area, particularly in developing methods to incorporate the tubes into plastics and metals. This particular application, like some of the applications discussed earlier, will, of course, require inexpensive bulk fabrication of nanotubes.

FURTHER READING

P. M. Ajayan, "Carbon Nanotubes," *Handbook of Nanostructured Materials and Nanotechnology,* H. S. Nalwa, ed., Academic Press, San Diego, 2000, Vol. 5, Chapter 6, p. 375.

P. G. Collins and P. Avouris, "Carbon Nanotubes," *Sci. Am.*, 62 (Dec. 2000).

M. S. Dresselhaus, G. Dresselhaus, and P. C. Eklund, *Science of Fullerenes and Carbon Nanotubes*, Academic Press, San Diego, 1995.

M. S. Dresselhaus, G. Dresselhaus, and R. Saito, "Nanotechnology in Carbon Materials," in *Nanotechnology*, G. Timp, ed., Springer-Verlag, 1998, Chapter 7, p. 285.

T. W. Ebbesen, "Carbon Nanotubes," *Phys. Today*, 26 (June, 1996).

R. Saito, G. Dresselhaus, and M. S. Dresselhaus, *Physical Properties of Carbon Nanotubes*, Imperial College Press, London, 1999.

Scientific American, Sept. 2001 issue (contains a number of articles on nanotechnology).

R. E. Smalley and B. I. Yakobson, "Future of Fullerenes," *Solid State Commun.* **107**, 597 (1998).

6

BULK NANOSTRUCTURED
MATERIALS

In this chapter the properties of bulk nanostructured materials are discussed. Bulk nanostructured materials are solids having a nanosized microstructure. The basic units that make up the solids are nanoparticles. The nanoparticles can be disordered with respect to each other, where their symmetry axes are randomly oriented and their spatial positions display no symmetry. The particles can also be ordered in lattice arrays displaying symmetry. Figure 6.1a illustrates a hypothetical two-dimensional ordered lattice of Al_{12} nanoparticles, and Fig. 6.1b shows a two-dimensional bulk disordered nanostructure of these same nanoparticles.

6.1. SOLID DISORDERED NANOSTRUCTURES

6.1.1. Methods of Synthesis

In this section we will discuss some of the ways that disordered nanostructured solids are made. One method is referred to as *compaction and consolidation*. As an example of such a process, let us consider how nanostructured Cu–Fe alloys are made. Mixtures of iron and copper powders having the composition $Fe_{85}Cu_{15}$ are ballmilled for 15 h at room temperature. The material is then compacted using a

---

Introduction to Nanotechnology, by Charles P. Poole Jr. and Frank J. Owens.
ISBN 0-471-07935-9. Copyright © 2003 John Wiley & Sons, Inc.

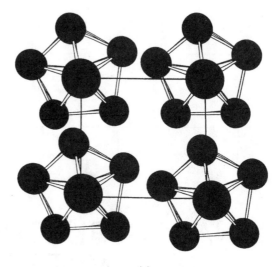

(a)

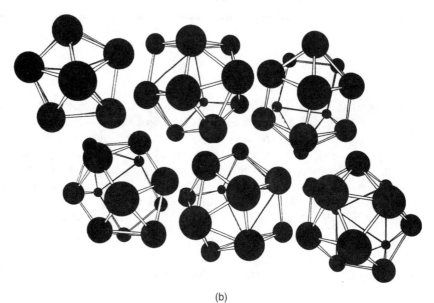

(b)

Figure 6.1. (a) Illustration of a hypothetical two-dimensional square lattice of Al_{12} particles, and (b) illustration of a two-dimensional bulk solid of Al_{12} where the nanoparticles have no ordered arrangement with respect to each other.

tungsten–carbide die at a pressure of 1 GPa for 24 h. This compact is then subjected to hot compaction for 30 min at temperatures in the vicinity of 400°C and pressure up to 870 MPa. The final density of the compact is 99.2% of the maximum possible density. Figure 6.2 presents the distribution of grain sizes in the material showing that it consists of nanoparticles ranging in size from 20 to 70 nm, with the largest

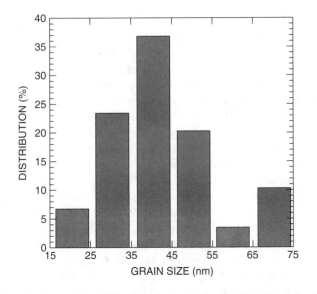

Figure 6.2. Distribution of sizes of Fe–Cu nanoparticles made by hot compaction methods described in the text. [Adapted from L. He and E. Ma, *J. Mater. Res.* **15**, 904 (2000).]

number of particles having 40 nm sizes. Figure 6.3 shows a stress–strain curve for this case. Its Young's modulus, which is the slope of the curve in the linear region, is similar to that of conventional iron. The deviation from linearity in the stress–strain curve shows there is a ductile region before fracture where the material displays elongation. The data show that fracture occurs at 2.8 GPa which is about 5 times the fracture stress of iron having larger grain sizes, ranging from 50 to 150 μm. Significant modifications of the mechanical properties of disordered bulk

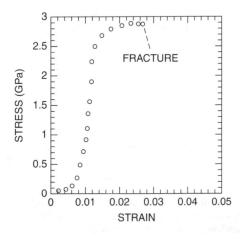

Figure 6.3. Stress–strain curve for bulk compacted nanostructured Fe–Cu material, showing fracture at a stress of 2.8 GPa. [Adapted from L. He and E. Ma, *J. Mater. Res.* **15**, 904 (2000).]

materials having nanosized grains is one of the most important properties of such materials. Making materials with nanosized grains has the potential to provide significant increases in yield stress, and has many useful applications such as stronger materials for automobile bodies. The reasons for the changes in mechanical properties of nanostructured materials will be discussed below.

Nanostructured materials can be made by rapid solidification. One method illustrated in Fig. 6.4 is called "chill block melt spinning." RF (radiofrequency) heating coils are used to melt a metal, which is then forced through a nozzle to form a liquid stream. This stream is continuously sprayed over the surface of a rotating metal drum under an inert-gas atmosphere. The process produces strips or ribbons ranging in thickness from 10 to 100 μm. The parameters that control the nano-structure of the material are nozzle size, nozzle-to-drum distance, melt ejection pressure, and speed of rotation of the metal drum. The need for light weight, high strength materials has led to the development of 85–94% aluminum alloys with other metals such as Y, Ni, and Fe made by this method. A melt spun alloy of Al–Y–Ni–Fe consisting of 10–30-nm Al particles embedded in an amorphous matrix can have a tensile strength in excess of 1.2 GPa. The high value is attributed to the presence of defect free aluminum nanoparticles. In another method of making nanostructured

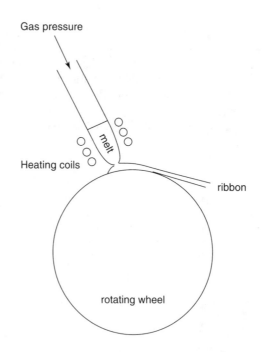

Figure 6.4. Illustration of the chill block melting apparatus for producing nanostructured materials by rapid solidification on a rotating wheel. (With permission from I. Chang, in *Handbook of Nanostructured Materials and Nanotechnology*, H. S. Nalwa, ed., Academic Press, San Diego, 2000, Vol. 1, Chapter 11, p. 501.)

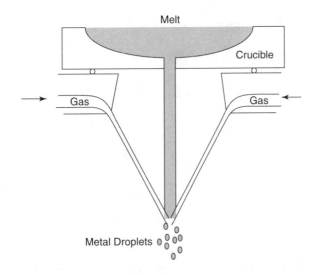

Figure 6.5. Illustration of apparatus for making droplets of metal nanoparticles by gas atomization. (With permission from I. Chang, in *Handbook of Nanostructured Materials and Nanotechnology*, H. S. Nalwa, ed., Academic Press, San Diego, 2000, Vol. 1, Chapter 11, p. 501.)

materials, called *gas atomization*, a high-velocity inert-gas beam impacts a molten metal. The apparatus is illustrated in Fig 6.5. A fine dispersion of metal droplets is formed when the metal is impacted by the gas, which transfers kinetic energy to the molten metal. This method can be used to produce large quantities of nanostructured powders, which are then subjected to hot consolidation to form bulk samples.

Nanostructured materials can be made by electrodeposition. For example, a sheet of nanostructured Cu can be fabricated by putting two electrodes in an electrolyte of $CuSO_4$ and applying a voltage between the two electrodes. A layer of nanostructured Cu will be deposited on the negative titanium electrode. A sheet of Cu 2 mm thick can be made by this process, having an average grain size of 27 nm, and an enhanced yield strength of 119 MPa.

6.1.2. Failure Mechanisms of Conventional Grain-Sized Materials

In order to understand how nanosized grains effect the bulk structure of materials, it is necessary to discuss how conventional grain-sized materials fail mechanically. A brittle material fractures before it undergoes an irreversible elongation. Fracture occurs because of the existence of cracks in the material. Figure 6.6 shows an example of a crack in a two-dimensional lattice. A "crack" is essentially a region of a material where there is no bonding between adjacent atoms of the lattice. If such a material is subjected to tension, the crack interrupts the flow of stress. The stress accumulates at the bond at the end of the crack, making the stress at that bond very high, perhaps exceeding the bond strength. This results in a breaking of the bond at the end of the crack, and a lengthening of the crack. Then the stress builds up on the

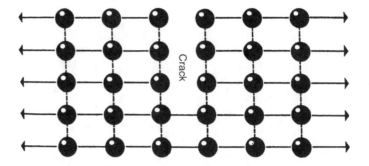

Figure 6.6. A crack in a two-dimensional rectangular lattice.

next bond at the bottom of the crack and it breaks. This process of crack propagation continues until eventually the material separates at the crack. A crack provides a mechanism whereby a weak external force can break stronger bonds one by one. This explains why the stresses that induce fracture are actually weaker than the bonds that hold the atoms of the metal together. Another kind of mechanical failure, is the *brittle-to-ductile transition*, where the stress–strain curve deviates from linearity, as seen in Fig. 6.3. In this region the material irreversibly elongates before fracture. When the stress is removed after the brittle to ductile transition the material does not return to its original length. The transition to ductility is a result of another kind of defect in the lattice called a *dislocation*. Figure 6.7 illustrates an edge dislocation in a two-dimensional lattice. There are also other kinds of dislocations such as a screw dislocation. Dislocations are essentially regions where lattice deviations from a regular structure extend over a large number of lattice spacings. Unlike cracks, the atoms in the region of the dislocation are bonded to each other, but the bonds are weaker than in the normal regions. In the ductile region one part of the lattice is able to slide across an adjacent part of the lattice. This occurs between sections of the lattice located at dislocations where the bonds between the atoms

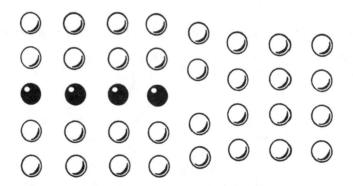

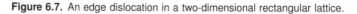

Figure 6.7. An edge dislocation in a two-dimensional rectangular lattice.

along the dislocation are weaker. One method of increasing the stress at which the brittle-to-ductile transition occurs is to impede the movement of the dislocations by introducing tiny particles of another material into the lattice. This process is used to harden steel, where particles of iron carbide are precipitated into the steel. The iron carbide particles block the movement of the dislocations.

6.1.3. Mechanical Properties

The intrinsic elastic modulus of a nanostructured material is essentially the same as that of the bulk material having micrometer-sized grains until the grain size becomes very small, less than 5 nm. As we saw in Chapter 5, Young's modulus is the factor relating stress and strain. It is the slope of the stress–strain curve in the linear region. The larger the value of Young's modulus, the less elastic the material. Figure 6.8 is a plot of the ratio of Young's modulus E in nanograined iron, to its value in conventional grain-sized iron E_0, as a function of grain size. We see from the figure that below $\sim$20 nm, Young's modulus begins to decrease from its value in conventional grain-sized materials.

The yield strength σ_y of a conventional grain-sized material is related to the grain size by the Hall–Petch equation

$$\sigma_y = \sigma_0 + Kd^{-(1/2)} \tag{6.1}$$

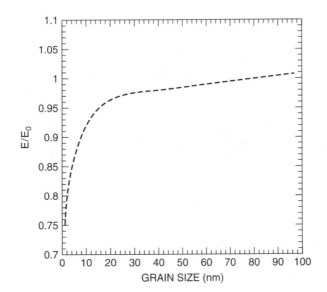

Figure 6.8. Plot of the ratio of Young's modulus E in nanograin iron to its value E_0 in conventional granular iron as a function of grain size.

where σ_o is the frictional stress opposing dislocation movement, K is a constant, and d is the grain size in micrometers. Hardness can also be described by a similar equation. Figure 6.9 plots the measured yield strength of Fe–Co alloys as a function of $d^{-(1/2)}$, showing the linear behavior predicted by Eq. (6.1). Assuming that the equation is valid for nanosized grains, a bulk material having a 50-nm grain size would have a yield *strength* of 4.14 GPa. The reason for the increase in yield strength with smaller grain size is that materials having smaller grains have more grain boundaries, blocking dislocation movement. Deviations from the Hall–Petch behavior have been observed for materials made of particles less than 20 nm in size. The deviations involve no dependence on particle size (zero slope) to decreases in yield strength with particle size (negative slope). It is believed that conventional dislocation-based deformation is not possible in bulk nanostructured materials with sizes less than 30 nm because mobile dislocations are unlikely to occur. Examination of small-grained bulk nanomaterials by transmission electron microscopy during deformation does not show any evidence for mobile dislocations.

Most bulk nanostructured materials are quite brittle and display reduced ductility under tension, typically having elongations of a few percent for grain sizes less than 30 nm. For example, conventional coarse-grained annealed polycrystalline copper is very ductile, having elongations of up to 60%. Measurements in samples with grain sizes less than 30 nm yield elongations no more than 5%. Most of these measurements have been performed on consolidated particulate samples, which have large residual stress, and flaws due to imperfect particle bonding, which restricts disloca-

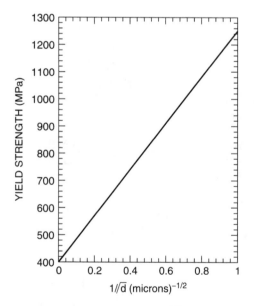

Figure 6.9. Yield *strength* of Fe–Co alloys versus $1/d^{1/2}$, where d is the size of the grain. [Adapted from C.-H. Shang et al., *J. Mater. Res.* **15**, 835 (2000).]

tion movement. However, nanostructured copper prepared by electrodeposition displays almost no residual stress and has elongations up to 30% as shown in Fig. 6.10. These results emphasize the importance of the choice of processing procedures, and the effect of flaws and microstructure on measured mechanical properties. In general, the results of ductility measurements on nanostructured bulk materials are mixed because of sensitivity to flaws and porosity, both of which depend on the processing methods.

6.1.4. Nanostructured Multilayers

Another kind of bulk nanostructure consists of periodic layers of nanometer thickness of different materials such as alternating layers of TiN and NbN. These layered materials are fabricated by various vapor-phase methods such as sputter deposition and chemical vapor-phase deposition. They can also be made by electrochemical deposition, which is discussed in Section 6.1.1. The materials have very large interface area densities. This means that the density of atoms on the planar boundary between two layers is very high. For example, a square centimeter of a 1-μm-thick multilayer film having layers of 2 nm thickness has an interface area of 1000 cm^2. Since the material has a density of about 6.5 g/cm^3, the interface area density is 154 m^2/g, comparable to that of typical heterogeneous catalysts (see Chapter 10). The interfacial regions have a strong influence on the properties of these materials. These layered materials have very high hardness,

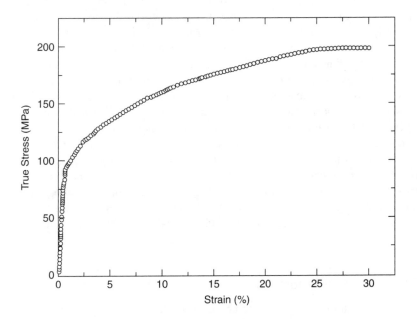

Figure 6.10. Stress–strain curve of nanostructured copper prepared by electrodeposition. [Adapted from L. Lu et al., *J. Mater. Res.* **15**, 270 (2000).]

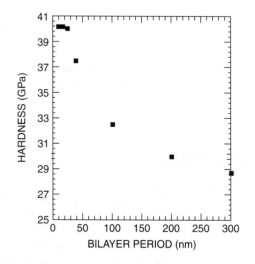

Figure 6.11. Plot of the hardness of TiN/NbN multilayer materials as a function of the thickness of the layers. (Adapted from B. M. Clemens, MRS Bulletin, Feb. 1999, p. 20.)

which depends on the thickness of the layers, and good wear resistance. Hardness is measured using an indentation load depth sensing apparatus which is commercially available, and is called a *nanoindenter*. A pyramidal diamond indenter is pressed into the surface of the material with a load, $L(h)$ and the displacement of the tip is measured. Hardness is defined as $L(h)/A(h)$ where $A(h)$ is the area of the indentation. Typically measurements are made at a constant load rate of ~ 20 mN/s.

Figure 6.11 shows a plot of the hardness of a TiN/NbN nanomultilayered structure as a function of the bilayer period (or thickness) of the layers, showing that as the layers get thinner in the nanometer range there is a significant enhancement of the hardness until ~ 30 nm, where it appears to level off and become constant. It has been found that a mismatch of the crystal structures between the layers actually enhances the hardness. The compounds TiN and NbN both have the same rock salt or NaCl structure with the respective lattice constants 0.4235 and 0.5151 nm, so the mismatch between them is relatively large, as is the hardness. Harder materials have been found to have greater differences between the shear modulus of the layers. Interestingly, multilayers in which the alternating layers have different crystal structures were found to be even harder. In this case dislocations moved less easily between the layers, and essentially became confined in the layers, resulting in an increased hardness.

6.1.5. Electrical Properties

For a collection of nanoparticles to be a conductive medium, the particles must be in electrical contact. One form of a bulk nanostructured material that is conducting consists of gold nanoparticles connected to each other by long molecules. This network is made by taking the gold particles in the form of an aerosol spray and subjecting them to a fine mist of a thiol such as dodecanethiol RSH, where R is $C_{12}H_{25}$. These alkyl thiols have an end group $-SH$ that can attach to a methyl $-CH_3$, and a methylene chain 8–12 units long that provides steric repulsion between the chains. The chainlike molecules radiate out from the particle. The encapsulated gold particles are stable in aliphatic solvents such as hexane. However, the addition of a small amount of dithiol to the solution causes the formation of a three-dimensional cluster network that precipitates out of the solution. Clusters of particles can also be deposited on flat surfaces once the colloidal solution of encapsulated nanoparticles has been formed. In-plane electronic conduction has been measured in two-dimensional arrays of 500-nm gold nanoparticles connected or linked to each other by conjugated organic molecules. A lithographically fabricated device allowing electrical measurements of such an array is illustrated in Fig. 6.12. Figure 6.13 gives a measurement of the current versus voltage for a chain without (line a) and with (line b) linkage by a conjugated molecule. Figure 6.14 gives the results of a measurement of a linked cluster at a number of different temperatures. The conductance G, which is defined as the ratio of the current I, to the voltage V, is the reciprocal of the resistance: $R = V/I = 1/G$. The data in Fig. 6.13 show that linking the gold nanoparticles substantially increases the conductance. The temperature dependence of the low-voltage conductance is given by

$$G = G_o \exp\left(\frac{-E}{k_B T}\right) \tag{6.2}$$

where E is the activation energy. The conduction process for this system can be modeled by a hexagonal array of single-crystal gold clusters linked by resistors, which are the connecting molecules, as illustrated in Fig. 6.15. The mechanism of

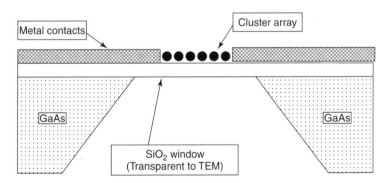

Figure 6.12. Cross-sectional view of a lithographically fabricated device to measure the electrical conductivity in a two-dimensional array of gold nanoparticles linked by molecules. (With permission from R. P Andres et al., in *Handbook of Nanostructured Materials and Nanotechnology*, H. S. Nalwa, ed., Academic Press, San Diego, 2000, Vol. 3, Chapter 4, p. 217.

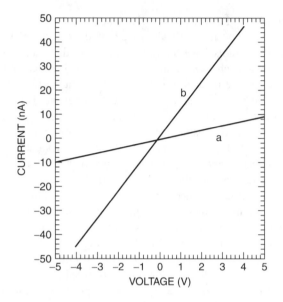

Figure 6.13. Room-temperature current–voltage relationship for a two-dimensional cluster array: without linkage (line a) and with the particles linked by a $(CN)_2C_{18}H_{12}$ molecule (line b). [Adapted from D. James et al., *Superlatt. Microstruct.* **18**, 275 (1995).]

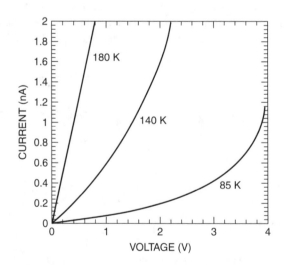

Figure 6.14. Measured current–voltage relationship for a two-dimensional linked cluster array at the temperatures of 85, 140, and 180 K. [Adapted from D. James et al., *Superlatt. Microstruct.* **18**, 275 (1995).]

(111) facets

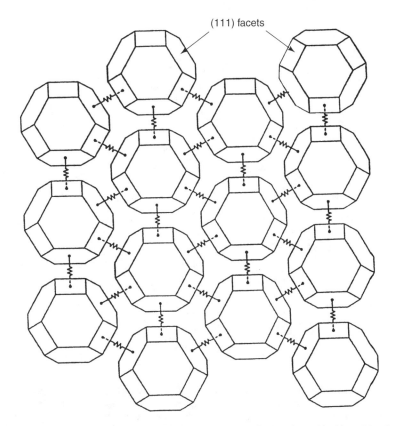

Figure 6.15. Sketch of a model to explain the electrical conductivity in an ideal hexagonal array of single-crystal gold clusters with uniform intercluster resistive linkage provided by resistors connecting the molecules. (With permission from R. P. Andres et al., in *Handbook of Nanostructured Materials and Nanotechnology*, H. S. Nalwa, ed., Academic Press, San Diego, 2000, Vol. 3, Chapter 4, p. 221.)

conduction is electron tunneling from one metal cluster to the next. Section 9.5 (of Chapter 9) discusses a similar case for smaller gold nanoparticles.

The tunneling process is a quantum-mechanical phenomenon where an electron can pass through an energy barrier larger than the kinetic energy of the electron. Thus, if a sandwich is constructed consisting of two similar metals separated by a thin insulating material, as shown in Fig. 6.16a, under certain conditions an electron can pass from one metal to the other. For the electron to tunnel from one side of the junction to the other, there must be available unoccupied electronic states on the other side. For two identical metals at $T = 0 \, \text{K}$, the Fermi energies will be at the same level, and there will be no states available, as shown in Fig. 6.16b, and tunneling cannot occur. The application of a voltage across the junction increases the electronic energy of one metal with respect to the other by shifting one Fermi level relative to the other. The number of electrons that can then move across the

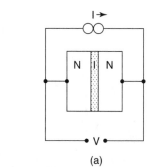

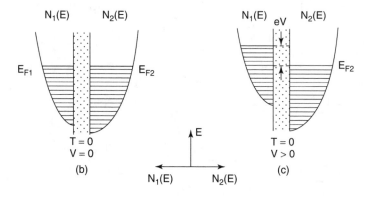

Figure 6.16. (a) Metal–insulator–metal junction; (b) density of states of occupied levels and Fermi level before a voltage is applied to the junction; (c) density of states and Fermi level after application of a voltage. Panels (b) and (c) plot the energy vertically and the density of states horizontally, as indicated at the bottom of the figure. Levels above the Fermi level that are not occupied by electrons are not shown.

junction from left to right (Fig. 6.16c) in an energy interval dE is proportional to the number of occupied states on the left and the number of unoccupied states on the right, which is

$$N_1(E - eV)f(E - eV)[N_2(E)(1 - f(E))] \tag{6.3}$$

where N_1 is the density of states in metal 1, N_2 is the density of states in metal 2, and $f(E)$ is the Fermi–Dirac distribution of states over energy, which is plotted in Fig. 9.8. The net flow of current I across the junction is the difference between the currents flowing to the right and the left, which is

$$I = K \int N_1(E - eV)N_2(E)[f(E - eV) - f(E)]dE \tag{6.4}$$

where K is the matrix element, which gives the probability of tunneling through the barrier. The current across the junction will depend linearly on the voltage. If the density of states is assumed constant over an energy-range eV (electron volt), then for small V and low T, we obtain

$$I = KN_1(E_f)N_2(E_f)eV \tag{6.5}$$

which can be rewritten in the form

$$I = G_{nn}V \tag{6.6}$$

where

$$G_{nn} = KN_1(E_f)N_2(E_f)e \tag{6.7}$$

and G_{nn} is identified as the conductance. The junction, in effect, behaves in an ohmic manner, that is, with the current proportional to the voltage.

6.1.6. Other Properties

While the emphasis of the previous discussion has been on the effect of nanosized microstructure on mechanical and electrical properties, many other properties of bulk nanostructured materials are also affected. For example, the magnetic behavior of bulk ferromagnetic material made of nanosized grains is quite different from the same material made with conventional grain sizes. Because of its technological importance relating to the possibility of enhancing magnetic information storage capability, this is discussed in more detail in Chapter 7.

In Chapter 4 we saw that the inherent reactivity of nanoparticles depends on the number of atoms in the cluster. It might be expected that such behavior would also be manifested in bulk materials made of nanostructured grains, providing a possible way to protect against corrosion and the detrimental effects of oxidation, such as the formation of the black silver oxide coating on silver. Indeed, there have been some advances in this area. The nanostructured alloy $Fe_{73}B_{13}Si_9$ has been found to have enhanced resistance to oxidation at temperatures between 200 and 400°C. The material consists of a mixture 30-nm particles of Fe(Si) and Fe_2B. The enhanced resistance is attributed to the large number of interface boundaries, and the fact that atom diffusion occurs faster in nanostructured materials at high temperatures. In this material the Si atoms in the FeSi phase segregate to interface boundaries where they can then diffuse to the surface of the sample. At the surface the Si interacts with the oxygen in the air to form a protective layer of SiO_2, which hinders further oxidation.

The melting temperature of nanostructured materials is also affected by grain size. It has been shown that indium containing 4-nm nanoparticles has its melting temperature lowered by 110 K.

In the superconducting phase there is a maximum current that a material can carry called the *critical current* I_C. When the current I exceeds that value, the superconducting state is removed, and the material returns to its normal resistance. It has been found that in the bulk granular superconductor Nb_3Sn decreasing the grain size of the sample can increase the critical current.

In Chapter 4 we saw that the optical absorption properties of nanoparticles, determined by transitions to excited states, depend on their size and structure. In principle, it should therefore be possible to engineer the optical properties of bulk nanostructured materials. A high-strength transparent metal would have many application possibilities. In the next sections we will discuss some examples of how nanostructure influences the optical properties of materials.

6.1.7. Metal Nanocluster Composite Glasses

One of the oldest applications of nanotechnology is the colored stained-glass windows in medieval cathedrals which are a result of nanosized metallic particles embedded in the glass. Glasses containing a low concentration of dispersed nanoclusters display a variety of unusual optical properties that have application potential. The peak wavelength of the optical absorption, which largely determines the color, depends on the size, and on the type of metal particle. Figure 6.17 shows an example of the effect of the size of gold nanoparticles on the optical absorption properties of an SiO_2 glass in the visible region. The data confirm that the peak of

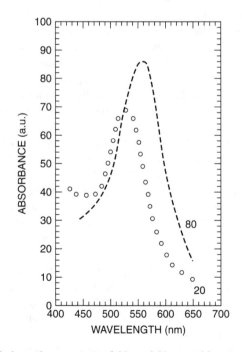

Figure 6.17. Optical absorption spectrum of 20- and 80-nm gold nanoparticles embedded in glass. (Adapted from F. Gonella et al., in *Handbook of Nanostructured Materials and Nanotechnology*, H. S. Nalwa, ed., Academic Press, San Diego, 2000, Vol. 4, Chapter 2, p. 85.)

the optical absorption shifts to shorter wavelengths when the nanoparticle size decreases from 80 to 20 nm. The spectrum is due to plasma absorption in the metal nanoparticles. At very high frequencies the conduction electrons in a metal behave like a plasma, that is, like an electrically neutral ionized gas in which the negative charges are the mobile electron, and the positive charges reside on the stationary background atoms. Provided the clusters are smaller than the wavelength of the incident visible light, and are well dispersed so that they can be considered non-interacting, the electromagnetic wave of the light beam causes an oscillation of the electron plasma that results in absorption of the light. A theory developed by Mie may be used to calculate the absorption coefficient versus the wavelength of the light. The absorption coefficient α of small spherical metal particles embedded in a nonabsorbing medium is given by

$$\alpha = \frac{18\pi N_s V n_0 \varepsilon_2^3 / \lambda}{[\varepsilon_1 + 2n_0^2]^2} + \varepsilon_2^2 \tag{6.8}$$

where N_s is the number of spheres of volume V, ε_1 and ε_2 are the real and imaginary parts of the dielectric constant of the spheres, n_0 is the refractive index of the insulating glass, and λ is the wavelength of the incident light.

Another technologically important property of metallic glass composites is that they display nonlinear optical effects, which means that their refractive indices depend on the intensity of the incident light. The glasses have an enhanced third-order susceptibility that results in an intensity dependent refractive index n given by

$$n = n_0 + n_2 I \tag{6.9}$$

where I is the intensity of the light beam. Nonlinear optical effects have potential application as optical switches, which would be a major component of photon-based computers. When metal particles are less than 10 nm in size, confinement effects become important, and these alter the optical absorption properties. Quantum confinement is discussed in Chapter 9.

The earliest methods for making composite metal glasses involve mixing metal particles in molten glasses. However, it is difficult to control the properties of the glasses, such as the aggregation of the particles. More controllable processes have been developed such as ion implantation. Essentially, the glasses are subjected to an ion beam consisting of atoms of the metal to be implanted, having energies in the range from 10 keV to 10 MeV. Ion exchange is also used to put metal particles into glasses. Figure 6.18 shows an experimental setup for an ion exchange process designed to put silver particles in glasses. Monovalent surface atoms such as sodium present near the surface of all glasses are replaced with other ions such as silver. The glass substrate is placed in a molten salt bath that contains the electrodes, and a voltage is applied across the electrodes with the polarity shown in Fig. 6.18. The sodium ion diffuses in the glass toward the negative electrode, and the silver diffuses from the silver electrolyte solution into the surface of the glass.

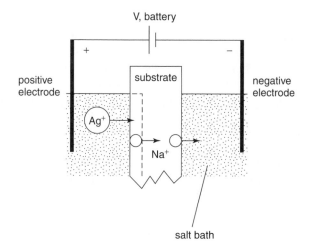

Figure 6.18. Electric field assisted ion exchange apparatus for doping glasses (substrate) with metals such as Ag^+ ions. [Adapted from G. De Marchi et al., *J. Non-Cryst. Solids* **196**, 79 (1996).]

6.1.8. Porous Silicon

When a wafer of silicon is subjected to electrochemical etching, the silicon wafer develops pores. Figure 6.19 is a scanning electron microscope picture of the (100) surface of etched silicon showing pores (dark regions) of micrometer dimensions.

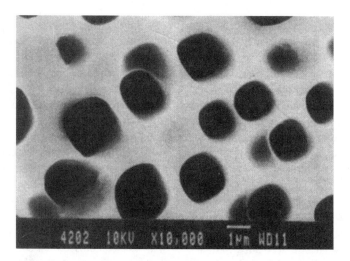

Figure 6.19. Scanning electron microscope (SEM) picture of the surface of n-doped etched silicon. The micrometer-sized pores appear as dark regions. [With permission from C. Levy-Clement et al., *J. Electrochem. Soc.* **141**, 958 (1994).]

This silicon is called *porous silicon* (PoSi). By controlling the processing conditions, pores of nanometer dimensions can be made. Research interest in porous silicon was intensified in 1990 when it was discovered that it was fluorescent at room temperature. *Luminescence* refers to the absorption of energy by matter, and its re-emission as *visible* or *near-visible light*. If the emission occurs within 10^{-8} s of the excitation, then the process is called *fluorescence*, and if there is a delay in the emission it is called *phosphorescence*. Nonporous silicon has a weak fluorescence between 0.96 and 1.20 eV in the region of the band gap, which is 1.125 eV at 300 K. This fluorescence is due to band gap transitions in the silicon. However, as shown in Fig. 6.20, porous silicon exhibits a strong photon-induced luminescence well above 1.4 eV at room temperature. The peak wavelength of the emission depends on the length of time the wafer is subjected to etching. This observation generated much excitement because of the potential of incorporating photoactive silicon using current silicon technology, leading to new display devices or optoelectronic coupled elements. Silicon is the element most widely used to make transistors, which are the on/off switching elements in computers.

Figure 6.21 illustrates one method of etching silicon. Silicon is deposited on a metal such as aluminum, which forms the bottom of a container made of poly-ethylene or Teflon, which will not react with the hydrogen fluoride (HF) etching solution. A voltage is applied between the platinum electrode and the Si wafer such that the Si is the positive electrode. The parameters that influence the nature of the pores are the concentration of HF in the electrolyte or etching solution, the

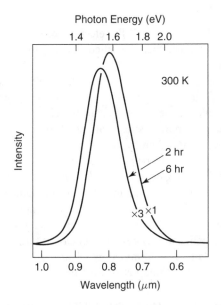

Figure 6.20. Photoluminescence spectra of porous silicon for two different etching times at room temperature. Note the change in scale for the two curves. [Adapted from L. T. Camham, *Appl. Phys. Lett.* **57**, 1046 (1990).]

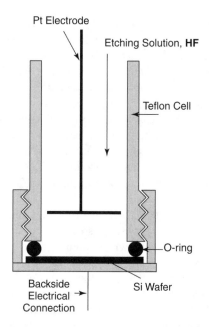

Figure 6.21. A cell for etching a silicon wafer in a hydrogen fluoride (HF) solution in order to introduce pores. (With permission from D. F. Thomas et al., in *Handbook of Nanostructured Materials and Nanotechnology*, H. S. Nalwa, ed., Academic Press, San Diego, 2000, Vol. 4, Chapter 3, p. 173.)

magnitude of current flowing through the electrolyte, the presence of a surfactant (surface-active agent), and whether the silicon is negatively (n) or positively (p) doped.

The Si atoms of a silicon crystal have four valence electrons, and are bonded to four nearest-neighbor Si atoms. If an atom of silicon is replaced by a phosphorus atom, which has five valence electrons, four of the electrons will participate in the bonding with the four neighboring silicon atoms. This will leave an extra electron available to carry current, and thereby contribute to the conduction process. This puts an energy level in the gap just below the bottom of the conduction band. Silicon doped in this way is called an *n-type semiconductor*. If an atom of aluminum, which has three valence electrons, is doped into the silicon lattice, there is a missing electron referred to as a *hole* in one of the bonds of the neighboring silicon atoms. This hole can also carry current and contribute to increasing the conductivity. Silicon doped in this manner is called a *p-type semiconductor*. It turns out that the size of the pores produced in the silicon is determined by whether silicon is n- or p-type. When p-type silicon is etched, a very fine network of pores having dimensions less than 10 nm is produced.

A number of explanations have been offered to explain the origin of the fluorescence of porous silicon, such as the presence of oxides on the surface of the pores that emit molecular fluorescence, surface defect states, quantum wires, quantum dots and the resulting quantum confinement, and surface states on quantum dots. Porous silicon also displays electroluminscence, whereby the luminescence is induced by the application of a small voltage across electrodes mounted on the silicon, and cathodoluminescence from bombarding electrons.

6.2. NANOSTRUCTURED CRYSTALS

In this section we discuss the properties of crystals made of ordered arrays of nanoparticles.

6.2.1. Natural Nanocrystals

There are some instances of what might be called "natural nanocrystals." An example is the 12-atom boron cluster, which has an icosahedral structure, that is, one with 20 faces. There are a number of crystalline phases of solid boron containing the B_{12} cluster as a subunit. One such phase with tetragonal symmetry has 50 boron atoms in the unit cell, comprising four B_{12} icosahedra bonded to each other by an intermediary boron atom that links the clusters. Another phase consists of B_{12} icosahedral clusters shown in Fig. 6.22 arranged in a hexagonal array. Of course there are other analogous nanocrystals such as the fullerene C_{60} compound, which forms the lattice shown in Fig. 5.7 (of Chapter 5).

6.2.2. Computational Prediction of Cluster Lattices

Viewing clusters as superatoms raises the intriguing possibility of designing a new class of solid materials whose constituent units are not atoms or ions, but rather clusters of atoms. Solids built from such clusters may have new and interesting properties. There have been some theoretical predictions of the properties of solids made from clusters such as $Al_{12}C$. The carbon is added to this cluster so that it has 40 electrons, which is a closed-shell configuration that stabilizes the cluster. This is necessary for building solids from clusters because clusters that do not have closed shells could chemically interact with each other to form a larger cluster. Calculations of the face-centered cubic structure $Al_{12}C$ predict that it would have a very small band gap, in the order of 0.05 eV, which means that it would be a semiconductor. The possibility of ionic solids made of KAl_{13} clusters has been considered. Since the electron affinity of Al_{13} is close to that of Cl, it may be possible for this cluster to form a structure similar to KCl. Figure 6.23 shows a possible body-centered structure for this material. Its calculated cohesive energy is 5.2 eV, which can be compared with the cohesive energy of KCl, which is 7.19 eV. This cluster solid is quite stable. These calculations indicate that new solids with clusters as their subunits are possible, and may have new and interesting properties; perhaps even new high-temperature superconductors could emerge. New ferromagnetic materials could result from solids made of clusters, which have a net magnetic moment.

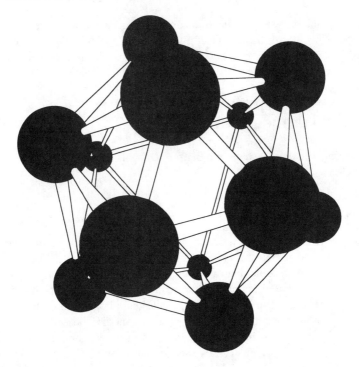

Figure 6.22. The icosohedral structure of a boron cluster containing 12 atoms. This cluster is the basic unit of a number of boron lattices.

6.2.3. Arrays of Nanoparticles in Zeolites

Another approach that has enabled the formation of latticelike structures of nano-particles is to incorporate them into zeolites. Zeolites such as the cubic mineral, faujasite, $(Na_2,Ca)(Al_2Si_4)O_{12} \cdot 8H_2O$, are porous materials in which the pores have a regular arrangement in space. The pores are large enough to accommodate small clusters. The clusters are stabilized in the pores by weak van der Waals interactions between the cluster and the zeolite. Figure 6.24 shows a schematic of a cluster assembly in a zeolite. The pores are filled by injection of the guest material in the molten state. It is possible to make lower-dimensional nanostructured solids by this approach using a zeolite material such as mordenite, which has the structure illustrated in Fig. 6.25. The mordenite has long parallel channels running through it with a diameter of 0.6 nm. Selenium can be incorporated into these channels, forming chains of single atoms. A trigonal crystal of selenium also has parallel chains, but the chains are sufficiently close together so that there is an interaction between them. In the mordenite this interaction is reduced significantly, and the electronic structure is different from that of a selenium crystal. This causes the optical absorption spectra of the selenium crystal and the selenium in mordenite to differ in the manner shown in Fig. 6.26.

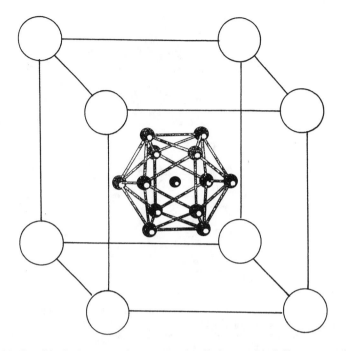

Figure 6.23. Possible body-centered structure of a lattice made of Al_{13} nanoparticles and potassium (large circles). [Adapted from S. N. Khanna and P. Jena, *Phys. Rev.* **51**, 13705 (1995).]

Figure 6.24. Schematic of cluster assemblies in zeolite pores. (With permission from S. G. Romanov et al., in *Handbook of Nanostructured Materials and Nanotechnology*, H. S. Nalwa, ed., Academic Press, San Diego, 2000, Vol. 4, Chapter 4, p. 236.)

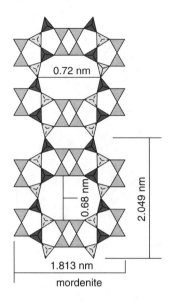

Figure 6.25. Illustration of long parallel channels in a crystal of mordenite, an orthrorhombic variety of zeolite $(Ca,Na_2,K_2)(Al_2Si_{10})O_{24} \cdot 7H_2O$. (Adapted from S. G. Romanov et al., in *Handbook of Nanostructured Materials and Nanotechnology*, H. S. Nalwa, ed., Academic Press, San Diego, 2001, Vol. 4, Chapter 4, p. 238.)

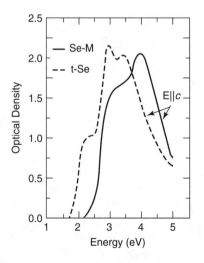

Figure 6.26. Optical absorption spectra of chains of selenium atoms in mordenite (solid line, Se–M) and in crystalline selenium (dashed line, t-Se) showing the shift in the peak of the absorption, and the change in shape. [With permission V. N. Bogomolov, *Solid State Commun.* **47**, 181, (1983).]

6.2.4. Crystals of Metal Nanoparticles

A two-phase water toluene reduction of $AuCl_4^-$ by sodium borohydride in the presence of an alkanethiol ($C_{12}H_{25}SH$) solution produces gold nanoparticles Au_m having a surface coating of thiol, and embedded in an organic compound. The overall reaction scheme is

$$AuCl_4^-(aq) + N(C_8H_{17})_4 + (C_6H_5Me) \rightarrow N(C_8H_{17})_4 + AuCl_4^-(C_6H_5Me) \quad (6.10)$$

$$mAuCl_4^-(C_6H_5Me) + n(C_{12}H_{25}SH)(C_6H_5Me) + 3me^- \rightarrow$$
$$4mCl^-(aq) + (Au_m)(C_{12}H_{25}SH)_n(C_6H_5Me) \quad (6.11)$$

Essentially, the result of the synthesis is a chemical compound denoted as c-Au:SR, where SR is $(C_{12}H_{25}SH)_n(C_6H_5Me)$, and Me denotes the methyl radical CH_3. When the material was examined by X-ray diffraction it showed, in addition to the diffuse peaks from the planes of gold atoms in the nanoparticles Au_m, a sequence of sharp peaks at low scattering angles indicating that the gold nanoparticles had formed a giant three-dimensional lattice in the SR matrix. The crystal structure was determined to be a body-centered cubic (BCC) arrangement. In effect, a highly ordered symmetric arrangement of large gold nanoparticles had spontaneously self-assembled during the chemical processing.

Superlattices of silver nanoparticles have been produced by aerosol processing. The lattices are electrically neutral, ordered arrangements of silver nanoparticles in a dense mantel of alkylthiol surfactant, which is a chain molecule, n-$CH_3(CH_2)_mSH$. The fabrication process involves evaporation of elemental silver above 1200°C into a flowing preheated atmosphere of high-purity helium. The flow stream is cooled over a short distance to about 400 K, resulting in condensation of the silver to nanocrystals. Growth can be abruptly terminated by expansion of the helium flow through a conical funnel accompanied by exposure to cool helium. The flowing nanocrystals are condensed into a solution of alkylthiol molecules. The material produced in this way has a superlattice structure with FCC arrangement of silver nanoparticles having separations of <3 nm. Metal nanoparticles are of interest because of the enhancement of the optical and electrical conductance due to confinement and quantization of conduction electrons by the small volume of the nanocrystal.

When the size of the crystal approaches the order of the de Broglie wavelength of the conduction electrons, the metal clusters may exhibit novel electronic properties. They display very large optical polarizabilities as discussed earlier, and nonlinear electrical conductance having small thermal activation energies. Coulomb blockade and Coulomb staircase current–voltage curves are observed. The Coulomb blockade phenomenon is discussed in Chapter 9. Reduced-dimensional lattices such as quantum dots and quantum wires have been fabricated by a subtractive approach that removes bulk fractions of the material leaving nanosized wires and dots. These nanostructures are discussed in Chapter 9.

6.2.5. Nanoparticle Lattices in Colloidal Suspensions

Colloidal suspensions consist of small spherical particles 10–100 nm in size suspended in a liquid. The interaction between the particles is *hard-sphere repulsion*, meaning that the center of the particles cannot get closer than the diameters of the particles. However it is possible to increase the range of the repulsive force between the particles in order to prevent them from aggregating. This can be done by putting an electrostatic charge on the particles. Another method is to attach soluble polymer chains to the particles, in effect producing a dense brush with flexible bristles around the particle. When the particles with these brush polymers about them approach each other, the brushes compress and generate a repulsion between the particles. In both charge and polymer brush suspensions the repulsion extends over a range that can be comparable to the size of the particles. This is called "soft repulsion." When such particles occupy over 50% of the volume of the material, the particles begin to order into lattices. The structure of the lattices is generally hexagonal close-packed, face-centered cubic, or body-centered cubic. Figure 6.27 shows X-ray densitometry measurements on a 3-mM salt solution containing 720-nm polystyrene spheres. The dashed-line plots are for the equations of state of the material, where the pressure P is normalized to the thermal energy $k_B T$, versus the fraction of particles in the fluid. The data show a gradual transition from a phase where the particles are disordered in the liquid to a phase where there is lattice ordering. In between there is a mixed region where there is both a fluid phase and a crystal phase. This transition is called the *Kirkwood–Alder transition*, and it can be altered by changing the concentration of the particles or the charge on them. At high concentrations or for short-range

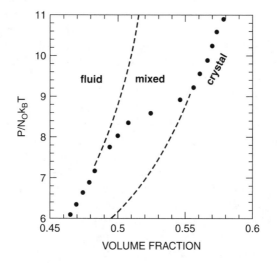

Figure 6.27. Equations of state (dashed curves) plotted as a function of fraction of 720-nm styrene spheres in a 3-mM salt solution. The constant N_0 is Avogadro's number. [Adapted from A. P. Gast and W. B. Russel, *Phys. Today* (Dec. 1998.)]

repulsions the lattice structure is face-centered cubic. Increasing the range of the repulsion, or lowering the concentration, allows the formation of the slightly less compact body-centered structure. Figure 6.28 shows the phase diagram of a system of soft spherical particles. The excluded volume is the volume that is inaccessible to one particle because of the presence of the other particles. By adjusting the fraction of particles, a structural phase transition between the face-centered and body-centered structures can be induced. The particles can also be modified to have attractive potentials. For the case of charged particles in aqueous solutions, this can be accomplished by adding an electrolyte to the solution. When this is done, abrupt aggregation occurs.

6.2.6. Photonic Crystals

A photonic crystal consists of a lattice of dielectric particles with separations on the order of the wavelength of visible light. Such crystals have interesting optical properties. Before discussing these properties, we will say a few words about the reflection of waves of electrons in ordinary metallic crystal lattices.

The wavefunction of an electron in a metal can be written in the free-electron approximation as

$$\Psi_{k[r]} = \left[\frac{1}{V}\right]^{1/3} e^{ik\cdot r} \tag{6.12}$$

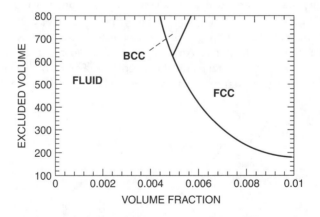

Figure 6.28. Phase diagram for soft spherical particles in suspension showing the fluid state, the body-centered cubic (BCC) and face-centered cubic (FCC) phases. The vertical axis (ordinate) represents the volume that is inaccessible to one particle because the presence of the others. The excluded volume has been scaled to the cube of the Debye screening length, which characterizes the range of the interaction. [Adapted from A. P. Gast and W. B. Russel, *Phys. Today* (Dec. 1998.)]

where V is the volume of the solid, the momentum $p = \hbar k$, and the wavevector k is related to the wavelength λ by the expression $k = 2\pi/\lambda$. In the nearly free-electron model of metals the valence or conduction electrons are treated as noninteracting free electrons moving in a periodic potential arising from the positively charged ion cores. Figure 6.29 shows a plot of the energy versus the wavevector for a one-dimensional lattice of identical ions. The energy is proportional to the square of the wavevector, $E = h^2 k^2 / 8\pi^2 m$, except near the band edge where $k = \pm \pi/a$. The important result is that there is an energy gap of width E_g, meaning that there are certain wavelengths or wavevectors that will not propagate in the lattice. This is a result of Bragg reflections. Consider a series of parallel planes in a lattice separated by a distance d containing the atoms of the lattice. The path difference between two waves reflected from adjacent planes is $2d \sin \Theta$, where Θ is the angle of incidence of the wavevector to the planes. If the path difference $2d \sin \Theta$ is a half-wavelength, the reflected waves will destructively interfere, and cannot propagate in the lattice, so there is an energy gap. This is a result of the lattice periodicity and the wave nature of the electrons.

In 1987 Yablonovitch and John proposed the idea of building a lattice with separations such that light could undergo Bragg reflections in the lattice. For visible light this requires a lattice dimension of about 0.5 μm or 500 nm. This is 1000 times larger than the spacing in atomic crystals but still 100 times smaller than the thickness of a human hair. Such crystals have to be artificially fabricated by methods such as electron-beam lithography or X-ray lithography. Essentially a photonic crystal is a periodic array of dielectric particles having separations on the order of 500 nm. The materials are patterned to have symmetry and periodicity in their dielectric constant. The first three-dimensional photonic crystal was fabricated by Yablonovitch for microwave wavelengths. The fabrication consisted of covering a block of a dielectric material with a mask consisting of an ordered array of holes and drilling through these holes in the block on three perpendicular facets. A technique of stacking micromachined wafers of silicon at consistent separations has been used to build the photonic structures. Another approach is to build the lattice out of isolated dielectric materials that are not in contact. Figure 6.30 depicts a two-dimensional photonic crystal made of dielectric rods arranged in a square lattice.

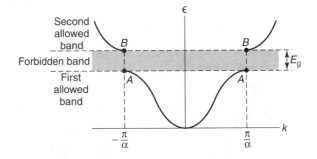

Figure 6.29. Curve of energy E plotted versus wavevector k for a one-dimensional line of atoms.

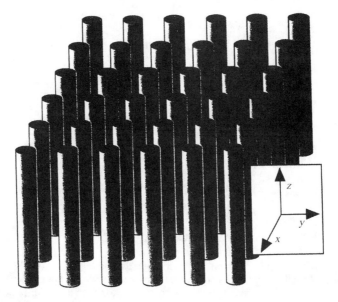

Figure 6.30. A two-dimensional photonic crystal made by arranging long cylinders of dielectric materials in a square lattice array.

The description of the behavior of light in photonic crystals involves solving Maxwell's equations in a periodic dielectric structure. The associated Helmholtz equation obtained for the case of no external current sources is

$$\nabla^2 H(r) + \varepsilon \left[\frac{\omega^2}{c^2} \right] H(r) = 0 \qquad (6.13)$$

where H is the magnetic field associated with the electromagnetic radiation, and ε is the relative dielectric constant of the components constituting the photonic crystal. This equation can be solved exactly for light in a photonic crystal primarily because there is little interaction between photons, and quite accurate predictions of the dispersion relationship are possible. The dispersion relationship is the dependence of the frequency or energy on the wavelength or k vector. Figure 6.31 shows a plot of the dispersion relationship of alumina rods (Al_2O_3, $\varepsilon = 8.9$), having the structure shown in Fig. 6.30, with a radius of 0.37 mm, and a length of 100 mm for the transverse magnetic modes. This corresponds to the vibration of the magnetic H vector of the electromagnetic wave. The separation between the centers of the rods is 1.87 mm. This lattice is designed for the microwave region, but the general properties would be similar at the smaller rod separations needed for visible light. The labels Γ and X refer to special symmetry points in k space for the square lattice. The results show the existence of a photonic band gap, which is essentially a range of frequencies where electromagnetic energy cannot propagate in the lattice. The light

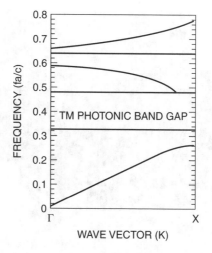

Figure 6.31. A part of the dispersion relationship of a photonic crystal mode, TM, of a photonic crystal made of a square lattice of alumina rods. The ordinate scale is the frequency f multiplied by the lattice parameter a divided by the speed of light c. [Adapted from J. D. Joannopoulos, *Nature* **386**, 143 (1997).]

power is intense below this band gap, and in analogy with the terms *conduction band* and *valence band*, this region is called the *dielectric band*. Above the forbidden gap the light power is low, and the region there is referred to as the "air band."

Now let us consider what would happen if a line defect were introduced into the lattice by removing a row of rods. The region where the rods have been removed would act like a wave guide, and there would now be an allowed frequency in the band gap, as shown in Fig. 6.32. This is analogous to p and n doping of

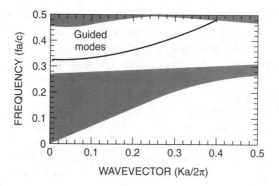

Figure 6.32. Effect of removing one row of rods from a square lattice of a photonic crystal, which introduces a level (guided mode) in the forbidden gap. The ordinate scale is the frequency f multiplied by the lattice parameter a divided by the speed of light c. [Adapted from J. D. Joannopoulos, *Nature* **386**, 143 (1997).]

semiconductors, which puts an energy level in the energy gap. A wave guide is like a pipe that confines electromagnetic energy, enabling it to flow in one direction. An interesting feature of this wave guide is that the light in the photonic crystal guide can turn very sharp corners, unlike light traveling in a fiberoptic cable. Because the frequency of the light in the guide is in the forbidden gap, the light cannot escape into the crystal. It essentially has to turn the sharp corner. Fiberoptic cables rely on total internal reflection at the inner surface of the cable to move the light along. If the fiber is bent too much, the angle of incidence is too large for total internal reflection, and light escapes at the bend.

A resonant cavity can be created in a photonic crystal by removing one rod, or changing the radius of a rod. This also puts an energy level into the gap. It turns out that the frequency of this level depends on the radius of the rod, as shown in Fig. 6.33. The air and dielectric bands discussed above are indicated on the figure. This provides a way to tune the frequency of the cavity. This ability to tune the light and concentrate it in small regions gives photonic crystals potential for use as filters and couplers in lasers. *Spontaneous emission* is the emission of light that occurs when an exited state decays to a lower energy state. It is an essential part of the process of producing lasing. The ability to control spontaneous emission is

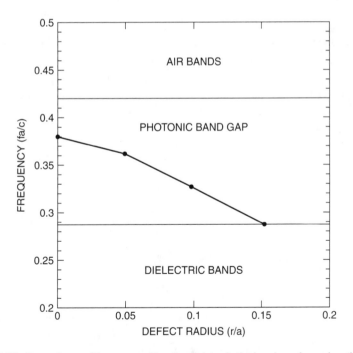

Figure 6.33. Dependence of frequency of localized states in the band gap formed on the radius *r* of a single rod in the square lattice. The ordinate scale is the frequency *f* multiplied by the lattice parameter *a* divided by the speed of light *c*. [Adapted from J. D. Joannopoulos, *Nature* **386**, 143 (1997).]

necessary to produce lasing. The rate at which atoms decay depends on the coupling between the atom and the photon, and the density of the electromagnetic modes available for the emitted photon. Photonic crystals could be used to control each of these two factors independently.

Semiconductor technology constitutes the basis of integrated electronic circuitry. The goal of putting more transistors on a chip requires further miniaturization. This unfortunately leads to higher resistances and more energy dissipation. One possible future direction would be to use light and photonic crystals for this technology. Light can travel much faster in a dielectric medium than an electron can in a wire, and it can carry a larger amount of information per second. The bandwidth of optical systems such as fiberoptic cable is terahertz in contrast to that in electron systems (with current flowing through wires), which is a few hundred kilohertz. Photonic crystals have the potential to be the basis of future optical integrated circuits.

FURTHER READING

S. A. Asher et al., *Mesoscopically Periodic Photonic Crystal Materials for Linear and Non Linear Optics and Chemical Sensing*, MRS Bulletin, Oct. 1998.

I. Chang, "Rapid Solidification Processing of Nanocrystalline Metallic Alloys," in *Handbook of Nanostructured Materials and Nanotechnology*, H. S. Nalwa, ed., Academic Press, San Deigo, 2000, Vol. 1, Chapter 11, p. 501.

A. L. Gast and W. B. Russel, "Simple Ordering in Complex Fluids," *Phys. Today* (Dec. 1998).

J. E. Gordon, *The New Science of Strong Materials*, Penguin Books, Middlesex, UK, 1968.

J. D. Joanpoulos, P. R. Villeineuve, and S. Fan, "Photonic Crystals," *Nature* **386**, 143 (1997).

C. C. Koch, D. G. Morris, K. Lu, and A. Inoue, *Ductility of Nanostructured Materials*, MRS Bulletin, Feb. 1999.

M. Marder and J. Fineberg, "How Things Break," *Phys. Today* (Sept. 1996).

R. L. Whetten et al., "Crystal Structure of Molecular Gold Nanocrystal Array," *Acc. Chem. Res.* **32**, 397 (1999).

7

NANOSTRUCTURED FERROMAGNETISM

7.1. BASICS OF FERROMAGNETISM

In this chapter we discuss the effect of nanostructure on the properties of ferromagnets, and how the size of nanograins that make up bulk magnet materials affects ferromagnetic properties. We will examine how control of nanoparticle size can be used to design magnetic materials for various applications. In order to understand the role of nanostructure on ferromagnetism, we will present a brief overview of the properties of ferromagnets. In Chapter 4 we saw that certain atoms whose energy levels are not totally filled have a net magnetic moment, and in effect behave like small bar magnets. The value of the magnetic moment of a body is a measure of the strength of the magnetism that is present. Atoms in the various transition series of the periodic table have unfilled inner energy levels in which the spins of the electrons are unpaired, giving the atom a net magnetic moment. The iron atom has 26 electrons circulating about the nucleus. Eighteen of these electrons are in filled energy levels that constitute the argon atom inner core of the electron configuration. The d level of the $N = 3$ orbit contains only 6 of the possible 10 electrons that would

Introduction to Nanotechnology, by Charles P. Poole Jr. and Frank J. Owens.
ISBN 0-471-07935-9. Copyright © 2003 John Wiley & Sons, Inc.

fill it, so it is incomplete to the extent of four electrons. This incompletely filled electron d shell causes the iron atom to have a strong magnetic moment.

When crystals such as bulk iron are formed from atoms having a net magnetic moment a number of different situations can occur relating to how the magnetic moments of the individual atoms are aligned with respect to each other. Figure 7.1 illustrates some of the possible arrangements that can occur in two dimensions. The point of the arrow is the north pole of the tiny bar magnet associated with the atom. If the magnetic moments are randomly arranged with respect to each other, as shown in Fig. 7.1a, then the crystal has a zero net magnetic moment, and this is referred to as the *paramagnetic state*. The application of a DC magnetic field aligns some of the moments, giving the crystal a small net moment. In a ferromagnetic crystal these moments all point in the same direction, as shown in Fig. 7.1b, even when no DC magnetic field is applied, so the whole crystal has a magnetic moment and behaves like a bar magnet producing a magnetic field outside of it. If a crystal is made of two types of atoms, each having a magnetic moment of a different strength (indicated in Fig. 7.1c by the length of the arrow) the arrangement shown in Fig. 7.1c can occur, and it is called *ferrimagnetic*. Such a crystal will also have a net magnetic moment, and behave like a bar magnet. In an antiferromagnet the moments are arranged in

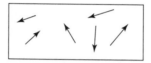

(a) PARAMAGNETIC

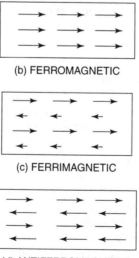

(b) FERROMAGNETIC

(c) FERRIMAGNETIC

(d) ANTIFERROMAGNETIC

Figure 7.1. Illustration of various arrangements of individual atomic magnetic moments that constitute paramagnetic (a), ferromagnetic (b), ferrimagnetic (c), and antiferromagnetic (d) materials.

an antiparallel scheme, that is, opposite to each other, as shown in Fig. 7.1d, and hence the material has no net magnetic moment. In the present chapter we are concerned mainly with ferromagnetic ordering.

Now let us consider the question of why the individual atomic magnets align in some materials and not others. When a DC magnetic field is applied to a bar magnet, the magnetic moment tends to align with the direction of the applied field. In a crystal each atom having a magnetic moment has a magnetic field about it. If the magnetic moment is large enough, the resulting large DC magnetic field can force a nearest neighbor to align in the same direction provided the interaction energy is larger than the thermal vibrational energy $k_B T$ of the atoms in the lattice. The interaction between atomic magnetic moments is of two types: the so-called exchange interaction and the dipolar interaction. The exchange interaction is a purely quantum-mechanical effect, and is generally the stronger of the two interactions.

In the case of a small particle such as an electron that has a magnetic moment, the application of a DC magnetic field forces its spin vector to align such that it can have only two projections in the direction of the DC magnetic field, which are $\pm \frac{1}{2}\mu_B$, where μ_B is the unit magnetic moment called the *Bohr magneton*. The wavefunction representing the state $+\frac{1}{2}\mu_B$ is designated α, and for $-\frac{1}{2}\mu_B$ it is β. The numbers $\pm\frac{1}{2}$ are called the *spin quantum numbers* m_s. For a two-electron system it is not possible to specify which electron is in which state. The Pauli exclusion principle does not allow two electrons in the same energy level to have the same spin quantum numbers m_s. Quantum mechanics deals with this situation by requiring that the wavefunction of the electrons be antisymmetric, that is, change sign if the two electrons are interchanged. The form of the wavefunction that meets this condition is $\frac{1}{2}^{-1/2}[\Psi_A(1)\Psi_B(2) - \Psi_A(2)\Psi_B(1)]$. The electrostatic energy for this case is given by the expression

$$E = \int \left[\frac{\frac{1}{2}e^2}{r_{12}}\right][\Psi_A(1)\Psi_B(2) - \Psi_A(2)\Psi_B(1)]^2 dV_1 \, dV_2 \qquad (7.1)$$

which involves carrying out a mathematical operation from the calculus called *integration*. Expanding the square of the wavefunctions gives two terms:

$$E = \int \left[\frac{e^2}{r_{12}}\right][\Psi_A(1)\Psi_B(2)]^2 dV_1 \, dV_2 - \int \left[\frac{e^2}{r_{12}}\right]\Psi_A(1)\Psi_B(1)\Psi_A(2)\Psi_B(2) dV_1 \, dV_2 \qquad (7.2)$$

The first term is the normal Coulomb interaction between the two charged particles. The second term, called the *exchange interaction*, represents the difference in the Coulomb energy between two electrons with spins that are parallel and antiparallel. It can be shown that under certain assumptions the exchange interaction can be written in a much simpler form as $J S_1 \cdot S_2$, where J is called the exchange integral, or exchange interaction constant. This is the form used in the Heisenberg model of magnetism. For a ferromagnet J is negative, and for an antiferromagnet it is positive. The exchange interaction, because it involves overlap of orbitals, is primarily a nearest-neighbor interaction, and it is generally the dominant interaction. The other interaction, which can occur in a lattice of magnetic ions, is called the *dipole–dipole interaction*, and has the form

$$\frac{\boldsymbol{\mu}_1 \cdot \boldsymbol{\mu}_2}{r^3} - 3(\boldsymbol{\mu}_1 \cdot \mathbf{r})\frac{\boldsymbol{\mu}_2 \cdot \mathbf{r}}{r^5} \tag{7.3}$$

where $\mathbf{r}$ is a vector along the line separating the two magnetic moments $\boldsymbol{\mu}_1$ and $\boldsymbol{\mu}_2$, and r is the magnitude of this distance.

The magnetization M of a bulk sample is defined as the total magnetic moment per unit volume. It is the vector sum of all the magnetic moments of the magnetic atoms in the bulk sample divided by the volume of the sample. It increases strongly at the Curie temperature T_c, the temperature at which the sample becomes ferromagnetic, and the magnetization continues to increase as the temperature is lowered further below T_c. It has been found empirically that far below the Curie temperature, the magnetization depends on temperature as

$$M(T) = M(0)(1 - cT^{3/2}) \tag{7.4}$$

where $M(0)$ is the magnetization at zero degrees Kelvin and c is a constant. The susceptibility χ of a sample is defined as the ratio of the magnetization at a given temperature to the applied field H, that is, $\chi = M/H$.

Generally for a bulk ferromagnetic material below the Curie temperature, the magnetic moment M is less than the moment the material would have if every atomic moment were aligned in the same direction. The reason for this is the existence of domains. *Domains* are regions in which all the atomic moments point in the same direction so that within each domain the magnetization is saturated; that is, it attains its maximum possible value. However, the magnetization vectors of different domains in the sample are not all parallel to each other. Thus the total sample has an overall magnetization less than the value for the complete alignment of all moments. Some examples of domain configurations are illustrated in Fig. 7.2a. They exist when the magnetic energy of the sample is lowered by the formation of domains.

Applying a DC magnetic field can increase the magnetic moment of a sample. This occurs by two processes. The first process occurs in weak applied fields when the volume of the domains which are oriented along the field direction increases. The second process dominates in stronger applied fields that force the domains to rotate toward the direction of the field. Both of these processes are illustrated in Fig. 7.2b. Figure 7.3, which shows the magnetization curve of a ferromagnetic material, is a plot of the total magnetization of the sample M versus the applied DC field strength H. In the MKS system the units of both H and M are amperes per meter; in the CGS system the units of M are emu/g (electromagnetic units per gram), and the units of H are oersteds. Initially as H increases, M increases until a saturation point M_s, is reached. When H is decreased from the saturation point, M does not decrease to the same value it had earlier when the field was increasing; rather, it is higher on the curve of the decreasing field. This is called *hysteresis*. It occurs

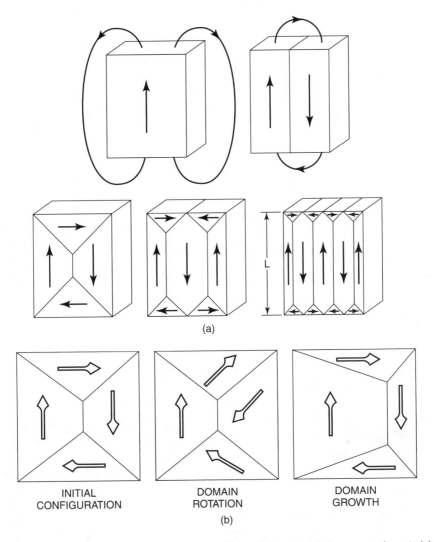

(a)

INITIAL
CONFIGURATION

DOMAIN
ROTATION

DOMAIN
GROWTH

(b)

Figure 7.2. Illustrations of (a) some examples of domain structure in ferromagnetic materials and (b) how the domains can change by the mechanisms of rotation and growth when a DC magnetic is applied.

because the domains that were aligned with the increasing field do not return to their original orientation when the field is lowered. When the applied field H is returned to zero, the magnet still has *magnetization*, referred to as the *remnant magnetization* M_r. In order to remove the remnant magnetization, a field H_c has to be applied in the direction opposite to the initial applied field, as shown in Fig. 7.3. This field, called the *coercive field*, causes the domains to rotate back to their original positions. The properties of the magnetization curve of a ferromagnet have a strong bearing on

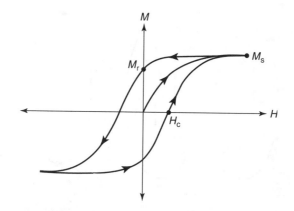

Figure 7.3. Plot of the magnetization M versus an applied magnetic field H for a hard ferromagnetic material, showing the hysteresis loop with the coercive field H_c, the remnant magnetization M_r, and the saturation magnetization M_s, as indicated.

the use of magnet materials, and there is same ongoing research to design permanent magnets with different shapes of magnetization curves.

7.2. EFFECT OF BULK NANOSTRUCTURING ON MAGNETIC PROPERTIES

The diverse applications of magnets require the magnetization curve to have different properties. Magnets used in transformers and rotating electrical machinery are subjected to rapidly alternating AC magnetic fields, so they repeat their magnetization curve many times a second, causing a loss of efficiency and a rise in the temperature of the magnet. The rise in temperature is due to frictional heating from domains as they continuously vary their orientations. The amount of loss during each cycle, meaning the amount of heat energy generated during each cycle around a hysteresis loop, is proportional to the area enclosed by the loop. In these applications small or zero coercive fields are required to minimize the enclosed area. Such magnets are called "soft magnetic materials." On the other hand, in the case of permanent magnets used as a part of high-field systems, large coercive fields are required, and the widest possible hysteresis loop is desirable. Such magnets are called "hard magnets." High-saturation magnetizations are also needed in permanent magnets.

Nanostructuring of bulk magnetic materials can be used to design the magnetization curve. Amorphous alloy ribbons having the composition $Fe_{73.5}Cu_1Nb_3Si_{13.5}B_9$ prepared by a roller method, and subjected to annealing at 673 to 923 K for one hour in inert-gas atmospheres, were composed of 10-nm iron grains in solid solutions. Such alloys had a saturation magnetization M_s of 1.24 T, a remnant magnetization M_r of 0.67 T, and a very small coercive field H_c of 0.53 A/m. Nanoscale amorphous alloy powders of $Fe_{69}Ni_9CO_2$ having grain sizes of 10–15 nm prepared by decomposition of solutions of $Fe(CO)_5$, $Ni(CO)_4$, and $Co(NO)(CoO)_3$ in the hydrocarbon solvent decalin ($C_{10}H_{18}$) under an inert-gas atmosphere showed almost no hysteresis in the magnetization curve. Figure 7.4 presents the magnetization curve for this material. A magnetic material with grain-sized single domain magnetic moments, which has no hysteresis at any temperature, is said to be *superparamagnetic*.

The strongest known permanent magnets are made of neodymium, iron, and boron. They can have remnant magnetizations as high as 1.3 T and coercive fields as high as 1.2 T. The effect of the size of the nanoparticle grain structure on $Nd_2Fe_{14}B$ has been investigated. The results, shown in Figs. 7.5 and 7.6, indicate that in this material the coercive field decreases significantly below ~40 nm and the remnant magnetization increases. Another approach to improving the magnetization curves of this material has been to make nanoscale compositions of hard $Nd_2Fe_{14}B$ and the soft α phase of iron. Measurements of the effect of the presence of the soft iron phase mixed in with the hard material confirm that the remnant field can be increased

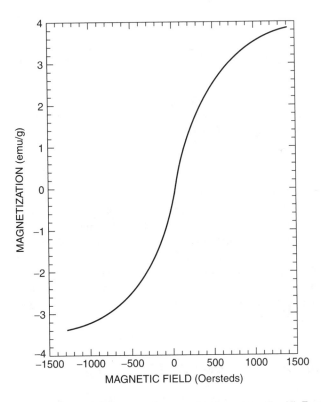

Figure 7.4. Reversible magnetization curve for nanosized powders of a Ni–Fe–Co alloy that exhibits no hysteresis. An oersted corresponds to 10^{-4} T (tesla). [Adapted from K. Shafi et al., *J. Mater. Res.* **15**, 332 (2000).]

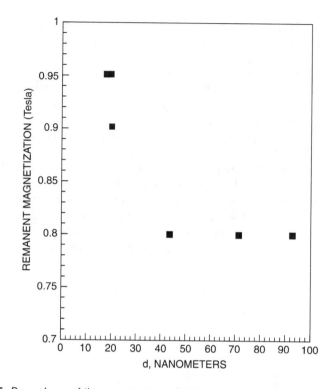

Figure 7.5. Dependence of the remnant magnetization M_r on the particle size d of the grains that form the structure of a Nd–B–Fe permanent magnet normalized to the value $M_s(90)$ for a 90-nm grain size. [Adapted from A. Manaf et al., *J. Magn. Magn. Mater.*, **101**, 360 (1991).]

by this approach. This is believed to be due to the exchange coupling between the hard and soft nanoparticles, which forces the magnetization vector of the soft phase to be rotated to the direction of the magnetization of the hard phase.

The size of magnetic nanoparticles has also been shown to influence the value M_s at which the magnetization saturates. Figure 7.7 shows the effect of particle size on the saturation magnetization of zinc ferrite, illustrating how the magnetization increases significantly below a grain size of 20 nm. Thus, decreasing the particle size of a granular magnetic material can considerably improve the quality of magnets fabricated from it.

7.3. DYNAMICS OF NANOMAGNETS

The study of magnetic materials, particularly of films made of nanomagnets, some-times called *mesoscopic magnetism*, is driven by the desire to increase storage space on magnetic storage devices such as hard drives in computers. The basic information

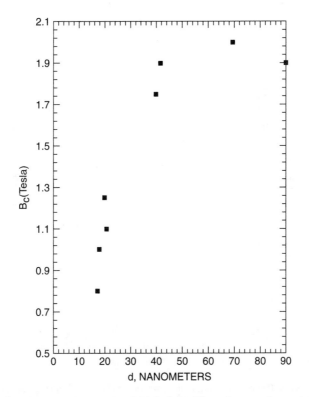

Figure 7.6. Dependence of the coercive field B_c (i.e., H_c) on the granular particle size d of a Nd–B–Fe permanent magnet. [Adapted from A. Manaf et al., *J. Magn. Magn. Mater.* **101**, 360 (1991).]

storage mechanism involves alignment of the magnetization in one direction of a very small region on the magnetic tape called a *byte*. To achieve a storage of 10 gigabytes (10^{10} bytes) per square inch, a single bit would be approximately 1 μm wide and 70 nm long. The film thickness could be about 30 nm. Existing magnetic storage devices such as hard drives are based on tiny crystals of cobalt chromium alloys. One difficulty that arises when bits are less than 10 nm in size is that the magnetization vector can be flipped by random thermal vibrations, in effect erasing the memory. One solution to this is to use nanosized grains, which have higher saturation magnetizations, and hence stronger interactions between the grains. A group at IBM has developed a magnetic nanograin, FePt, which has a much higher magnetization. The FePt particles were made in a heated solution of platinum acetylacetonate and iron carbonyl with a reducing agent added. Oleic acid was also added as a surfactant to prevent aggregation of the particles by coating them with it. The solution is then spread on a substrate and allowed to evaporate, leaving behind the coated particles on the substrate. The resulting thin films are then baked at 560°C

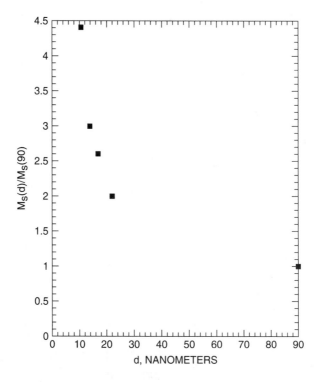

Figure 7.7. Dependence of the saturation magnetization M_s of zinc ferrite on the granular particle size d normalized to the value $M_s(90)$ for a 90-nm grain. [Adapted from C. N. Chinnasamy, *J. Phys. Condens. Matter* **12**, 7795 (2000).]

for 30 min, forming a carbonized hard crust containing 3-nm particles of FePt. This size of magnetic nanoparticle would result in a storage density of 150 gigabytes per square inch, which is about 10 times higher than commercially available magnetic storage units.

When length scales of magnetic nanoparticles become this small, the magnetic vectors become aligned in the ordered pattern of a single domain in the presence of a DC magnetic field, eliminating the complication of domain walls and regions having the magnetization in different directions. The Stone–Wohlfarth (SW) model has been used to account for the dynamical behavior of small nanosized elongated magnetic grains. Elongated grains are generally the type used in magnetic storage devices. The SW model postulates that in the absence of a DC magnetic field ellipsoidal magnetic particles can have only two stable orientations for their magnetization, either up or down with respect to the long axis of the magnetic particles, as illustrated in Fig. 7.8. The energy versus orientation of the vectors is a symmetric double-well potential with a barrier between the two orientations. The particle may flip its orientation by

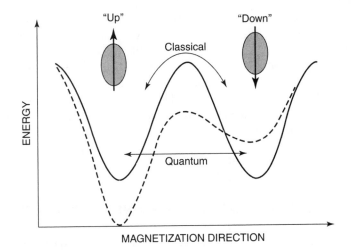

Figure 7.8. Sketch of double-well potential showing the energy plotted against the orientation of the magnetization for up and down orientations of magnetic nanoparticles in the absence (————) and presence (– – – –) of an applied magnetic field. [With permission from D. D. Awschalom and D. P. DiVincenzo, *Phys. Today* 44 (April 1995).]

thermal activation, due to an Arrhenius process, where the probability P for reorientation is given by

$$P \sim \exp\left(\frac{-E}{k_B T}\right) \tag{7.5}$$

where E is the height of the energy barrier between the two orientations. The particle can also flip its orientation by a much lower probability process called *quantum-mechanical* tunneling. This can occur when the thermal energy $k_B T$ of the particle is much less than the barrier height. This process is a purely quantum-mechanical effect resulting from the fact that solution to the wave equation for this system predicts a small probability for the up state of the magnetization to change to the down state. If a magnetic field is applied, the shape of the potential changes, as shown by the dashed line in Fig. 7.8, and one minimum becomes unstable at the coercive field.

The SW model provides a simple explanation for many of the magnetic properties of small magnetic particles, such as the shape of the hysteresis loop. However, the model has some limitations. It overestimates the strength of the coercive field because it allows only one path for reorientation. The model assumes that the magnetic energy of a particle is a function of the collective orientation of the spins of the magnetic atoms in the particle and the effect of the applied DC magnetic field. This implies that the magnetic energy of the particle depends on its volume. However, when particles are in the order of 6 nm in size, most of their atoms are on the surface, which means that they can have very different magnetic properties than larger grain particles. It has been shown that treating the surfaces of nanoparticles of α-Fe that are 600 nm long and 100 nm wide with various chemicals can produce variations in the coercive field by as much as 50%, underlining the importance of the surface of nanomagnetic particles in determining the magnetic properties of the grain. Thus the dynamical behavior of very small magnetic particles is somewhat more complicated than predicted by the SW model, and remains a subject of continuing research.

7.4. NANOPORE CONTAINMENT OF MAGNETIC PARTICLES

Another area of ongoing research in nanomagnetism involves developing magnetic materials by filling porous substances with nanosized magnetic particles. In fact, there are actually naturally occurring materials having molecular cavities filled with nanosized magnetic particles. Ferritin is a biological molecule, 25% iron by weight, which consists of a symmetric protein shell in the shape of a hollow sphere having an inner diameter of 7.5 nm and an outer diameter of 12.5 nm. The molecule plays the role in biological systems as a means of storing Fe^{3+} for an organism. One quarter of the iron in the human body is in ferritin, and 70% is in hemoglobin. The ferritin cavity is normally filled with a crystal of iron oxide $5Fe_2O_3 \cdot 9H_2O$. The iron oxide can be incorporated into the cavity from solution, where the number of iron atoms per protein is controlled from a few to a few thousand per protein molecule. The magnetic properties of the molecule depend on the number and kind of particles in the cavity, and the system can be engineered to be ferromagnetic or antiferromagnetic. The blocking temperature T_B is the temperature below which thermally assisted hopping, between different magnetic orientations, becomes frozen out. Figure 7.9 shows that the blocking temperature decreases with a decrease in the number of iron atoms in the cavity. Ferritin also displays magnetic quantum tunneling at very low temperatures. In zero magnetic field and at the very low temperature of 0.2 K the magnetization tunnels coherently back and forth between two minima. This effect makes its appearance as a resonance line in frequency-dependent magnetic susceptibility data. Figure 7.10 shows the results of a measurement of the resonant frequency of this susceptibility versus the number of iron atoms per molecule. We see that the frequency decreases from 3×10^8 Hz for 800 atoms to 10^6 Hz for 4600 atoms. The resonance disappears when a DC magnetic field is applied, and the symmetry of the double walled potential is broken.

Zeolites are crystalline silicates with intrinsic pores of well-defined shape, Fig. 6.24 (of Chapter 6) gives a schematic of a zeolite structure. These materials can be used as a matrix for the confinement of magnetic nanoparticles. Measurements of the temperature dependence of the susceptibility of iron particles incorporated into the pores exhibited paramagnetic behavior, with the magnetic susceptibility χ obeying the Curie law, $\chi = C/T$, where C is a constant, but there was no evidence of ferromagnetism.

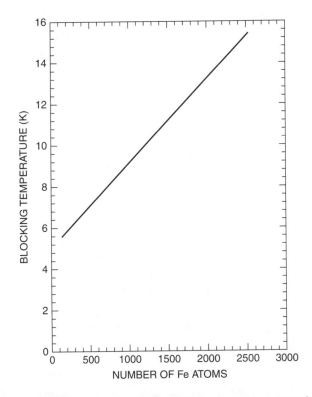

Figure 7.9. Plot of blocking temperature T_B versus the number of iron atoms in the cavity of ferritin. [Adapted from D. D. Awschalom and D. P. DiVincenzo, *Phys. Today* (April 1995).]

7.5. NANOCARBON FERROMAGNETS

As discussed in Chapter 5, the presence of iron or cobalt particles is necessary for the nucleation and growth of carbon nanotubes fabricated by pyrolysis. In the preparation of aligned carbon nanotubes by pyrolyzing iron II phthalacyanide (FePc) it has been demonstrated that nanotube growth involves two iron nanoparticles. A small iron particle serves as a nucleus for the tube to grow on, and at the other end of the tube larger particles enhance the growth. Aligned nanotubes are prepared on quartz glass by pyrolysis of FePc in an argon–hydrogen atmosphere.

Figure 7.11 shows a scanning electron microscope (SEM) image of iron particles on the tips of aligned nanotubes, and Fig. 7.12 shows the magnetization curve at 5 K and 320 K for the magnetic field parallel to the tubes showing that the hystersis is larger at 5 K. Figures 7.13 and 7.14 show plots of the temperature dependence of the coercive field H_c, and ratio of remnant magnetization M_r to saturation magnetization M_s. We see that the coercive field increases by more than a factor of 3 when the temperature is reduced from room temperature (300 K) to liquid helium temperature

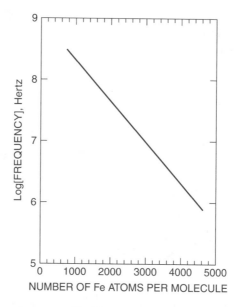

Figure 7.10. Resonance frequency of ferritin as a function of the number of atoms in the cavity of the molecule. [Adapted from D. D. Awschalom and D. P. DiVincenzo, *Phys. Today* (April 1995).]

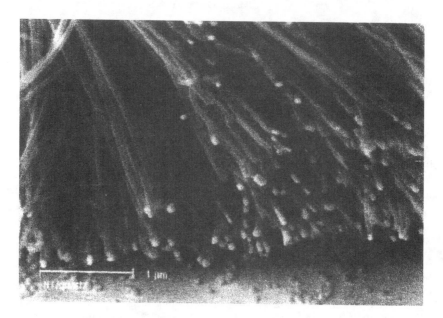

Figure 7.11. Scanning electron microscope image of iron particles (light spots) on the tips of aligned carbon nanotubes. [With permission from Z. Zhang et al., *J. Magn. Magn. Mater.* **231**, L9 (2001).]

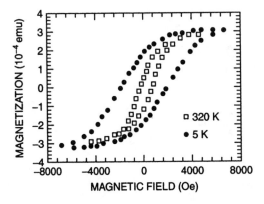

Figure 7.12. Magnetization curve hystersis loops for iron particles on the tips of aligned nanotubes at the temperatures of 5 and 320 K, for a magnetic field *H* applied parallel to the tubes. An oersted corresponds to 10^{-4} T. [Adapted from Z. Zhang et al., *J. Magn. Magn. Mater.* **231**, L9 (2001).]

(4 K). These iron particles at the tips of aligned nanotubes could be the basis of high-density magnetic storage devices. The walls of the tubes can provide nonmagnetic separation between the iron nanoparticles, ensuring that the interaction between neighboring nanoparticles is not too strong. If the interaction is too strong, the fields required to flip the orientation would be too large.

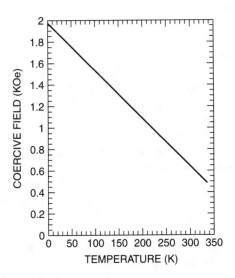

Figure 7.13. Plot of coercive field H_c versus temperature *T* for iron particles on the tips of aligned nanotubes. A kilooersted corresponds to 0.1 T. [Adapted from Z. Zhang et al., *J. Magn. Magn. Mater.* **231**, L9 (2001).]

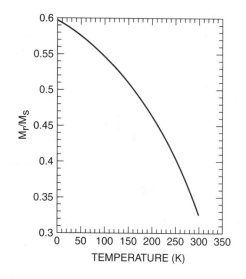

Figure 7.14. Ratio of remnant magnetization M_r to saturation magnetization M_s versus temperature T for iron nanoparticles on the tips of aligned carbon nanotubes. [Adapted from Z. Zhang et al., *J. Magn. Magn. Mater.* **231**, L9 (2001).]

Efforts to synthesize a nonpolymeric organic ferromagnet have received increased interest because of the possibility that such a magnet engineered by chemical modification of the molecules of the system would be lighter and an insulator. The C_{60} molecule discussed in Chapter 5 has a very high electron affinity, which means that it strongly attracts electrons. On the other hand, the molecule $C_2N_2(CH_3)_8$ or tetrakis (dimethylamino ethylene) (TDAE) is a strong electron donor, which means it readily gives its electron to another molecule. When C_{60} and TDAE are dissolved in a solvent of benzene and toluene, a precipitate of a new material forms that consists of the TDAE molecule complexed with C_{60}, and this 1:1 complex has a monoclinic crystal structure. A measured large increase in the susceptibility at 16 K indicated this material had become ferromagnetic, and up until relatively recently this was the highest Curie temperature for an organic ferromagnet.

When solid C_{60} is subjected to a 6 GPa pressure at 1000 K, a new crystal structure of C_{60} forms. In this structure the C_{60} molecules lie in parallel planes, and are bonded to each other in a way that forms a hexagonal array in the planes. The structure is very similar to that of graphite, with the carbon atoms replaced by C_{60} molecules. Although formed under high pressure, the structure is stable at ambient conditions. Magnetization measurements indicate that this new form of C_{60} becomes ferromagnetic at a remarkable 500 K. However, there is some skepticism in the scientific community about this result because the C_{60} hexagonal structure has no unpaired electrons, which are necessary for the formation of ferromagnetism. The researchers who made the observations suggest that the material has defects such as disrupted chemical bonds involving neighboring C_{60} molecules. Such broken bonds could be the source of conduction electrons, which might bring about itinerant ferromagnetism. Itinerant ferromagnetism corresponds to ferromagnetism from a spin that can move through the lattice. The result has not yet been confirmed by other researchers.

7.6. GIANT AND COLOSSAL MAGNETORESISTANCE

Magnetoresistance is a phenomenon where the application of a DC magnetic field changes the resistance of a material. The phenomenon has been known for many years in ordinary metals, and is due to the conduction electrons being forced to move in helical trajectories about an applied magnetic field. The effect becomes evident only when the magnetic field is strong enough to curve the electron trajectory within a length equal to its mean free path. The mean free path is the average distance an electron travels in a metal when an electric field is applied before it undergoes a collision with atoms, defects, or impurity atoms. The resistance of a material is the result of the scattering of electrons out of the direction of current flow by these collisions. The magnetoresistance effect occurs in metals only at very high magnetic fields and low temperatures. For example, in pure copper at 4 K a field of 10 T produces a factor of 10 change in the resistance.

Because of the large fields and low temperatures, magnetoresistance in metals originally had few potential application possibilities. However, that changed in 1988 with the discovery of what is now called *giant magnetoresistance* (GMR) in materials synthetically fabricated by depositing on a substrate alternate layers of nanometer thickness of a ferromagnetic material and a nonferromagnetic metal. A schematic of the layered structure and the alternating orientation of the magnetization in the ferromagnetic layer is shown in Fig. 7.15a. The effect was first observed in films made of alternating layers of iron and chromium, but since then other layered materials composed of alternating layers of cobalt and copper have been made that display much higher magnetoresistive effects. Figure 7.16 shows the effect of a DC magnetic field on the resistance of the iron–chromium multilayered system. The magnitude of the change in the resistance depends on the thickness of the iron layer, as shown in Figure 7.17, and it reaches a maximum at a thickness of 7 nm.

The effect occurs because of the dependence of electron scattering on the orientation of the electron spin with respect to the direction of magnetization. Electrons whose spins are not aligned along the direction of the magnetization M are scattered more strongly than those with their spins aligned along M. The application of a DC magnetic field parallel to the layers forces the magnetization of all the magnetic layers to be in the same direction. This causes the magnetizations pointing opposite to the direction of the applied magnetic field to become flipped. The conduction electrons with spins aligned opposite to the magnetization are more strongly scattered at the metal–ferromagnet interface, and those aligned along the field direction are less strongly scattered. Because the two spin channels are in parallel, the lower-resistance channel determines the resistance of the material.

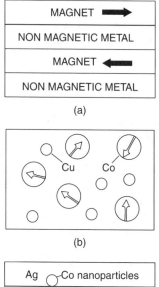

(a)

(b)

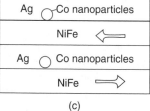

(c)

Figure 7.15. Three arrangements for producing colossal magnetoresistance: (a) layers of nonmagnetic material alternating with oppositely magnetized (arrows) ferromagnetic layers; (b) randomly oriented ferromagnetic cobalt nanoparticles (large circles) in a nonmagnetic copper matrix (small circles); (c) hybrid system consisting of cobalt nanoparticles in a silver (Ag) matrix sandwiched between nickel–iron (NiFe) magnetic layers, with alternating magnetizations indicated by arrows.

The magnetoresistance effect in these layered materials is a sensitive detector of DC magnetic fields, and is the basis for the development of a new, more sensitive reading head for magnetic disks. Prior to this, magnetic storage devices have used induction coils to both induce an alignment of the magnetization in a small region of the tape (write mode), and to sense the alignment of a recorded area (read mode). The magnetoresistive reading head is considerably more sensitive than the inductive coil method.

Materials made of single-domain ferromagnetic nanoparticles with randomly oriented magnetizations embedded in a nonmagnetic matrix also display giant magnetoresistance. Figure 7.15b shows a schematic of this system. The magneto-resistance in these materials, unlike the layered materials, is isotropic. The application of the DC magnetic field rotates the magnetization vector of the ferromagnetic

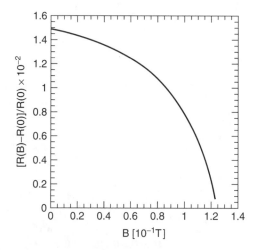

Figure 7.16. Dependence of the electrical resistance $R(B)$, relative to its value $R(0)$ in zero field, of a layered iron–chromium system on a magnetic field B applied parallel to the surface of the layers. [Adapted from R. E. Camley, *J. Phys. Condens. Matter* **5**, 3727 (1993).]

nanoparticles parallel to the direction of the field, which reduces the resistance. The magnitude of the effect of the applied magnetic field on the resistance increases with the strength of the magnetic field and as the size of the magnetic nanoparticles decreases. Figure 7.18 shows a representative measurement at 100 K of a film

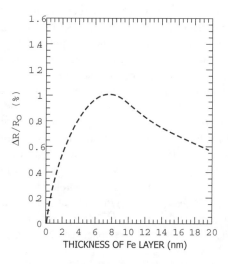

Figure 7.17. Dependence of the change of magnetoresistance ΔR on the thickness of the iron magnetic layer in a constant DC magnetic field for the Fe–Cr layered system. [Adapted from R. E. Camley, *J. Phys. Condens. Matter* **5**, 3727 (1993).]

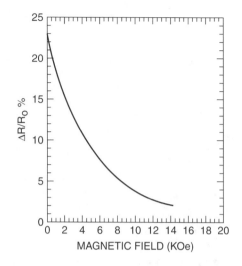

Figure 7.18. Dependence of the change of magnetoresistance ΔR versus the applied magnetic field for a thin film of Co nanoparticles in a copper matrix. A kilooersted corresponds to 0.1 T [Adapted from A. E. Berkowitz, *Phys. Rev. Lett.* **68**, 3745 (1992).]

consisting of Co nanoparticles in a copper matrix. Hybrid systems consisting of nanoparticles in metal matrices sandwiched between metal magnetic layers, as illustrated schematically in Fig. 7.15c, have also been developed and exhibit similar magnetoresistance properties.

Materials have been discovered having larger magnetoresistive effects than the layered materials, and this phenomenon in them is called *colossal magnetoresistance* (CMR). These materials also have a number of application possibilities, such as in magnetic recording heads, or as sensing elements in magnetometers. The perovskite-like material $LaMnO_3$ has manganese in the Mn^{3+} valence state. If the La^{3+} is partially replaced with ions having a valence of 2+, such as Ca, Ba, Sr, Pb, or Cd, some Mn^{3+} ions transform to Mn^{4+} to preserve the electrical neutrality. The result is a mixed valence system of Mn^{3+}/Mn^{4+}, with the presence of many mobile charge carriers. This mixed valence system has been shown to exhibit very large magneto-resistive effects. The unit cell of the crystal is sketched in Fig. 7.19. The particular system $La_{0.67}Ca_{0.33}MnO_x$ displays more than a thousandfold change in resistance with the application of a 6-T DC magnetic field. Figure 7.20 shows how the normalized resistance, called the *resistivity* (normalized magnetoresistance) of a thin film of the material exhibits a pronounced decrease with increasing values of the DC magnetic field. The temperature dependence of the resistivity also displays the unusual behavior shown in Fig. 7.21 as the temperature is lowered through the Curie point. Although the effect of nanostructuring on these materials has not been extensively studied, it is expected to have a pronounced influence on the magnitude of the magnetoresistive effect.

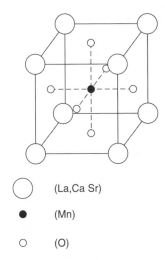

    ◯   (La,Ca Sr)

    ●   (Mn)

    ○   (O)

Figure 7.19. Crystal structure of $LaMnO_3$, which displays colossal magneto resistance when the La site is doped with Ca or Sr. (With permission from F. J. Owens and C. P. Poole Jr., *Electromagnetic Absorption in the Copper Oxide Superconductors*, Kluwer/Plenum, 1999, p. 89.)

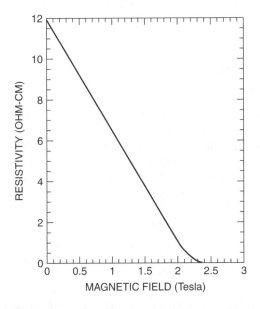

Figure 7.20. Dependence of the resistivity (normalized magnetoresistance) of La–Ca–Mn–O on an applied magnetic field in the neighborhood of the Curie temperature at 250 K. (With permission from F. J. Owens and C. P. Poole Jr., *Electromagnetic Absorption in the Copper Oxide Superconductors*, Kluwer/Plenum, 1999, p. 90.)

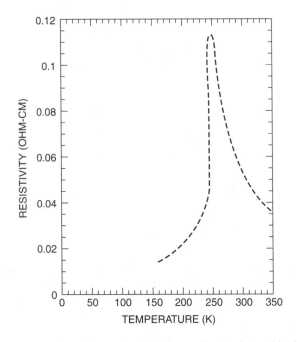

Figure 7.21. Temperature dependence of resistivity (normalized resistance) in sintered samples of La–Ca–Mn–O in zero magnetic field. (With permission from F. J. Owens and C. P. Poole, Jr. *Electromagnetic Absorption in the Copper Oxide Superconductors*, Kluwer/Plenum, 1999, p. 90.)

7.7. FERROFLUIDS

Ferrofluids, also called *magnetofluids*, are colloids consisting typically of 10-nm magnetic particles coated with a surfactant to prevent aggregation, and suspended in a liquid such as transformer oil or kerosene. The nanoparticles are single-domain magnets, and in zero magnetic field, at any instant of time, the magnetization vector of each particle is randomly oriented so the liquid has a zero net magnetization. When a DC magnetic field is applied, the magnetizations of the individual nanoparticles all align with the direction of the field, and the fluid acquires a net magnetization. Typically ferrofluids employ nanoparticles of magnetite, Fe_3O_4. Figure 7.22 shows the magnetization curve for a ferrofluid made of 6-nm Fe_3O_4 particles exhibiting almost immeasurable hysteresis. Ferrofluids are soft magnetic materials that are superparamagnetic. Interestingly, suspensions of magnetic particles in fluids have been used since the 1940s in magnetic clutches, but the particles were larger, having micrometer dimensions. Application of a DC magnetic field to this fluid causes the fluid to congeal into a solid mass, and in the magnetic state the material is not a liquid. A prerequisite for a ferrofluid is that the magnetic particles have nanometer sizes. Ferrofluids have a number of interesting properties, such as magnetic-field-dependent anisotropic optical properties.

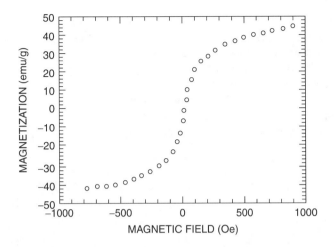

Figure 7.22. Magnetization curve for a ferrofluid made of magnetite, Fe_3O_4, nanoparticles showing the soft (nonhysteretic) magnetic behavior. An oersted corresponds to 10^{-4} T [Adapted from D. K Kim, *J. Magn. Magn. Mater.* **225**, 30 (2001).]

Analogous properties are observed in liquid crystals, which consist of long molecules having large electric dipole moments, which can be oriented by the application of an electric field in the fluid phase. Electric-field-modulated birefringence or double refraction of liquid crystals is widely used in optical devices, such as liquid crystal displays in digital watches, and screens of portable computers. This suggests a potential application of ferrofluids employing magnetic field induced bifringence. To observe the behavior, the ferrofluid is sealed in a glass cell having a thickness of several micrometers. When a DC magnetic field is applied parallel to

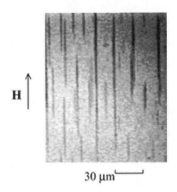

Figure 7.23. Picture taken through an optical microscope of chains of magnetic nanoparticles formed in a film of a ferrofluid when the DC magnetic field is parallel to the plane of the film. [With permission from H. E. Hornig et al., *J. Phys. Chem. Solids* **62**, 1749 (2001).]

the surface and the film is examined by an optical microscope, it is found that some of the magnetic particles in the fluid agglomerate to form needle like chains parallel to the direction of the magnetic field. Figure 7.23 depicts the chains viewed through an optical microscope. As the magnetic field increases more particles join the chains, and the chains become broader and longer. The separation between the chains also decreases. Figures 7.24a and 7.24b show plots of the chain separation and the chain width as a function of the DC magnetic field. When the field is applied perpendicular to the face of the film the ends of the chains arrange themselves in the pattern shown

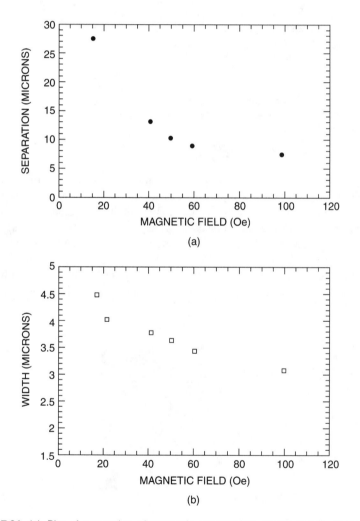

Figure 7.24. (a) Plot of separation of magnetic nanoparticle chains versus strength of a magnetic field applied parallel to the surface of the film; (b) plot of the width of the chains as a function of DC magnetic field strength. An oersted corresponds to 10^{-4} T. [Adapted from H. E. Hornig et al., *J. Phys. Chem. Solids* **62**, 1749 (2001).]

10 μm

Figure 7.25. Optical microscope picture of the ends of chains of magnetic nanoparticles in a ferrofluid film when the DC magnetic field is perpendicular to the surface of the film. The field strength is high enough to form the hexagonal lattice configuration. [With permission from H. E. Hornig et al., *J. Phys. Chem. Solids* **62**, 1749 (2001).]

in Fig. 7.25, which is also a picture taken through an optical microscope. Initially at low fields the ends of the chains are randomly distributed in the plane of the fluid. As the field increases a critical field is reached where the chain ends become ordered in a two-dimensional hexagonal array as shown in Fig. 7.25. This behavior is analogous to the formation of the vortex lattice in type II superconductors.

The formation of the chains in the ferrofluid film when a DC magnetic field is applied makes the fluid optically anisotropic. Light, or more generally electromagnetic waves, have oscillatory magnetic and electric fields perpendicular to the direction of propagation of the beam. Light is linearly polarized when these vibrations are confined to one plane perpendicular to the direction of propagation, rather than having random transverse directions. When linearly polarized light is incident on a magnetofluid film to which a DC magnetic field is applied, the light emerging from the other side of the film is elliptically polarized. Elliptically polarized light occurs when the E and H vibrations around the direction of propagation are confined to two mutually perpendicular planes, and the vibrations in each plane are out of phase. This is called the *Cotton–Mouton effect*. The experimental arrangement to investigate this effect is shown in Fig. 7.26. A He–Ne laser beam linearly polarized by a polarizer is incident on the magnetofluid film. A DC magnetic field is applied parallel to the plane of the film. To examine the polarization of the light emerging from the film, another polarizer called an *analyzer* is placed between the film and the light detector, which is a *photomultiplier*. The intensity of the transmitted light is measured as a function of the orientation of the

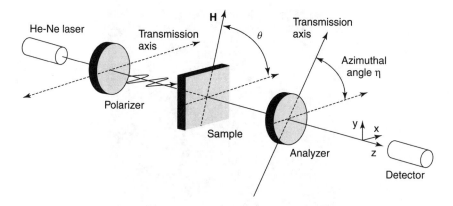

Figure 7.26. Experimental arrangement for measuring optical polarization effects in a ferrofluid film that has a DC magnetic field *H* applied parallel to its surface. [With permission from H. E. Hornig et al., *J. Phys. Chem. Solids* **62**, 1749 (2001).]

polarizing axis of the analyzer given by the angle η in the figure. Figure 7.27 shows that the transmitted light intensity depends strongly on the angle η. These effects could be the basis of optical switches where the intensity of transmitted light is switched on and off using a DC magnetic field, or the orientation of a polarizer.

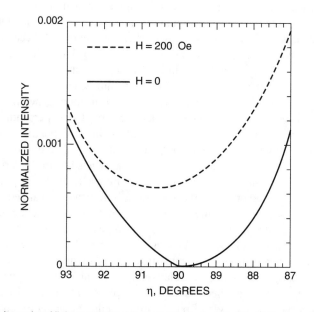

Figure 7.27. Intensity of light transmitted through the analyzer in Fig. 7.26 versus the azimuthal angle η of the light beam in zero field, and in a 200-Oe (0.02-T) applied magnetic field *H*. [Adapted from H. E. Hornig et al., *J. Phys. Chem. Solids* **62**, 1749 (2001).]

Ferrofluids can also form magnetic field tunable diffraction gratings. Diffraction is the result of interference of two or more light waves of the same wavelength traveling paths of slightly different lengths before arriving at a detector such as a photographic film. When the pathlength differs by half a wavelength, the waves destructively interfere, resulting in a dark band on the film. When the path lengths differ by a wavelength, then the waves constructively interfere, producing a bright band on the film. A diffraction grating consists of small slits separated by distances of the order of the wavelength of the incident light. We saw above that when a DC magnetic field of sufficient strength is applied perpendicular to a magnetofluid film, an equilibrium two-dimensional hexagonal lattice is formed with columns of nanoparticles occupying the lattice sites. This structure can act as a two-dimensional optical diffraction grating that diffracts incoming visible light. Figure 7.28 shows a black and white picture of the chromatic (colored) rings of light and darkness resulting from the diffraction and interference when a focused parallel beam of white light is passed through a magnetic fluid film that has a magnetic field applied perpendicular to it. The diffraction pattern is determined by the equation

$$d \sin \Theta = n\lambda \qquad (7.6)$$

where d is the distance between the chains of nanoparticles, Θ is the angle between the outgoing light and the direction normal to the film, n is an integer, and λ is the wavelength of the light. We saw earlier that the distance between the chains d

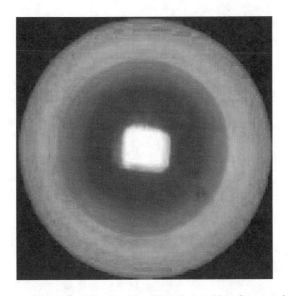

Figure 7.28. Chromatic rings resulting from the diffraction and interference of a beam of white light incident on a ferrofluid film in a perpendicular DC applied magnetic field. [With permission from H. E. Hornig et al., *J. Phys. Chem. Solids* **62**, 1749 (2001).]

depends on the strength of the applied DC magnetic field. In effect, we have a tunable diffraction grating, which can be adjusted to a specific wavelength by changing the strength of the DC magnetic field.

Ferrofluids have a number of present commercial uses. They are employed as contaminant exclusion seals on hard drives of personal computers, and vacuum seals for high-speed high vacuum motorized spindles. In this latter application the ferrofluid is used to seal the gap between the rotating shaft and the pole piece support structure, as illustrated in Fig. 7.29. The seal consists of a few drops of ferrofluid in the gap between the shaft and a cylindrical permanent magnet that forms a collar around it. The fluid forms an impermeable O-ring around the shaft, while allowing rotation of the shaft without significant friction. Seals of this kind have been utilized in a variety of applications. Ferrofluids are used in audiospeakers in the voice gap of the driver to dampen moving masses. Nature even employs ferrofluids. For example, it is believed that ferrofluids play a role in the directional sensing of trout fish. Cells near the nose of the trout are thought to contain a ferrofluid of magnetite nanoparticles. When the trout changes its orientation with respect to the Earth's magnetic field, the direction of magnetization in the ferrofluid of the cells changes, and this change is processed by the brain to give the fish orientational information.

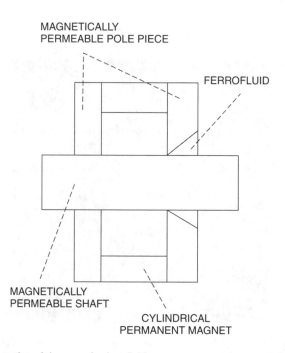

Figure 7.29. Illustration of the use of a ferrofluid as a vacuum seal on a rotating magnetically permeable shaft mounted on the pole pieces of a permanent magnet.

FURTHER READING

D. D. Awschalom and D. P. DiVincenzo, "Complex Dynamics of Mesoscopic Magnets," *Phys. Today* 43 (April 1995).

D. D. Awschalom and S. von Molnar, "Physical Properties of Nanometer-scale Magnets," in *Nanotechnology*, G. Timp, ed., Springer-Verlag, Heidelberg, 1999, Chapter 12.

R. E. Camley and R. L. Stamps, "Magnetic Multilayers," *J. Phys. Condens. Matter* **15**, 3727 (1993).

H. E. Horng, C. Hong, S. Y. Yang, and H. C. Yand, "Novel Properties and Applications in Magnetic Fluids," *J. Phys. Chem. Solids* **62**, 1749 (2001).

H. Kronmuller, "Recent Developments in High Tech Magnetic Materials," *J. Magn. Magn. Mater.* **140**, 25 (1995).

K. Ounadjela and R. L. Stamps, "Mesoscopic Magnetism in Metals," in *Handbook of Nanostructured Materials and Nanotechnology*, H. S. Nalwa, ed., Academic Press, San Diego, 2000, Vol. 2, Chapter 9, p. 429.

R. E. Rosenweig, "Magnetic Fluids," *Sci. Am.* 136 (Oct. 1982).

J. L. Simonds, "Magnetoelectronics," *Phys. Today* 26 (April 1996).

8

OPTICAL AND VIBRATIONAL SPECTROSCOPY

8.1. INTRODUCTION

One of the main ways to study and characterize nanoparticles is by the use of spectroscopic techniques. We begin this chapter with some introductory comments on the nature of spectroscopy, and then discuss various ways in which it can enhance our understanding of nanoparticles.

There are several types of spectroscopy. Light or radiation incident on a material with intensity I_0 can be transmitted (I_T), absorbed (I_A), or reflected (I_R) by it, as shown in Fig. 8.1, and these three intensities are related by the conservation of intensity expression

$$I_0 = I_T + I_A + I_R \qquad (8.1)$$

Ordinarily the incoming or incident light I_0 is scanned by gradually varying its frequency $v = \omega/2\pi$ or wavelength λ, where $\lambda v = c$, and $c = 2.9979 \times 10^8 \sim 3 \times 10^8$ m/s is the velocity of light. In a typical case an incident photon or "particle" of light with the energy $E = \hbar \omega$ ($\hbar = h/2\pi$ is the reduced Planck's constant) induces a transition of an electron from a lower energy state to a higher energy state, and as a

Introduction to Nanotechnology, by Charles P. Poole Jr. and Frank J. Owens.
ISBN 0-471-07935-9. Copyright © 2003 John Wiley & Sons, Inc.

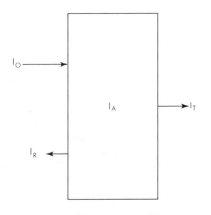

Figure 8.1. Sketch of incident electromagnetic wave intensity I_0 being partially absorbed I_A, partially reflected I_R and partially transmitted I_T by a sample.

result light is absorbed. In transmission spectroscopy the reflected signal is neglected, and the absorption is determined by a decrease in the transmitted intensity I_T as a function of the scanning frequency ω (or λ), while in reflection spectroscopy the transmission is neglected and the absorption is determined from the change I_R in the reflected light. The former is used for transparent samples and the latter, for opaque ones. Thus spectroscopic measurements can be made by gradually scanning E, ω, or λ of the incoming light beam and measuring its effect on I_T (or I_R), whose amplitude is recorded during the scan. More modern equipment makes use of charge-coupled devices (CCDs) as light detectors. These are arrays of metal oxide semi-conductors (MOS) that consist of a p-type silicon layer, a silicon dioxide layer, and a metal plate. Incident photons generate minority carriers, and the current is propor-tional to the intensity of the light, and the time of exposure. These devices make several rapid scans of the wavelength range, and in conjunction with computer pro-cessing, they enable the recording of the complete spectrum in a relatively short time.

This chapter discusses investigations of nanomaterials using spectroscopic techniques in the infrared and Raman regions of the spectrum (frequencies from 10^{12} to 4×10^{14} Hz, wavelengths λ from 300 to 1 μm), as well as visible and ultraviolet spectroscopy (frequencies from 4×10^{14} to 1.5×10^{15}, λ from 0.8 to 0.2 μm).

Another type of spectroscopy is *emission spectroscopy*. An incident photon $\hbar \omega_0$ raises an electron from its ground state E_{gnd} to an excited energy level E_{exc}, the electron undergoes a radiationless transition to an intermediate energy state, and then it returns to its ground-state level, emitting in the process a photon $\hbar \omega_{lum}$, that can be detected, as shown in Fig. 8.2. If the emission takes place immediately, it is called *fluorescence*, and if it is delayed as a result of the finite lifetime of the intermediate metastable state E_{met}, it is called *phosphorescence*. Both types of emission paths are referred to as *luminescence*, and the overall process of light absorption followed by emission is called *photoemission*. Emission spectroscopy can be studied by varying the frequency of the incident exciting light, by studying the frequency distribution of

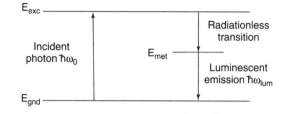

Figure 8.2. Energy-level diagram showing an incident photon $\hbar\omega_0$ raising an electron from its ground state E_{gnd} to an excited state E_{exc}, and a subsequent radiationless transition to a long-lived metastable state E_{met} followed by luminescent emission of a photon $\hbar\omega_{lum}$.

the emitted light, or by combining both techniques. Luminescent spectra will be examined for all these variations. Light emission can also be induced by gradually heating a sample, and the resulting thermal luminescence manifests itself by the emission of light over a characteristic temperature range. This emission is called a *glow peak*.

8.2. INFRARED FREQUENCY RANGE

8.2.1. Spectroscopy of Semiconductors; Excitons

These observations on spectroscopy that we have made are of a general nature, and apply to all types of ultraviolet, visible, Raman, and infrared spectroscopy. Semiconductors are distinguished by the nature and the mechanisms of the processes that bring about the absorption or emission of light. Incident light with photon energies less than the bandgap energy E_g passes through the sample without absorption, and higher-energy photons can raise electrons from the valence band to the conduction band, leaving behind holes in the valence band. Figure 8.3 presents a plot of the optical absorption coefficient for bulk GaAs, and we see that the onset of the absorption occurs at the bandgap edge where the photon energy $\hbar\omega$ equals E_g. The data tabulated in Table B.7 show that the temperature coefficient dE_g/dT of the energy gap is negative for all III–V and II–VI semiconductors, which means that the onset of absorption undergoes what is called a *blue shift* to higher energies as the temperature is lowered, as shown on the figure. The magnitude of the absorption, measured by the value of the absorption coefficient, also becomes stronger at lower temperatures, as shown in Fig. 8.3.

Another important contributor to the spectroscopy of semiconductors is the presence in the material of weakly bound excitons called *Mott–Wannier excitons*. This type of exciton is a bound state of an electron from the conduction band and a hole from the valence band attracted to each other by the Coulomb interaction $e^2/4\pi e^2\varepsilon_0 r^2$, and having a hydrogen atom like system of energy levels called a *Rydberg series*, as explained in Section 2.3.3. The masses m_e and m_h, of the electron and the hole, respectively, in a zinc blende semiconductor are both much less than

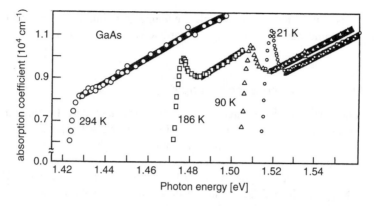

Figure 8.3. GaAs excitonic absorption spectra near the bandgap for several temperatures. The experimental points for the spectra are given, and the dark lines represent theoretical fits to the data. [From M. D. Sturge, *Phys. Rev.* **127**, 768 (1962); see also P. Y. Yu and M. Cardona, *Fundamentals of Semiconductors*, 3rd ed., Springer, Berlin, 2001, p. 287.]

the free-electron mass m_0, as is clear from the data listed in Table B.8. As a result, the effective mass $m^* = m_e m_h/(m_e + m_h)$ of the exciton is significantly less that the mass m_0 of a free electron. The material itself has a dielectric constant ε listed in Table B.11, which is appreciably larger than the value ε_0 in free space, and the result is a system of energy levels that is related to the ground-state hydrogen atom energy 13.6 eV through Eq. (2.18):

$$E = \frac{13.6 m^*/m_0}{(\varepsilon/\varepsilon_0)^2 n^2} \text{ eV} \tag{8.2}$$

where the quantum number n takes on the values $n = 1,2,3,4,\ldots$, with the value $n = 1$ for the lowest energy or ground state. It is clear from Eq. (8.2) that both the mass ratio m^*/m_0 and the dielectric constant ratio $\varepsilon/\varepsilon_0$ have the effect of decreasing the exciton energy considerably below that of a hydrogen atom, as shown in Fig. 2.20. In bulk semiconductors the absorption spectra from excitons are generally too weak to be observed at room temperature, but can be seen at low temperature. Extensive ionization of excitons at room temperature weakens their absorption. The resulting temperature dependence is illustrated by the series of spectra in Fig. 8.3, which display excitonic absorption near the band edge that becomes more prominent as the temperature is lowered.

Thus far we have discussed the optical absorption of the bulk semiconductor GaAs, and spectroscopic studies of its III–V sister compounds have shown that they exhibit the same general type of optical absorption. When nanoparticles are studied by optical spectroscopy, it is found that there is a shift toward higher energies as the size of the particle is decreased; this so-called blue shift is accompanied by an enhancement of the intensity, and the exciton absorption becomes more pronounced. The optical spectra displayed in Fig. 4.20 for CdSe illustrate this trend for the particle sizes 4 and 2 nm. Thus lowering the temperature influences the spectra very much in the same way as decreasing the particle size affects them.

8.2.2. Infrared Surface Spectroscopy

The general principles of infrared (IR) spectroscopy, including Fourier transform infrared spectroscopy (FTIR), were explained in Section 3.4.1. These spectroscopic techniques measure the absorption of radiation by high frequency (i.e., optical branch) phonon vibrations, and they are also sensitive to the presence of particular chemical groups such as hydroxyl ($-OH$), methyl ($-CH_3$), imido ($-NH$), and amido ($-NH_2$). Each of these groups absorbs infrared radiation at a characteristic frequency, and the actual frequency of absorption varies somewhat with the environment. We discuss some results based on work of Baraton (2000).

As an example, Fig. 8.4 shows the FTIR spectrum of titania (TiO_2), which exhibits IR absorption lines from the groups OH, CO, and CO_2. Titania is an important catalyst, and IR studies help elucidate catalytic mechanisms of processes that take place on its surface. This material has the anatase crystal structure at room temperature, and can be prepared with high surface areas for use in catalysis. It is a common practice to activate surfaces of catalysts by cleaning and exposure to particular gases in oxidizing or reducing atmospheres at high temperatures to prepare sites where catalytic reactions can take place. The spectrum of Fig. 8.4 was obtained after adsorbing carbon monoxide (CO) at 500°C on an activated titania nanopowder surface, and subtracting the spectrum of the initial activated surface before the adsorption. The strong carbon dioxide (CO_2) infrared absorption lines in the spectrum show that the adsorbed carbon monoxide had been oxidized to carbon

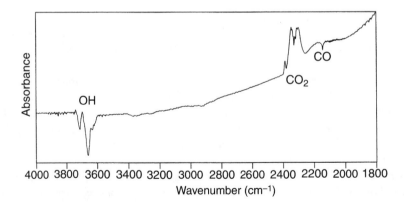

Figure 8.4. Fourier transform infrared (FTIR) spectrum of activated titania nanopowder with carbon monoxide (CO) adsorbed on the surface. The spectrum of the initial activated titania has been subtracted. The negative (downward) adsorption in the OH region indicates the replacement of hydroxyl groups by CO_2 on the surface. [From M.-I. Baraton and L. Merhari, *Nanostruct. Mater.* **10**, 699 (1998).]

dioxide on the surface. Note that the OH absorption signal is in the negative (downward) direction. This means that the initial activated titania surface, whose spectrum had been subtracted, had many more OH groups on it than the same surface after CO adsorption. Apparently OH groups originally present on the surface have been replaced by CO_2 groups. Also the spectrum exhibits structure in the range from 2100 to 2400 cm^{-1} due to the vibrational–rotational modes of the CO and CO_2. In addition, the gradually increasing absorption for decreasing wavenumber shown at the right side of the figure corresponds to a broad spectral band that arises from electron transfer between the valence and conduction bands of the n-type titania semiconductor.

To learn more about an infrared spectrum, the technique of isotopic substitution can be employed. We know from elementary physics that the frequency of a simple harmonic oscillator ω of mass m and spring constant C is proportional to $(C/m)^{1/2}$, which means that the frequency ω, and the energy E given by $E = \hbar\omega$, both decrease with an increase in the mass m. As a result, isotopic substitution, which involves nuclei of different masses, changes the IR absorption frequencies of chemical groups. Thus the replacements of ordinary hydrogen ^1H by the heavier isotope deuterium ^2D (0.015% abundant), ordinary carbon ^{12}C by ^{13}C (1.11% abundant), ordinary ^{14}N by ^{15}N (0.37% abundant), or ordinary ^{16}O by ^{17}O (0.047% abundant) all increase the mass, and hence decrease the infrared absorption frequency. The decrease is especially pronounced when deuterium is substituted for ordinary hydrogen since the mass ratio $m_D/m_H = 2$, so the absorption frequency is expected to decrease by the factor $\sqrt{2} \cong 1.414$. The FTIR spectrum of boron nitride (BN) nanopowder after deuteration (H/D exchange), presented in Fig. 8.5, exhibits this $\sqrt{2}$ shift. The figure shows the initial spectrum (tracing a) of the BN nanopowder after activation at 875 K, (tracing b) of the nanopowder after subsequent deuteration, and (tracing c) after subtraction of the two spectra. It is clear that the deuteration converted the initial $B-OH$, $B-NH_2$, and B_2-NH groups on the surface to $B-OD$, $B-ND_2$, and B_2-ND, respectively, and that in each case the shift in wavenumber (i.e., frequency) is close to the expected $\sqrt{2}$. The overtone bands that vanish in the subtraction of the spectra are due to harmonics of the fundamental BN lattice vibrations, which are not affected by the H/D exchange at the surface. Boron nitride powder is used commercially for lubrication. Its hexagonal lattice, with planar B_3N_3 hexagons, resembles that of graphite.

A close comparison of the FTIR spectra from gallium nitride GaN nanoparticles illustrated in Fig. 8.6 with the boron nitride nanoparticle spectra of Fig. 8.5 show how the various chemical groups $-OH$, $-NH_2$, and $-NH$ and their deuterated analogues have similar vibrational frequencies, but these frequencies are not precisely the same. For example, the frequency of the $B-ND_2$ spectral line of Fig. 8.5 is somewhat lower than that of the $Ga-ND_2$ line of Fig. 8.6, a small shift that results from their somewhat different chemical environments. The H/D exchange results of Fig. 8.6 show that all of the $Ga-OH$ and $Ga-NH_2$ are on the surface, while only some of the NH groups were exchanged. Notice that the strong GaH absorption band near 21,000 cm^{-1} was not appreciably disturbed by the H/D exchange, suggesting that it arises from hydrogen atoms bound to gallium inside the bulk.

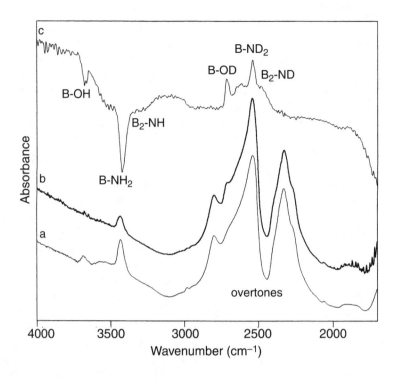

Figure 8.5. FTIR spectra of boron nitride nanopowder surfaces after activation at 875 K (tracing a), after subsequent deuteration (tracing b), and (c) difference spectrum of *a* subtracted from *b* (tracing c). [From M.-I. Baraton and L. Merhari, P. Quintard, V. Lorezenvilli, *Langmuir*, **9**, 1486 (1993).]

An example that demonstrates the power of infrared spectroscopy to elucidate surface features of nanomaterials is the study of γ-alumina (Al_2O_3), a catalytic material that can have a large surface area, up to 200–$300 \, m^2/g$, due to its highly porous morphology. It has a defect spinel structure, and its large oxygen atoms form a tetragonally distorted face-centered cubic lattice (see Section 2.1.2). There are one octahedral (VI) and two tetrahedral (IV) sites per oxygen atom in the lattice, and aluminum ions located at these sites are designated by the notations $_{VI}Al^{3+}$ and $_{IV}Al^{3+}$, respectively, in Fig. 8.7. There are a total of five configurations assumed by adsorbed hydroxyl groups that bond to aluminum ions at the surface, and these are sketched in the figure. The first two, types Ia and Ib, involve the simple cases of OH bonded to tetrahedrally and octahedrally coordinated aluminum ions, respectively. The remaining three cases involve the hydroxyl radical bound simultaneously to two or three adjacent trivalent aluminum ions. The frequency shifts assigned to these five surface species, which are listed in the figure ($v(OH)$), are easily distinguished by infrared spectroscopy.

The FTIR spectra from γ-alumina nanopowder before and after deuteration presented in Fig. 8.8 display broad absorption bands with structure arising from

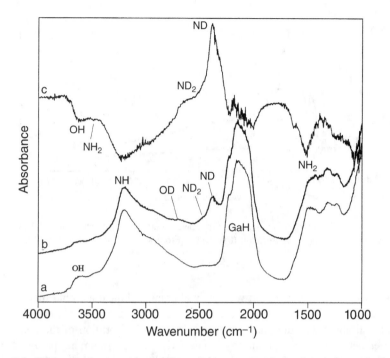

Figure 8.6. FTIR reflection spectra of gallium nitride nanopowder surfaces after activation at 500°C (tracing a), after subsequent deuteration (tracing b), and difference spectrum of *a* subtracted from *b* (tracing c). [From M.-I. Baraton, G. Carlson, and K. E. Gonsalves, *Mater. Sci. Eng.* **B50**, 42 (1997).]

the OH and OD groups, respectively, on alumina activated at the temperature 600°C, and spectrum (a) of Fig. 8.9 provides an expanded view of the OD region of Fig. 8.8 for γ-alumina activated at 500°C. The analysis of the positions and relative amplitudes of component lines of these spectra provide information on the distribution of aluminum ions in the octahedral and tetrahedral sites of the atomic layer at

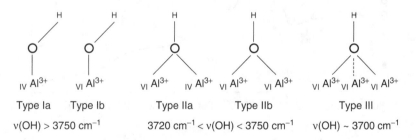

Figure 8.7. Five possible configurations of adsorbed hydroxyl groups bonded at tetrahedral ($_{IV}Al^{3+}$) and octahedral ($_{VI}Al^{3+}$) sites of a γ-alumina surface. [From M.-I. Baraton, in Nalwa (2000), Vol. 2. Chapter 2, p. 116.]

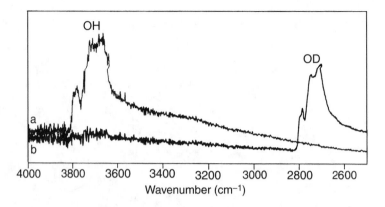

Figure 8.8. FTIR spectra of a γ-alumina nanopowder surfaces after activation at 600°C (tracing a), and after subsequent deuteration (tracing b). [From M.-I. Baraton, in Nalwa (2000), Vol. 2, Chapter 2, p. 115.]

the surface. The differences between the γ-alumina and the θ-alumina spectra of Fig. 8.9 provide evidence that the two aluminas differ in their allocations of Al^{3+} to octahedral and tetrahedral sites. In further work the adsorption of the heterocyclic six-membered ring compound pyridine, C_5H_5N, on γ-alumina provided FTIR spectra of the pyridine coordinated to different Al^{3+} Lewis acid sites. These infrared spectral results help clarify the coordination state of aluminum ions in the surface layer where Lewis acid sites play an important role in the catalytic activity.

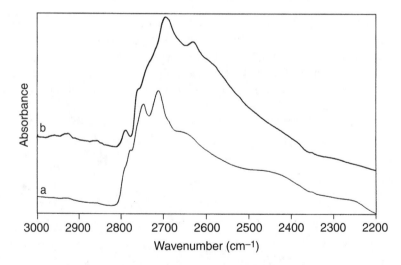

Figure 8.9. Details of FTIR surface spectra in the deuterated hydroxyl (OD) absorption region of γ-Al$_2$O$_3$ (tracing a) and θ-Al$_2$O$_3$ (tracing b) nanopowders after activation at 500°C followed by deuteration. [From M.-I. Baraton, in Nalwa (2000), Vol. 2, Chapter 2 p. 117.]

8.2.3. Raman Spectroscopy

The general principles of Raman spectroscopy were explained in Section 3.4.1. Raman scattering measures the frequency shift $\Delta\omega = \omega_{phonon} = |\omega_{inc} - \omega_{scat}|$ between the incident ω_{inc} and scattered ω_{scat} light frequencies when ω_{phonon} is an optical phonon mode vibration. When ω_{phonon} is an acoustic phonon, the process is called *Brillouin scattering*, discussed in the next section. Figure 2.10 makes it clear that the optical mode corresponds to high-frequency lattice vibrations, and the acoustic mode conforms to lattice vibrations at much lower frequencies: $\omega_{acous} \ll \omega_{opt}$. The scattered frequency has the value $\omega_{scat} = \omega_{inc} \pm \omega_{phonon}$ where the negative sign in the preceding expression for $\Delta\omega$ corresponds to a Stokes line, and the positive sign denotes an anti-Stokes line, as explained in Section 3.4.1. These two types of scattering that entail a change in frequency of the emitted photon are called *inelastic*. When there is no frequency shift (i.e., $\Delta\omega = 0$), then the scattering is the elastic Rayleigh type that takes place in X-ray diffraction. Here and in the next section we describe some Raman and Brillouin spectra from Milani and Bottani (2000).

The Raman spectrum of bulk crystalline germanium exhibits a narrow absorption line, $\cong 3\,cm^{-1}$ wide, arising from the $\Gamma_{25}{}^{+}$ optical phonon mode at the frequency of $300\,cm^{-1}$, as indicated in the inset to Fig. 8.10. When Ge is deposited on a silica (SiO_2) film, the Raman spectrum is featureless except for a broad shoulder near $270\,cm^{-1}$, as shown in the lowest spectrum of this figure. Annealing causes the shoulder to disappear, and the 300-cm^{-1} crystalline silica peak to appear. Figure 8.11 illustrates how this peak broadens and shifts to lower frequencies with decreasing particle size. Its width decreases with an increase in temperature and annealing time, as shown in Fig. 8.12, and these data provide an estimated particle size between 6

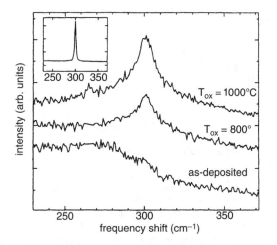

Figure 8.10. Raman spectra of Ge films on SiO_2 substrates as deposited and after oxidation at 800 and 1000°C. [From P. Milani and C. E. Bottani, in Nalwa (2000), Vol. 2, Chapter 4, p. 243.]

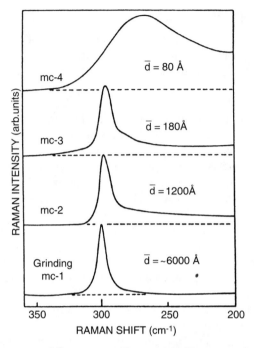

Figure 8.11. Raman spectra of Ge nanocrystallites produced by gas condensation showing the broadening and shift to lower wavenumbers as the particle size decreases. The lowest spectrum was obtained from ground bulk germanium. [From S. Hayashi, M. Ito, and H. Kanamori, *Solid State Commun.* **44**, 75 (1982).]

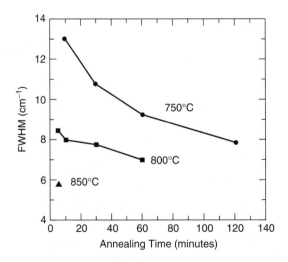

Figure 8.12. Plot of full width at half maximum height (FWHM) of a Ge Raman line against annealing time. [From D. C. Paine, C. Caragiantis, T. Y. Kim, Y. Shigesato, and T. Ishahara, *Appl. Phys. Lett.* **62**, 2842 (1993).]

and 13 nm, which increases with an increase in either the temperature or the annealing time, in accordance with the results plotted in Fig. 8.13.

Silicon nanoparticles exhibit the same behavior as their germanium counterparts for the Γ_{25}^{+} optical phonon mode, which in this case is centered at 521 cm^{-1}, as indicated in Fig. 8.14. This figure illustrates how the Raman line from fine-grained polycrystalline Si broadens and shifts to lower wavenumbers from the single-crystal spectrum. The normalized scans in Fig. 8.15 provide the evolution of the Raman line for spherical Si nanoparticles as the particle diameter decreases from infinity in bulk material to 3 nm. The dependence of the position of the peak absorption on the microcrystallite size is reported in Fig. 8.16 for annealed and unannealed samples of Si. The broadening and shift of the Si and Ge Raman lines to lower frequencies as the particle size decreases has been attributed to phonon confinement effects in the nanocrystals.

Raman spectra have been widely used to study carbon in its various crystallographic or allotropic forms. Diamond with the tetrahedrally bonded crystal structure sketched in Fig. 2.8a and graphite whose structure consists of stacked planar hexagonal sheets of the type sketched in Fig. 5.14, are the two traditional allotropic forms of diamond. More recently fullerenes such as C_{60} and nanotubes, which are discussed at length in Chapter 5, have been discovered, and are alternate allotropic forms of carbon. The Raman spectrum of diamond has a sharp line at 1332 cm^{-1}, while graphite has infrared-active vibrations at 867 and 1588 cm^{-1}, as well as Raman-active vibrational modes at 42, 1581, and 2710 cm^{-1}. The Raman scans of Fig. 8.17 show (a) the very narrow diamond line at 1332 cm^{-1}, and (b) the narrow graphite stretching mode line at 1581 cm^{-1}, which is called the *G band*. Micro-

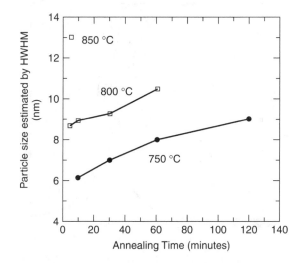

Figure 8.13. Plot of particle size estimated from the full width at half-maximum height (FWHM) of a Ge Raman line versus the annealing time at three temperatures. [From D. C. Paine, C. Caragiantis, T. Y. Kim, Y. Shigesato, and T. Ishahara, *Appl. Phys. Lett.* **62**, 2842 (1993).]

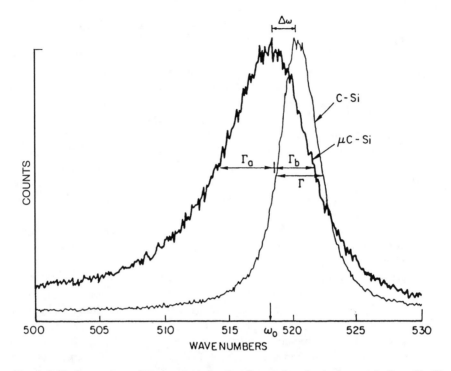

Figure 8.14. Comparison of Γ_{25}^{+} optical vibration Raman lines in single crystal silicon (C–Si) and fine-grained polycrystalline Si (μC–Si). [From P. M. Fauchet and I. H. Campbell, *Crit. Rev. Solid State Mater. Sci.* **14**, S79 (1988).]

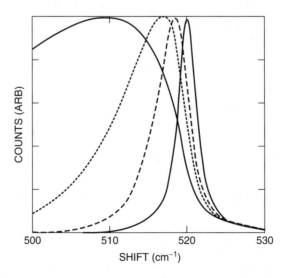

Figure 8.15. Shift to lower wavenumbers (cm^{-1}) and broadening of the Γ_{25}^{+} Raman line for spherical particles that decrease in diameter from right to left in the sequence: bulk, 10, 6, and 3 nm. [From P. M. Fauchet and I. H. Campbell, *Crit. Rev. Solid State Mater. Sci.* **14**, S79 (1988).]

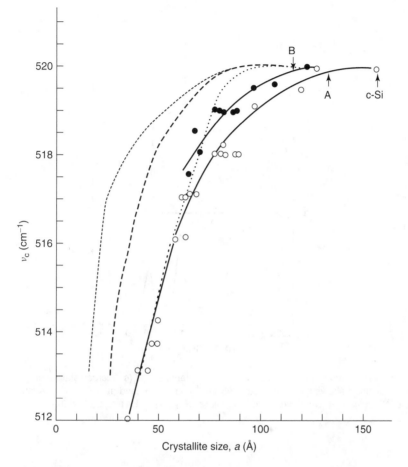

Figure 8.16. Dependence of the Raman peak frequency of the Γ_{25}^+ optical phonon line of annealed (filled circles) and unannealed (open circles) microcrystalline silicon. Broken and dotted curves represent calculated frequencies perpendicular and parallel to a (111) Si slab, and the curve with crosses was calculated for dispersion in the (111) direction. The solid curves drawn through the data points for the annealed and unannealed material are guides to the eye. [From Z. Iqbal and S. Veprek, *J. Phys. C: Solid State Phys.* **15**, 377 (1982).]

cystalline graphite (c) exhibits much broader Raman lines, as discussed below. Figure 8.18 presents the infrared and Raman spectra of the solid fullerene C_{60}, with each line labeled with its wavenumber (cm^{-1}) value. This figure provides an example of the fact that some vibrational normal modes are infrared-active, and others are Raman-active.

Raman spectroscopy is sensitive to deviations from the highly ordered diamond and graphite structures responsible for the narrow lines in Figs. 8.17a and 8.17b, respectively. Figure 8.17c shows that microcrystalline graphite exhibits a broadened G band at $\cong 1580\,cm^{-1}$, and a similarly broadened absorption at $\cong 1355\,cm^{-1}$ that

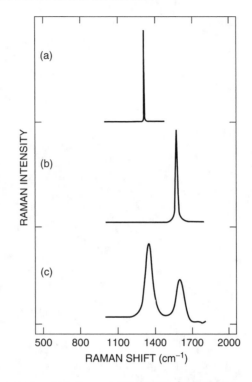

Figure 8.17. Raman spectra of (a) diamond, (b) graphite, and (c) microcrystalline graphite. The latter spectrum shows the D band at $1355\,cm^{-1}$ and the G band at $1580\,cm^{-1}$. [Adapted from R. E. Shroder, J. R. Nemanich, and T. J. Glass, *Phys. Rev.* **B41**, 3738 (1990).]

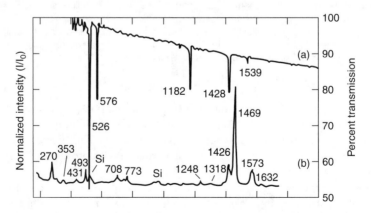

Figure 8.18. Infrared (a) and Raman (b) spectra of solid C_{60} with the lines labeled by their wavenumbers. [From A. M. Rao et al., *Science* **259**, 955 (1993).]

is referred to as the *D band*. The *D* band has been attributed to phonons that are Raman-active because of the finite size of the microcrystals. Figure 8.19 presents a series of Raman spectra from various forms of crystalline graphite, as well as spectra from four samples of amorphous carbon with graphitic features. These spectra can be used to assess the degree of disorder and the nature of the local chemical bonding. The log–log plot of Fig. 8.20 shows how the nanocrystal size L_a is related to the intensity ratio I_D/I_G of the *D* band to the *G* band. The linearity extends over two orders of magnitude, from $L_a = 3\,nm$ to $L_a = 300\,nm$. The nanocrystallite sizes for the data plotted in Fig. 8.20 had been independently determined from X-ray scattering data.

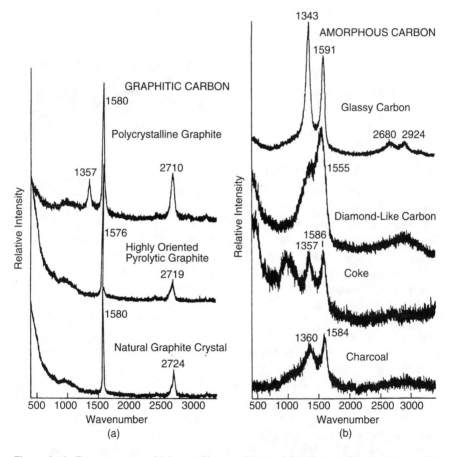

Figure 8.19. Raman spectra of (a) crystalline graphites and (b) noncrystalline, mainly graphitic, carbons. The *D* band appears near $1355\,cm^{-1}$ and the *G* band, near $1580\,cm^{-1}$. [From D. S. Knight and W. B. White, *J. Mater, Sci.* **4**, 385 (1989).]

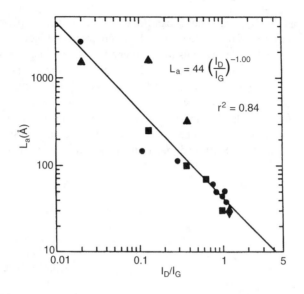

Figure 8.20. Plot of the relation between the graphite particle size L_a and the Raman D-band to G-band intensity ratio I_D/I_G on a log–log scale. The straight line is a least-squares fit to the data points that provided the linear relationship $L_a = 4.4\ I_G/I_D$, where L_a is expressed in namometers $(10\ Å = 1\ nm)$. [From D. S. Knight and W. B. White, *J. Mater Sci.* **4**, 385 (1989).]

8.2.4. Brillouin Spectroscopy

Brillouin scattering is a type of Raman scattering in which the difference frequency $\Delta\omega = \omega_{phonon} = (\omega_{inc} - \omega_{scat})$ corresponds to the acoustic branch of the phonon dispersion curves, with frequencies in the gigahertz $(\times 10^9\ Hz)$ range, as was explained in Section 3.4.1. The negative and positive signs in the expression above for ω_{phonon} correspond to Stokes and anti-Stokes lines, respectively.

Brillouin scattering has been used to study carbon films, and Fig. 8.21 compares the spectra of thick and thin films. The thick-film result (a) provides a bulk material response, namely, a strong central peak at zero frequency about 10 GHz wide, and a broad peak near 17 GHz attributed to longitudinal acoustic (LA) phonons. This latter frequency is consistent with the elastic moduli of carbon, which are measures of the stretching capability of solid carbon and its chemical bonds. The dotted line experimental spectrum of the 100 nm thick film at the top of Fig. 8.21b exhibits three peaks which come at positions close to the solid line theoretical spectrum

Figure 8.21. Brillouin spectra of (a) thick carbon film showing a Lorentzian fit to the data and (b) 100 nm thin carbon film. The upper experimentally measured spectrum of (b) is compared to the lower calculated spectrum, which does not take into account the scattering due to surface and structural irregularities that broaden the experimental spectrum. [From P. Milani and C. E. Bottani, in Nalwa (2000), Vol. 2, Chapter 4, p. 262.]

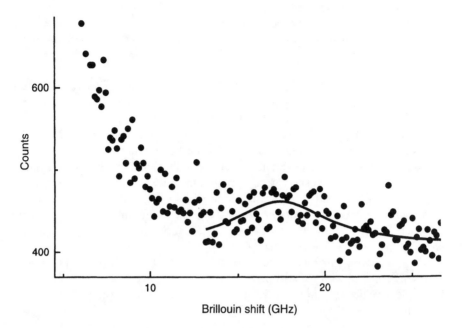

(a)

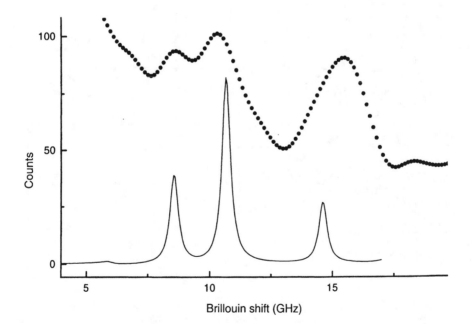

(b)

plotted below the experimental one. The peaks of the experimental spectrum are much broader due to the roughness and structural irregularities of the surface. The theoretical fit provided values for the shear modulus $\mu \cong 4.0$ GPa, the bulk modulus $\cong 3.7$ GPa, and Poisson's ratio $\cong 1.0$, consistent with the known value of the graphite elastic constant C_{44}.

Acoustic phonons of nanoparticles exhibit Brillouin scattering that depends on particle size, and an example of this is the results shown in Fig. 8.22 for Ag nanoparticles. Figure 8.22a presents the spectra for particle diameters $d = 2.7, 4.1$, and 5.2 nm, and Fig. 8.22b gives the dependence of the wavenumber (cm^{-1}) on the reciprocal of the particle diameter, $1/d$. The latter figure also plots the theoretically expected results for torsional and spheroidal vibrational modes with angular momenta $l = 1, 2, 3$. A related work carried out with nucleated cordierite glass ($Mg_2Al_4S_5O_{18}$) provided Brillouin scattering peaks with positions that were linear with the reciprocal of the diameter $1/d$ ($1/D$ on the figure) in the range of particle diameters from 15 to 40 nm, as shown in Fig. 8.23. Small-angle neutron scattering provided the nanoparticle diameters plotted in this figure.

Thus we have seen from Brillouin scattering data that acoustic modes shift to higher frequencies with reduced particle size, and we have seen from Raman data that optical modes shift to lower frequencies as the particle size is reduced.

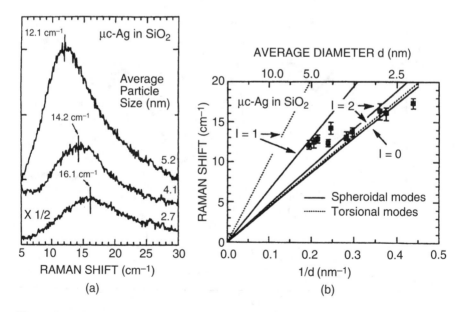

Figure 8.22. Low-frequency Raman shifts in the Brillouin scattering of Ag nanoparticles embedded in SiO$_2$: spectra for particle sizes 2.7, 4.1, and 52 nm and (b) peak position as a function of inverse particle diameter. Theoretical calculations for spheroidal (solid lines) and torsional (dashed lines) modes with angular momenta $l = 0$, 1 and 2 are indicated. [From E. Duval, A. Boukenter, and B. Champagnon, *Phys. Rev. Lett.* **56**, 2052 (1986).]

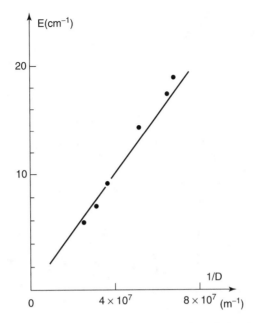

Figure 8.23. Dependence of the Brillouin spectral peaks of cordierite glass nanoparticles on their inverse particle diameter $1/D$ in the range $D = 15$–40 nm determined by small-angle neutron scattering. [From M. Fijii, T. Nagareda, S. Hayashi, and K. Yamamoto, *Phys. Rev.* **B44**, 6243 (1991).]

8.3. LUMINESCENCE

8.3.1. Photoluminescence

The technique of photoluminescence excitation (PLE) has become a standard one for obtaining information on the nature of nanostructures such as quantum dots, which are discussed in the next chapter. In bulk materials the luminescence spectrum often resembles a standard direct absorption spectrum, so there is little advantage to studying the details of both. High photon excitation energies above the band gap can be the most effective for luminescence studies of bulk materials, but it has been found that for the case of nanoparticles the efficiency of luminescence decreases at high incoming photon energies. Nonradiative relaxation pathways can short-circuit the luminescence at these high energies, and it is of interest to investigate the nature of these pathways. Various aspects of luminescence spectroscopy covered in the review by Chen (2000) are examined here.

The photoluminescence excitation technique involves scanning the frequency of the excitation signal, and recording the emission within a very narrow spectral range. Figure 8.24 illustrates the technique for the case of ~5.6-nm CdSe quantum dot nanoparticles. The solid line in Fig. 8.24a plots the absorption spectrum in the range from 2.0 to 3.1 eV, and the superimposed dashed line shows the photoluminescence

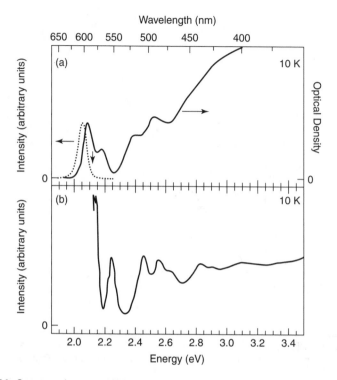

Figure 8.24. Spectra taken at 10 K for 5.6 nm diameter CdSe quantum dots: (a) absorption spectrum (solid line) and photoluminescence spectrum (dashed line) obtained with excitation at 2.655 eV (467 nm); (b) photoluminescence spectrum obtained with the emission position marked by the downward pointing arrow on the upper plot. [From D. J. Norris, and M. G. Bawendi, *Phys. Rev.* **B53**, 16338 (1996).]

response that appears near 2.05 eV. The sample was then irradiated with a range of photon energies of 2.13–3.5 eV, and the luminescence spectrum emitted at the photon energy of 2.13 eV is shown plotted in Fig. 8.24b as a function of the excitation energy. The downward-pointing arrow on Fig. 8.24a indicates the position of the detected luminescence. It is clear from a comparison of the absorption and luminescence spectra of this figure that the photoluminescence (b) is much better resolved.

The excitation spectra of nanoparticles of CdSe with a diameter of 3.2 nm exhibit the expected band-edge emission at 2.176 eV at the temperature 77 K, and they also exhibit an emission signal at 1.65 eV arising from the presence of deep traps, as explained in Section 2.3.1. Figure 8.25 compares the PLE spectra for the band-edge and deep-trap emissions with the corresponding absorption, and we see that the band-edge emission is much better resolved. This is because, as is clear from Fig. 4.20, each particle size emits light at a characteristic frequency so the PLE spectrum reflects the emission from only a small fraction of the overall particle size distribution. Shallow traps that can be responsible for band-edge emission have the same particle

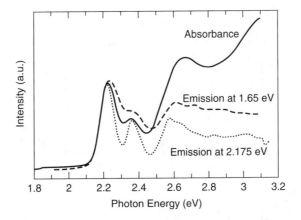

Figure 8.25. Spectra for CdSe nanoparticles of diameter 3.2 nm, showing absorption spectrum (solid line), excitation spectrum for emission at the 2.175-eV band-edge fluorescence maximum (dark dashed line), and excitation spectrum for emission at the 1.65-eV deep-trap level (light dashed line). [From W. Hoheisel, V. L. Colvin, C. S. Johnson, and A. P. Alivisatos, *J. Chem. Phys.* **101**, 8455 (1994).]

size dependence in their spectral response. This considerably reduces the inhomogeneous broadening, and the result is a narrowed, nearly homogeneous spectrum. The emission originating from the deep traps does not exhibit this same narrowing, which explains the low resolution of the 1.65 eV-emission spectrum of Fig. 8.25.

We mentioned above that there is a blue shift, that is, a shift of spectral line positions to higher energies as the size of a nanoparticle decreases. This is dramatically illustrated by the photoluminescence emission spectra presented in Fig. 8.26 arising from seven quantum dot samples ranging in size from ∼1.5 nm for the top spectrum to ∼4.3 nm for the bottom spectrum. We see that the band edge gradually shifts to higher energies, and the distances between the individual lines also gradually increase with the decrease in particle size. Another way to vary spectral parameters is to excite the sample with a series of photon energies and record the fluorescence spectrum over a range of energies, and this produces the series of spectra illustrated in Fig. 8.27. On this figure the peak of the fluorescence spectrum shifts to higher energies as the excitation photon energy increases. We also notice from the absorption spectrum, presented at the bottom of the figure for comparison purposes, that for all photon excitation energies the fluorescence maximum is at lower energies than the direct absorption maximum.

8.3.2. Surface States

As nanoparticles get smaller and smaller, the percentage of atoms on the surface becomes an appreciable fraction of the total number of atoms. For example, we see from Table 2.1 that a 5.7-nm-diameter FCC nanoparticle formed from atoms with a typical diameter of $d = 0.3$ nm (shell 10 in the table) has 28% of its 2869 atoms on

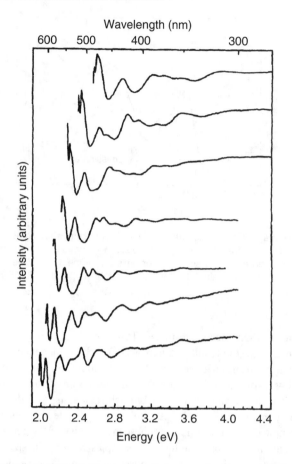

Figure 8.26. Normalized photoluminescence excitation spectra for seven CdSe quantum dots ranging in size from ~1.5 nm (top spectrum) to ~4.3 nm (bottom spectrum). [From D. J. Norris and M. G. Bawendi, *Phys. Rev.* **B53**, 16338 (1996).]

the surface, and a smaller (2.1-nm) nanoparticle (shell 4) has 63% of its 147 atoms on the surface. Irregularities of the surface topology can provide electron and hole traps during optical excitation. The presence of trapped electron–hole pairs bleaches the exciton absorption, but this absorption recovers when the trapped electron–hole pairs decay away. We will describe how this complex process has been studied by time-resolved laser spectroscopy, which furnishes us with details about how the initial excitation energy passes through various intermediate states before finally being dissipated.

The surface states of two nanoparticles of CdS with dimensions of 3.4 and 4.3 nm, respectively, were studied by fluorescence spectroscopy. We see from the resulting spectra presented in Fig. 8.28 that they each exhibit a sharp fluorescence at 435 and 480 nm, respectively, arising from excitons, and a broad fluorescence

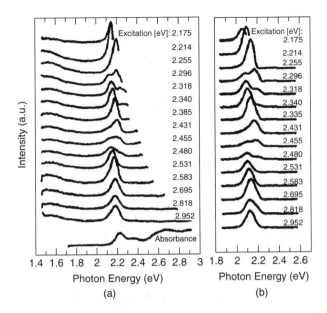

Figure 8.27. Fluorescence spectra for 3.2-nm-diameter CdSe nanocrystals for various indicated excitation energies at 77 K: (a) experimental and (b) simulated spectra. For comparison purposes, the experimental absorption spectrum is shown at the bottom left. [From W. Hoheisel, V. L. Colvin, C. S. Johnson, and A. P. Alivisatos, *J. Chem. Phys.* **101**, 8455 (1994).]

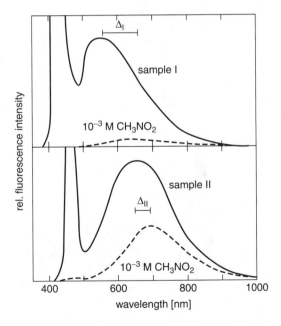

Figure 8.28. Fluorescence spectra for two samples of CdS nanoparticles obtained before (solid lines) and after (dashed lines) the addition of 10^{-3} M nitromethane CH_3NO_2. [From A. Hasselbarth, E. Eychmuller, and H. Well, *Chem. Phys. Lett.* **203**, 271 (1993).]

emission at longer wavelengths. We also see from this figure that the addition of nitromethane (CH_3NO_2) quenches the fluorescence by bringing about a shift toward longer wavelengths, plus an appreciable decrease in the magnitude of the broad band, and in addition it practically eliminates the sharp exciton emission. The temperature dependence of the excitonic and trapped carrier recombination fluorescence bands from CdS nanoparticles both exhibit a decrease in intensity and a shift toward longer wavelengths when the temperature is raised from 4 to 259 K, as illustrated in Fig. 8.29. These spectral data suggest that the hole traps lie much deeper (i.e., have much lower energies) than do electron traps.

To try and elucidate the mechanisms involved in the exciton relaxation and the detrapping of electrons, the time dependencies of the excitonic fluorescence and the trapped fluorescence were determined at a series of temperatures from 4 to 269 K, and the results are displayed in Figs. 8.30 and 8.31, respectively. Both types of decay were found to have a complicated multiexponential behavior, with the rates of decay changing as the processes proceeded. The decay time was shortest for the excitonic emission at intermediate temperatures, requiring the time $\tau_{1/2}$ of less than 10 ns for the decay to reach half of its initial intensity at 121 K. In contrast to this, the trapped fluorescence decayed much more slowly, being particularly slow at intermediate temperatures, with the rate constant $\tau_{1/2} \sim 100$ nsec at 70 K. It had been determined independently that the trapping of electrons in CdS nanoparticles is extremely fast, in the picosecond (ps) time range requiring 10^{-13} s or less time to complete the trapping, so all the trapped electrons are in place before there is an appreciable onset of the fluorescence.

To probe into spectral changes that take place during the initial extremely short picosecond timescale ($1000\,ps = 1\,ns$, or $1\,ps = 10^{-3}\,ns = 10^{-15}\,s$), the data from decay curves of the type presented in Figs. 8.30 and 8.31 were used to reconstruct

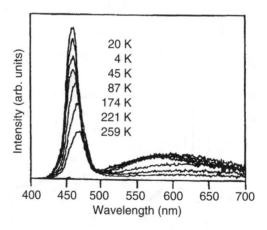

Figure 8.29. Fluorescence spectra of CdS nanoparticles (sample II of Fig 8.28) recorded at a series of temperatures from 4 to 259 K, using $\lambda = 360$ nm excitation. [From A. Eychmuller, A. Hasselbarth, L. Katsicas, and H. Weller, *Ber. Bunsen-Ges. Phys. Chem.* **95**, 79 (1991).]

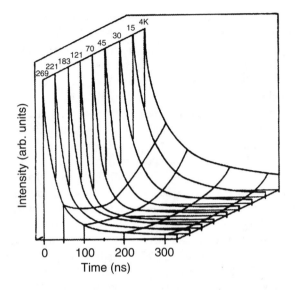

Figure 8.30. Decay curves for exciton fluorescence of CdS nanoparticles (sample II of Fig 8.28). [From A. Eychmuller et al., *Ber Bunsen-Ges. Phys. Chem.* **95**, 79 (1991).]

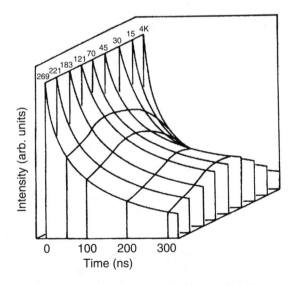

Figure 8.31. Decay curves for the trapped fluorescence of CdS nanoparticles (sample II of Fig 8.28). [From A. Eychmuller et al., *Ber. Bunsen-Ges. Phys. Chem.* **95**, 79 (1991).]

luminescence spectra at various times during the early stages of the emission process, and the results are presented in Fig. 8.32. The four spectra at the top of the figure cover the timespan from 0.05 to 1 ns, and they demonstrate that there is a gradual shift of the ~556 nm spectral line peak toward longer wavelengths during the first nanosecond of the emission, with the spectral features remaining stable during the remainder of the decay. The initial extremely fast component of the decay, for times less than 0.05 ns, arises from resonant emission, and the subsequent fast component that underwent the wavelength shift $\Delta\lambda \sim 2$ nm shown in Fig. 8.32 was attributed to longitudinal optical (LO) phonon vibrations.

The model sketched in Fig. 8.33 has been proposed to explain these results. The initial 400-nm laser excitation produces electron–hole pairs that either form free excitons or become trapped at surface states. Some of the free excitons decay rapidly by the emission of a ~1.87-eV photon, and others quickly become trapped and then decay almost as rapidly with the emission of a 1.85-eV photon. The electron–hole pairs trapped at surface states decay much more slowly, either radiatively by the emission of photons in the range from 1.77 to 1.83 eV, or nonradiatively. The rapid decays occur over a picosecond timescale, and the slower decays over a nanosecond timescale. This model provides a reasonable explanation of the dynamics of the nanoparticle luminescence that we have been discussing.

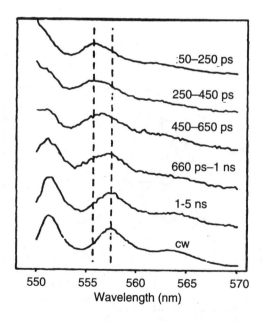

Figure 8.32. Time resolved CdS luminescence spectra of the 75 cm^{-1} shift toward longer wavelengths of the 556-nm line during the first nanosecond after the onset of the emission. The excitation was at the wavelength $\lambda = 549$ nm. The intensities of the spectra were adjusted to facilitate lineshape comparison. [From M. G. Bawendi, P. J. Carroll, W. L. Wilson, and E. L. Brus, *J. Chem. Phys.* **96**, 946 (1992).]

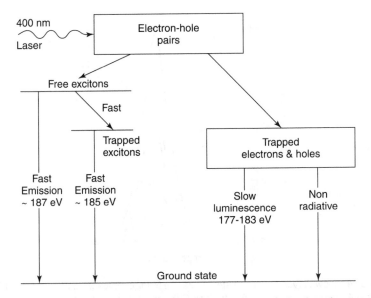

Figure 8.33. Sketch of a model to explain the luminescence emission from laser-generated electron–hole pairs in medium sized CdSe nanocrystals. [Adapted from P. Lefebvre, H. Matthieu, J. Allegre, T. Richard, A. Combettes-Roos, M. Pauthe, and W. Granier, *Semicond. Sci. Tech.* **12**, 598 (1997).]

8.3.3. Thermoluminescence

Another spectral technique that can provide information on surface states, detrapping, and other processes involved in light emission from nanoparticles is *thermolumines-cence*, the emission of light brought about by heating. Sometimes electron–hole pairs produced by irradiating a sample do not recombine rapidly, but become trapped in separate metastable states with prolonged lifetimes. The presence of traps is especially pronounced in small nanoparticles where a large percentage of the atoms are at the surface, many with unsatisfied chemical bonds and unpaired electrons. Heating the sample excites lattice vibrations that can transfer kinetic energy to electrons and holes held at traps, and thereby release them, with the accompaniment of emitted optical photons that constitute the thermal luminescence.

To measure thermoluminescence, the energy needed to bring about the release of electrons and holes from traps is provided by gradually heating the sample, and recording the light emission as a function of temperature, as shown in Fig. 8.34 for CdS residing in the cages of the material zeolite-Y, which will be discussed in the next section. The energy corresponding to the maximum emission, called the *glow peak*, is the energy needed to bring about the detrapping, and it may be considered as a measure of the depth of the trap. This energy, however, is generally insufficient to excite electrons from their ground states to excited states. For example, at room temperature (300 K) the thermal energy $k_B T = 25.85$ meV is far less than typical gap energies E_g, although it is comparable to the ionization energies of many donors and

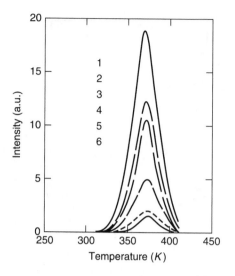

Figure 8.34. Glow curves of CdS clusters in zeolite-Y for CdS loadings of 1, 3, 5, and 20 wt% (curves 1–4, respectively). Curve 5 is for bulk CdS, and curve 6 is for a mechanical mixture of CdS with zeolite-Y powder. [From W. Chen, Z. G. Wang, and L. Y. Lin, *J. Lumin.* **71**, 151 (1997).]

acceptors in semiconductors (see Table B.10). It is quite common for trap depths to be in the range of thermal energies.

8.4. NANOSTRUCTURES IN ZEOLITE CAGES

An example of the efficacy of thermoluminescence to provide information on nanostructures is provided by studies of cadmium sulfide (CdS) clusters introduced into the cages of zeolite-Y. This material, which is found in nature as the mineral faujasite $(Na_2,Ca)(Al_2Si_4)O_{12} \cdot 8H_2O$, is cubic in structure with lattice constant $a = 2.474$ nm. It has a porous network of silicate (SiO_4) and aluminate (AlO_4) tetrahedra that form cube–octahedral cages ~ 0.5 nm in diameter, called *sodalite cages*, since they resemble those found in the mineral sodalite $Na_4Al_3Si_3O_{12}Cl$. The Al and Si atoms are somewhat randomly distributed in their assigned lattice sites. The sodalite cages have entrance windows ~ 0.25 nm in diameter. There are also larger supercages with diameter ~ 1.3 nm, and ~ 0.75-nm windows. Figure 8.35 shows a sketch of the structure with tetrahedrally bonded Cd_4S_4 cubic clusters occupying the sodalite cages, and with the supercage in the center empty.

As the CdS is introduced into the sodalite, it initially tends to enter the sodalite cages shown in Fig. 8.35, but it can also form clusters in the supercages, especially for higher loading. In addition, clusters of CdS in near-neighbor pores can connect to form larger effective cluster sizes. Thus, as the loading increases, the average cluster size also increases. The ultraviolet (200–400 nm) and blue–green–yellow range

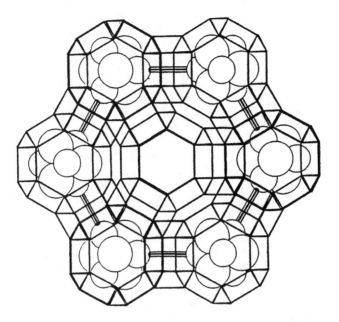

Figure 8.35. Sketch of zeolite-Y structure showing six sodalite cages (diameter ∼0.5 nm) occupied by Cd_4S_4 tetrahedral clusters, and one empty supercage (diameter ∼1.3 nm) in the center. (From H. Herron and Y. Wang, in *Nanomaterials, Synthesis, Properties and Applications*, A. S. Edelstein, ed., IOP, Bristol, UK, 1996, p. 73.)

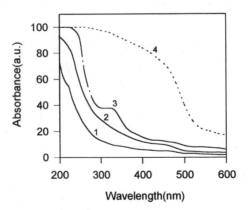

Figure 8.36. Reflectance spectra of CdS clusters in zeolite-Y, with CdS loadings of 1, 3, 5, and 20 wt% corresponding to curves 1, 2, 3, and 4, respectively. [From W. Chen et al., *J. Lumin.* **71**, 151 (1997).]

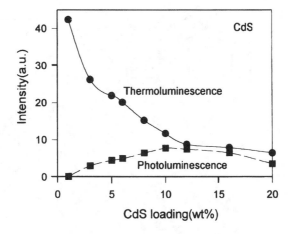

Figure 8.37. Dependence of thermoluminescent and phosphorescent intensities of CdS clusters in zeolite-Y on the CdS loading. [From W. Chen et al., *J. Lumin.* **71**, 151 (1997).]

(400–600 nm) optical absorption spectra shown in Fig. 8.36 exhibit a shift to the red (toward longer wavelengths) as the loading increases from 1 to 5% in zeolite-Y, as expected from the increase in the average cluster size with greater loading. At 20% loading the spectrum corresponds to that of bulk CdS, suggesting that some bulk phase has formed outside the zeolite pores. The photoluminescence is low for low CdS loading, increases in intensity as CdS is added, then falls off again for high loadings, as shown in Fig. 8.37. In contrast to this, the thermoluminescence glow curve intensity is highest for low loading, and decreases in intensity as the loading increases, as indicated by the data presented in Figs. 8.34 and 8.37. This is explained by the presence of trapped carriers introduced into the CdS clusters during sample processing. These carriers are detrapped by the thermal energy added to the sample near the temperature 375 K of the thermoluminescence peak in Fig. 8.34, the temperature where the thermal energy $k_B T$ equals the trap depth. The smaller clusters associated with low loading have more surface states, and hence more electrons to detrap and contribute to the glow peak. The increase in quantum confinement characteristic of smaller clusters also contributes to the increase in recombination probability, with the resulting enhanced thermoluminescence.

FURTHER READING

M.-I. Baraton, "Fourier Transform Infrared Surface Spectrometry of Nano-Sized Particles," in Nalwa (2000), Vol. 2, Chapter 2, p. 89.

W. Chen, "Fluorescence, Thermoluminescence, and Photostimulated Luminescence of Nanoparticles," in Nalwa (2000), Vol. 4, Chapter 5, p. 325.

P. Milani and C. E. Bottani, "Vibrational Spectroscopy of Mesoscopic Structures," in Nalwa (2000), Vol. 2, Chapter 4, p. 213.

H. S. Nalwa, ed., *Handbook of Nanostructured Materials and Nanotechnology*, Vol 2, *Spectroscopy and Theory*; Vol. 4, *Optical Properties*, Academic Press, San Diego, 2000.

9

QUANTUM WELLS, WIRES, AND DOTS

9.1. INTRODUCTION

When the size or dimension of a material is continuously reduced from a large or macroscopic size, such as a meter or a centimeter, to a very small size, the properties remain the same at first, then small changes begin to occur, until finally when the size drops below 100 nm, dramatic changes in properties can occur. If one dimension is reduced to the nanorange while the other two dimensions remain large, then we obtain a structure known as a *quantum well*. If two dimensions are so reduced and one remains large, the resulting structure is referred to as a *quantum wire*. The extreme case of this process of size reduction in which all three dimensions reach the low nanometer range is called a *quantum dot*. The word *quantum* is associated with these three types of nanostructures because the changes in properties arise from the quantum-mechanical nature of physics in the domain of the ultrasmall. Figure 9.1 illustrates these processes of diminishing the size for the case of rectilinear geometry, and Fig. 9.2 presents the corresponding reductions in curvilinear geometry. In this chapter we are interested in all three of these ways to decrease the size. In other words, we probe the dimensionality effects that occur when one, two, or all three dimensions becomes small. Of particular interest is how the electronic properties are

Introduction to Nanotechnology, by Charles P. Poole Jr. and Frank J. Owens.
ISBN 0-471-07935-9. Copyright © 2003 John Wiley & Sons, Inc.

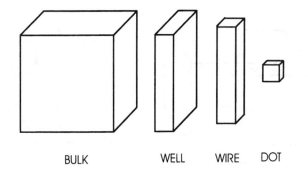

Figure 9.1. Progressive generation of rectangular nanostructures.

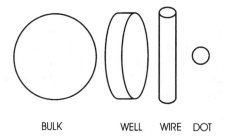

Figure 9.2. Progressive generation of curvilinear nanostructures.

altered by these changes. Jacak et al. (1998) have surveyed the field of quantum nanostructures.

9.2. PREPARATION OF QUANTUM NANOSTRUCTURES

One approach to the preparation of a nanostructure, called the *bottom–up* approach, is to collect, consolidate, and fashion individual atoms and molecules into the structure. This is carried out by a sequence of chemical reactions controlled by catalysts. It is a process that is widespread in biology where, for example, catalysts called *enzymes* assemble amino acids to construct living tissue that forms and supports the organs of the body. The next chapter explains how nature brings about this variety of self-assembly.

The opposite approach to the preparation of nanostructures is called the *top–down method*, which starts with a large-scale object or pattern and gradually reduces its dimension or dimensions. This can be accomplished by a technique called *lithography* which shines radiation through a template on to a surface coated with a radiation-sensitive resist; the resist is then removed and the surface is chemically treated to produce the nanostructure. A typical resist material is the polymer polymethyl methacrylate $[C_5O_2H_8]_n$ with a molecular weight in the range from

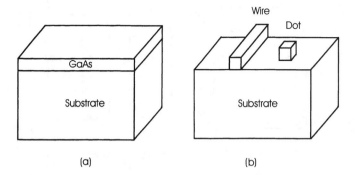

Figure 9.3. (a) Gallium arsenide quantum well on a substrate; (b) quantum wire and quantum dot formed by lithography.

10^5 to 10^6 Da (see Section 11.2.1). The lithographic process is illustrated by starting with a square quantum well located on a substrate, as shown in Fig. 9.3a. The final product to be produced from the material (e.g., GaAs) of the quantum well is either a quantum wire or a quantum dot, as shown in Fig. 9.3b. The steps to be followed in this process are outlined in Fig. 9.4.

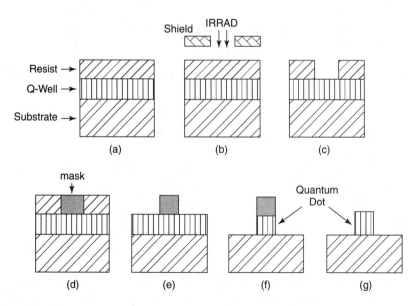

Figure 9.4. Steps in the formation of a quantum wire or quantum dot by electron-beam lithography: (a) initial quantum well on a substrate, and covered by a resist; (b) radiation with sample shielded by template; (c) configuration after dissolving irradiated portion of resist by developer; (d) disposition after addition of etching mask; (e) arrangement after removal of remainder of resist; (f) configuration after etching away the unwanted quantum-well material; (g) final nanostructure on substrate after removal of etching mask. [See also T. P. Sidiki and C. M. S. Torres, in Nalwa (2000), Vol. 3, Chapter 5, p. 250.]

The first step of the lithographic procedure is to place a radiation-sensitive resist on the surface of the sample substrate, as shown in Fig. 9.4a. The sample is then irradiated by an electron beam in the region where the nanostructure will be located, as shown in Fig. 9.4b. This can be done by using either a radiation mask that contains the nanostructure pattern, as shown, or a scanning electron beam that strikes the surface only in the desired region. The radiation chemically modifies the exposed area of the resist so that it becomes soluble in a developer. The third step in the process (Fig. 9.4c) is the application of the developer to remove the irradiated portions of the resist. The fourth step (Fig. 9.4d) is the insertion of an etching mask into the hole in the resist, and the fifth step (Fig. 9.4e) consists in lifting off the remaining parts of the resist. In the sixth step (Fig. 9.4f) the areas of the quantum well not covered by the etching mask are chemically etched away to produce the quantum structure shown in Fig. 9.4f covered by the etching mask. Finally the etching mask is removed, if necessary, to provide the desired quantum structure (Fig. 9.4g), which might be the quantum wire or quantum dot shown in Fig. 9.3b.

In Chapter 1 we mentioned the most common process called *electron-beam lithography*, which makes use of an electron beam for the radiation. Other types of lithography employ neutral atom beams (e.g., Li, Na, K, Rb, Cs), charged ion beams (e.g., Ga^+), or electromagnetic radiation such as visible light, ultraviolet light, or X rays. When laser beams are utilized, frequency doublers and quadruplers can bring the wavelength into a range (e.g., $\lambda \sim 150 \, nm$) that is convenient for quantum-dot fabrication. Photochemical etching can be applied to a surface activated by laser light.

The lithographic technique can be used to make more complex quantum structures than the quantum wire and quantum dot shown in Fig. 9.3b. For example one might start with a multiple quantum-well structure of the type illustrated in Fig. 9.5, place the resist on top of it, and make use of a mask film or template with six circles cut out of it, as portrayed at the top of Fig. 9.5. Following the lithographic procedure outlined in Fig. 9.4, one can produce the 24-quantum-dot array consisting of six columns, each containing four stacked quantum dots, that is sketched in

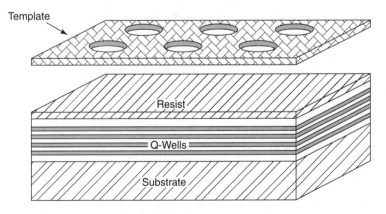

Figure 9.5. Four-cycle multiple-quantum-well arrangement mounted on a substrate and covered by a resist. A radiation shielding template for lithography is shown at the top.

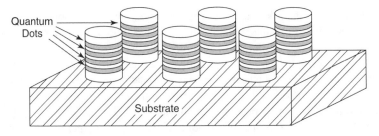

Figure 9.6. Quantum-dot array formed by lithography from the initial configuration of Fig. 9.5. This 24-fold quantum-dot array consists of six columns of four stacked quantum dots.

Fig 9.6. As an example of the advantages of fabricating quantum dot arrays, it has been found experimentally that the arrays produce a greatly enhanced photoluminescent output of light. Figure 9.7 shows a photoluminescence (PL) spectrum from a quantum-dot array that is much more than 100 times stronger than the spectrum obtained from the initial multiple quantum wells. The principles behind the photoluminescence technique are described in Section 8.3.1. The main peak of the spectrum of Fig. 9.7 was attributed to a localized exciton (LE), as explained in Section 9.4. This magnitude of enhancement shown in the figure has been obtained from initial samples containing, typically, a 15-period superlattice (SL) of alternating

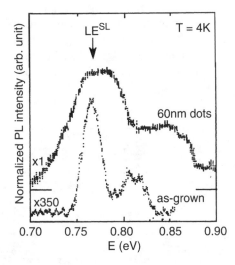

Figure 9.7. Photoluminescence spectrum of an array of 60-nm-diameter quantum dots formed by lithography, compared with the spectrum of the initial as-grown multiple quantum well. The intense peak at 0.7654 eV is attributed to localized excitons (LE) in the superlattice (SL). The spectra were taken at temperature 4 K. [From T. P. Sidiki and C. M. S. Torres, in Nalwa (2000), Vol. 3, Chapter 5, p. 251.]

3-nm-thick layers of Si and $Si_{0.7}Ge_{0.3}$ patterned into quantum-dot arrays consisting of 300-nm-high columns with 60 nm diameters and 200 nm separations.

9.3. SIZE AND DIMENSIONALITY EFFECTS

9.3.1. Size Effects

Now that we have seen how to make nanostructures, it is appropriate to say something about their sizes relative to various parameters of the system. If we select the type III–V semiconductor GaAs as a typical material, the lattice constant from Table B.1 (of Appendix B) is $a = 0.565$ nm, and the volume of the unit cell is $(0.565)^3 = 0.180$ nm^3. The unit cell contains four Ga and four As atoms. Each of these atoms lies on a face-centered cubic (FCC) lattice, shown sketched in Fig. 2.3, and the two lattices are displaced with respect to each other by the amount $\frac{1}{4}\frac{1}{4}\frac{1}{4}$ along the unit cell body diagonal, as shown in Fig. 2.8. This puts each Ga atom in the center of a tetrahedron of As atoms corresponding to the grouping GaAs$_4$, and each arsenic atom has a corresponding configuration AsGa$_4$. There are about 22 of each atom type per cubic nanometer, and a cube-shaped quantum dot 10 nm on a side contains 5.56×10^3 unit cells.

The question arises as to how many of the atoms are on the surface, and it will be helpful to have a mathematical expression for this in terms of the size of a particle with the zinc blende structure of GaAs, which has the shape of a cube. If the initial cube is taken in the form of Fig. 2.6 and nanostructures containing n^3 of these unit cells are built up, then it can be shown that the number of atoms N_S on the surface, the total number of atoms N_T, and the size or dimension d of the cube are given by

$$N_S = 12n^2 \tag{9.1}$$

$$N_T = 8n^3 + 6n^2 + 3n \tag{9.2}$$

$$d = na = 0.565n \tag{9.3}$$

here $a = 0.565$ nm is the lattice constant of GaAs, and the lattice constants of other zinc blende semiconductors are given in Table B.1. These equations, (9.1)–(9.3), represent a cubic GaAs nanoparticle with its faces in the x–y, y–z, and z–x planes, respectively. Table 9.1 tabulates N_S, N_T, d, and the fraction of atoms on the surface N_S/N_T, for various values of n. The large percentage of atoms on the surface for small n is one of the principal factors that differentiates properties of nanostructures from those of the bulk material. An analogous table could easily be constructed for cylindrical quantum structures of the types illustrated in Figs. 9.2 and 9.6.

Comparing Table 9.1, which pertains to a diamond structure nanoparticle in the shape of a cube, with Table 2.1, which concerns a face-centered cubic structure nanoparticle with an approximately spherical shape, it is clear that the results are qualitatively the same. We see from the comparison that the FCC nanoparticle has a greater percentage of its atoms on the surface for the same total number of atoms in

Table 9.1. Number of atoms on the surface N_S, number in the volume N_V, and percentage of atoms N_S/N_V on the surface of a nanoparticle[a]

n	Size na (nm)	Total Number of Atoms	Number of Surface Atoms	Percent of Atoms on Surface
2	1.13	94	48	51.1
3	1.70	279	108	38.7
4	2.26	620	192	31.0
5	2.83	1165	300	25.8
6	3.39	1962	432	22.0
10	5.65	8630	1200	13.9
15	8.48	2.84×10^4	2700	9.5
25	14.1	1.29×10^5	7500	5.8
50	28.3	1.02×10^6	3.0×10^4	2.9
100	56.5	8.06×10^6	1.2×10^5	1.5

[a]The nanoparticle has a diamond lattice structure in the shape of a cube n unit cells on a side, having a width na, where a is the unit cell dimension. Column 2 gives sizes for GaAs, which has $a = 0.565$ nm.

the particle. This is expected because according to the way the calculation was carried out, only one of the two types of atoms in the GaAs structure contributes to the surface.

A *charge carrier* in a conductor or semiconductor has its forward motion in an applied electric field periodically interrupted by scattering off phonons and defects. An electron or hole moving with a drift velocity v will, on the average, experience a scattering event every τ seconds, and travel a distance l called the mean free path between collisions, where

$$l = v\tau \tag{9.4}$$

This is called *intraband scattering* because the charge carrier remains in the same band after scattering, such as the valence band in the case of holes. Mean free paths in metals depend strongly on the impurity content, and in ordinary metals typical values might be in the low nanometer range, perhaps from 2 to 50 nm. In very pure samples they will, of course, be much longer. The resistivity of a polycrystalline conductor or semiconductor composed of microcrystallites with diameters significantly greater than the mean free path resembles that of a network of interconnected resistors, but when the microcrystallite dimensions approach or become less than l, the resistivity depends mainly on scattering off boundaries between crystallites. Both types of metallic nanostructures are common.

Various types of defects in a lattice can interrupt the forward motion of conduction electrons, and limit the mean free path. Examples of zero-dimensional defects are missing atoms called *vacancies*, and extra atoms called interstitial atoms located between standard lattice sites. A vacancy–interstitial pair is called a *Frenkel defect*. An example of a one-dimensional dislocation is a lattice defect at an edge, or a partial line of missing atoms. Common two-dimensional defects are a boundary between grains, and a stacking fault arising from a sudden change in the stacking arrangement of close-packed planes. A vacant space called a *pore*, a cluster of vacancies, and a precipitate of a second phase are three-dimensional defects. All of these can bring about the scattering of electrons, and thereby limit the electrical conductivity. Some nanostructures are too small to have internal defects.

Another size effect arises from the level of doping of a semiconductor. For typical doping levels of 10^{14} to 10^{18} donors/cm^3 a quantum-dot cube 100 nm on a side would have, on the average, from 10^{-1} to 10^3 conduction electrons. The former figure of 10^{-1} electrons per cubic centimeter means that on the average only 1 quantum dot in 10 will have one of these electrons. A smaller quantum-dot cube only 10 nm on a side would have, on the average 1 electron for the 10^{18} doping level, and be very unlikely to have any conduction electrons for the 10^{14} doping level. A similar analysis can be made for quantum wires and quantum wells, and the results shown in Table 9.2 demonstrate that these quantum structures are typically characterized by very small numbers or concentrations of electrons that can carry current. This results in the phenomena of single-electron tunneling and the Coulomb blockade discussed below.

9.3.2. Conduction Electrons and Dimensionality

We are accustomed to studying electronic systems that exist in three dimensions, and are large or macroscopic in size. In this case the conduction electrons are delocalized, and move freely throughout the entire conducting medium such as a copper wire. It is clear that all the wire dimensions are very large compared to the distances between atoms. The situation changes when one or more dimensions of the copper becomes so small that it approaches several times the spacings between the atoms in the lattice. When this occurs, the delocalization is impeded, and the electrons experience confinement. For example, consider a flat plate of copper that is 10 cm long, 10 cm wide, and only 3.6 nm thick. This thickness corresponds to the length of only 10 unit cells, which means that 20% of the atoms are in unit cells at the surface of the copper. The conduction electrons would be delocalized in the plane of the plate, but confined in the narrow dimension, a configuration referred to as a *quantum well*. A *quantum wire* is a structure such as a copper wire that is long in

Table 9.2. Conduction electron content of smaller size (on left) and larger size (on right) quantum structures containing donor concentrations of 10^{14}–10^{18} cm^{-3}

Quantum Structure	Size	Electron Content	Size	Electron Content
Bulk material	—	10^{14}–10^{18} cm^{-3}	—	10^{14}–10^{18} cm^{-3}
Quantum well	10 nm thick	1–$10^4 \mu m^{-2}$	100 nm thick	10–$10^5 \mu m^{-2}$
Quantum wire	10×10-nm cross section	10^{-2}–$10^2 \mu m^{-1}$	100 nm × 100 nm cross section	1–$10^4 \mu m^{-1}$
Quantum dot	10 nm on a side	10^{-4}–1	100 nm on a side	10^{-1}–10^3

Table 9.3. Delocalization and confinement dimensionalities of quantum nanostructures

Quantum Structure	Delocalization Dimensions	Confinement Dimensions
Bulk conductor	3 (x, y, z)	0
Quantum well	2 (x, y)	1 (z)
Quantum wire	1 (z)	2 (x, y)
Quantum dot	0	3 (x, y, z)

one dimension, but has a nanometer size as its diameter. The electrons are delocalized and move freely along the wire, but are confined in the transverse directions. Finally, a *quantum dot*, which might have the shape of a tiny cube, a short cylinder, or a sphere with low nanometer dimensions, exhibits confinement in all three spatial dimensions, so there is no delocalization. Figures 9.1 and 9.2, as well as Table 9.3, summarize these cases.

9.3.3. Fermi Gas and Density of States

Many of the properties of good conductors of electricity are explained by the assumption that the valence electrons of a metal dissociate themselves from their atoms and become delocalized conduction electrons that move freely through the background of positive ions such as Na^+ or Ag^+. On the average they travel a mean free path distance l between collisions, as mentioned in Section 9.3.1. These electrons act like a gas called a *Fermi gas* in their ability to move with very little hindrance throughout the metal. They have an energy of motion called *kinetic energy*, $E = \frac{1}{2}mv^2 = p^2/2m$, where m is the mass of the electron, v is its speed or velocity, and $\mathbf{p} = m\mathbf{v}$ is its momentum. This model provides a good explanation of Ohm's law, whereby the voltage V and current I are proportional to each other through the resistance R, that is, $V = IR$.

In a quantum-mechanical description the component of the electron's momentum along the x direction p_x has the value $p_x = \hbar k_x$, where $\hbar = h/2\pi$, h is Planck's universal constant of nature, and the quantity k_x is the x component of the wavevector $\mathbf{k}$. Each particular electron has unique k_x, k_y, and k_z values, and we saw in Section 2.2.2 that the k_x, k_y, k_z values of the various electrons form a lattice in k space, which is called *reciprocal space*. At the temperature of absolute zero, the electrons of the Fermi gas occupy all the lattice points in reciprocal space out to a distance k_F from the origin $k = 0$, corresponding to a value of the energy called the Fermi energy E_F, which is given by

$$E_F = \frac{\hbar^2 k_F^2}{2m} \tag{9.5}$$

We assume that the sample is a cube of side L, so its volume V in ordinary coordinate space is $V = L^3$. The distance between two adjacent electrons in k space is $2\pi/L$, and at the temperature of absolute zero all the conduction electrons are equally spread out inside a sphere of radius k_F, and of volume $4\pi k_F^3/3$ in k space, as was explained in Section 2.2.2. This equal density is plotted in Fig. 9.8a for the temperature absolute zero or 0 K, and in Fig. 9.8b we see that deviations from equal density occur near the Fermi energy E_F level at higher temperatures.

The number of conduction electrons with a particular energy depends on the value of the energy and also on the dimensionality of the space. This is because in one dimension the size of the Fermi region containing electrons has the length $2k_F$, in two dimensions it has the area of the Fermi circle πk_F^2, and in three dimensions it has the volume of the Fermi sphere $4\pi k_F^3/3$. These expressions are listed in column 3 of Table A.1 (of Appendix A). If we divide each of these Fermi regions by the size of the corresponding k-space unit cell listed in column 2 of this table, and make use of Eq. (9.5) to eliminate k_F, we obtain the dependence of the number of electrons N on the energy E given on the left side of Table 9.4, and shown plotted in Fig. 9.9. The slopes of the lines $N(E)$ shown in Fig. 9.9 provide the density of states $D(E)$, which is defined more precisely by the mathematical derivative $D(E) = dN/dE$, corresponding to the expression $dN = D(E)dE$. This means that the number of electrons dN with an energy E within the narrow range of energy $dE = E_2 - E_1$ is proportional to the density of states at that value of energy. The resulting formulas for $D(E)$ for the various dimensions are listed in the middle column of Table 9.4, and are shown plotted in Fig. 9.10. We see that the density of states decreases with increasing energy for one dimension, is constant for two dimensions, and increases with increasing energy for three dimensions. Thus the density of states has quite a different behavior for the three cases. These equations and plots of the density of states are very important in determining the electrical, thermal, and other properties

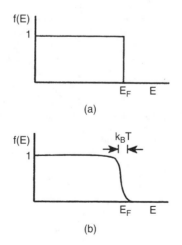

(a)

(b)

Figure 9.8. Fermi–Dirac distribution function $f(E)$, indicating equal density in k space, plotted for the temperatures (a) $T = 0$ and (b) $0 < T \ll T_F$. (From C. P. Poole, Jr., *Handbook of Physics*, Wiley, New York, 1998, p. 138.)

Table 9.4. Number of electrons N and density of states $D(E) = dN(E)/dE$ as a function of the energy E for conduction electrons delocalized in one, two, and three spatial dimensions[a]

Number of Electrons N	Density of States $D(E)$	Delocalization Dimensions
$N = K_1 E^{1/2}$	$D(E) = \frac{1}{2} K_1 E^{-1/2}$	1
$N = K_2 E$	$D(E) = K_2$	2
$N = K_3 E^{3/2}$	$D(E) = \frac{3}{2} K_3 E^{1/2}$	3

[a]The values of the constants K_1, K_2, and K_3 are given in Table A.2 (of Appendix A).

of metals and semiconductors, and make it clear why these features can be so dependent on the dimensionality. Examples of how various properties of materials depend on the density of states are given in Section 9.3.6.

9.3.4. Potential Wells

In the previous section we discussed the delocalization aspects of conduction electrons in a bulk metal. These electrons were referred to as *free electrons*, but perhaps *unconfined electrons* would be a better word for them. This is because when the size of a conductor diminishes to the nanoregion, these electrons begin to experience the effects of confinement, meaning that their motion becomes limited by the physical size of the region or domain in which they move. The influence of

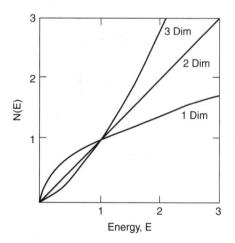

Figure 9.9. Number of electrons $N(E)$ plotted as a function of the energy E for conduction electrons delocalized in one (quantum wire), two (quantum well), and three dimensions (bulk material).

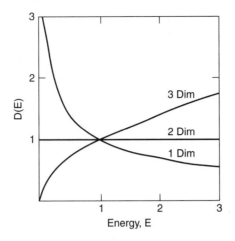

Figure 9.10. Density of states $D(E) = dN(E)/dE$ plotted as a function of the energy E for conduction electrons delocalized in one (Q-wire), two (Q-well), and three (bulk) dimensions.

electrostatic forces becomes more pronounced, and the electrons become restricted by a potential barrier that must be overcome before they can move more freely. More explicitly, the electrons become sequestered in what is called a *potential well*, an enclosed region of negative energies. A simple model that exhibits the principal characteristics of such a potential well is a square well in which the boundary is very sharp or abrupt. Square wells can exist in one, two, three, and higher dimensions; for simplicity, we describe a one-dimensional case.

Standard quantum-mechanical texts show that for an infinitely deep square potential well of width a in one dimension, the coordinate x has the range of values $-\frac{1}{2}a \leq x \leq \frac{1}{2}a$ inside the well, and the energies there are given by the expressions

$$E_n = \left[\frac{\pi^2\hbar^2}{2ma^2}\right]n^2 \tag{9.6a}$$

$$= E_0 n^2 \tag{9.6b}$$

which are plotted in Fig. 9.11, where $E_0 = \pi^2\hbar^2/2ma^2$ is the ground-state energy and the quantum number n assumes the values $n = 1, 2, 3, \ldots$. The electrons that are present fill up the energy levels starting from the bottom, until all available electrons are in place. An infinite square well has an infinite number of energy levels, with ever-widening spacings as the quantum number n increases. If the well is finite, then its quantized energies E_n all lie below the corresponding infinite well energies, and there are only a limited number of them. Figure 9.12 illustrates the case for a finite well of potential depth $V_0 = 7E_0$ which has only three allowed energies. No matter how shallow the well, there is always at least one bound state E_1.

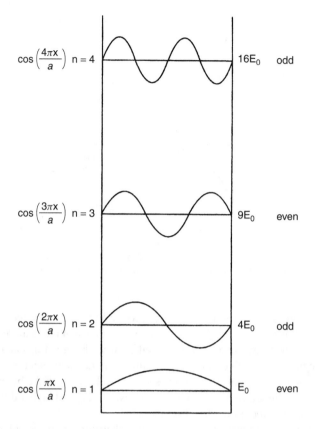

Figure 9.11. Sketch of wavefunctions for the four lowest energy levels ($n = 1$–4) of the one-dimensional infinite square well. For each level the form of the wavefunction is given on the left, and its parity (even or odd) is indicated on the right (From C. P. Poole, Jr., *Handbook of Physics*, Wiley, New York, 1998, p. 289.)

The electrons confined to the potential well move back and forth along the direction x, and the probability of finding an electron at a particular value of x is given by the square of the wavefunction $|\psi_n(x)|^2$ for the particular level n where the electron is located. There are even and odd wavefunctions $\psi_n(x)$ that alternate for the levels in the one-dimensional square well, and for the infinite square well we have the unnormalized expressions

$$\psi_n = \cos(n\pi x/a) \qquad n = 1, 3, 5, \ldots \qquad \text{even parity} \qquad (9.7)$$
$$\psi_n = \sin(n\pi x/a) \qquad n = 2, 4, 6, \ldots \qquad \text{odd parity} \qquad (9.8)$$

These wavefunctions are sketched in Fig. 9.11 for the infinite well. The property called *parity* is defined as even when $\psi_n(-x) = \psi_n(x)$, and it is odd when $\psi_n(-x) = -\psi_n(x)$.

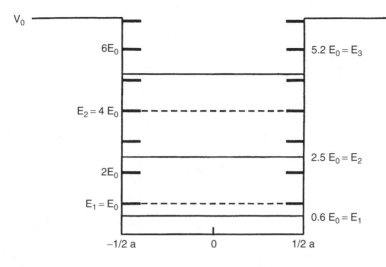

Figure 9.12. Sketch of a one-dimensional square well showing how the energy levels E_n of a finite well (right side, solid horizontal lines) lie below their infinite well counterparts (left side, dashed lines). (From C. P. Poole, Jr., *Handbook of Physics*, Wiley, New York, 1998, p. 285.)

Another important variety of potential well, is one with a curved cross section. For a circular cross section of radius a in two dimensions, the potential is given by $V = 0$ in the range $0 \leq \rho \leq a$, and has the value V_0 at the top and outside, where $\rho = (x^2 + y^2)^{1/2}$, and $\tan \phi = y/x$ in polar coordinates. The particular finite well sketched in Fig. 9.13 has only three allowed energy levels with the values E_1, E_2, and E_3. There is also a three-dimensional analog of the circular well in which the potential is zero for the radial coordinate r in the range $0 \leq r \leq a$, and has the value V_0 outside, where $r = (x^2 + y^2 + z^2)^{1/2}$. Another type of commonly used potential well is the *parabolic well*, which is characterized by the potentials $V(x) = \frac{1}{2}kx^2$, $V(\rho) = \frac{1}{2}k\rho^2$ and $V(r) = \frac{1}{2}kr^2$ in one, two, and three dimensions, respectively, and Fig. 9.14 provides a sketch of the potential in the one-dimensional case.

Another characteristic of a particular energy state E_n is the number of electrons that can occupy it, and this depends on the number of different combinations of quantum numbers that correspond to this state. From Eq. (9.6) we see that the one-dimensional square well has only one allowed value of the quantum number n for each energy state. An electron also has a spin quantum number m_s, which can take on two values, $m_s = +\frac{1}{2}$ and $m_s = -\frac{1}{2}$, for spin states up and down, respectively, and for the square well both spin states $m_s = \pm\frac{1}{2}$ have the same energy. According to the Pauli exclusion principle of quantum mechanics, no two electrons can have the same set of quantum numbers, so each square well energy state E_n can be occupied by two electrons, one with spin up, and one with spin down. The number of combinations of quantum numbers corresponding to each spin state is called its *degeneracy*, and so the degeneracy of all the one-dimensional square well energy levels is 2.

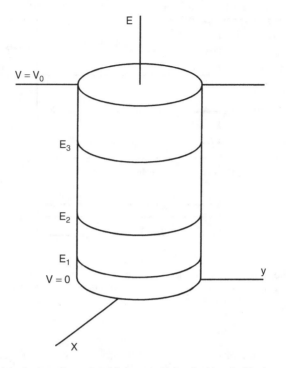

Figure 9.13. Sketch of a two-dimensional finite potential well with cylindrical geometry and three energy levels.

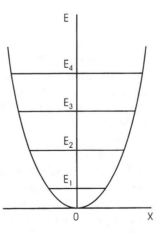

Figure 9.14. Sketch of a one-dimensional parabolic potential well showing the positions of the four lowest energy levels.

The energy of a two-dimensional infinite rectangular square well

$$E_n = \left(\frac{\pi^2 \hbar^2}{2ma^2}\right)(n_x^2 + n_y^2) = E_0 n^2 \tag{9.9}$$

depends on two quantum numbers, $n_x = 0, 1, 2, 3, \ldots$ and $n_y = 0, 1, 2, 3, \ldots$, where $n^2 = n_x^2 + n_y^2$. This means that the lowest energy state $E_1 = E_0$ has two possibilities, namely, $n_x = 0, n_y = 1$, and $n_x = 1, n_y = 0$, so the total degeneracy (including spin direction) is 4. The energy state $E_5 = 25E_0$ has more possibilities since it can have, for example, $n_x = 0, n_y = 5$, or $n_x = 3$ and $n_y = 4$, and so on, so its degeneracy is 8.

9.3.5. Partial Confinement

In the previous section we examined the confinement of electrons in various dimensions, and we found that it always leads to a qualitatively similar spectrum of discrete energies. This is true for a broad class of potential wells, irrespective of their dimensionality and shape. We also examined, in Section 9.3.3, the Fermi gas model for delocalized electrons in these same dimensions and found that the model leads to energies and densities of states that differ quite significantly from each other. This means that many electronic and other properties of metals and semiconductors change dramatically when the dimensionality changes. Some nanostructures of technological interest exhibit both potential well confinement and Fermi gas delocalization, confinement in one or two dimensions, and delocalization in two or one dimensions, so it will be instructive to show how these two strikingly different behaviors coexist.

In a three-dimensional Fermi sphere the energy varies from $E = 0$ at the origin to $E = E_F$ at the Fermi surface, and similarly for the one- and two-dimensional analogs. When there is confinement in one or two directions, the conduction electrons will distribute themselves among the corresponding potential well levels that lie below the Fermi level along confinement coordinate directions, in accordance with their respective degeneracies d_i, and for each case the electrons will delocalize in the remaining dimensions by populating Fermi gas levels in the delocalization direction of the reciprocal lattice. Table 9.5 lists the formulas for the energy dependence of the number of electrons $N(E)$ for quantum dots that exhibit total confinement, quantum wires and quantum wells, which involve partial confinement, and bulk material, where there is no confinement. The density of states formulas $D(E)$ for these four cases are also listed in the table. The summations in these expressions are over the various confinement well levels i.

Figure 9.15 shows plots of the energy dependence $N(E)$ and the density of states $D(E)$ for the four types of nanostructures listed in Table 9.5. We see that the number of electrons $N(E)$ increases with the energy E, so the four nanostructure types vary only qualitatively from each other. However, it is the density of states $D(E)$ that determines the various electronic and other properties, and these differ dramatically for each of the three nanostructure types. This means that the nature of the

Table 9.5. Number of electrons *N(E)* and density of states *D(E)* = *dN(E)*/*dE* as a function of the energy *E* for electrons delocalized/confined in quantum dots, quantum wires, quantum wells, and bulk material[a]

			Dimensions	
Type	Number of Electrons $N(E)$	Density of States $D(E)$	Delocalized	Confined
Dot	$N(E) =$ $K_0 \sum d_i \Theta(E - E_{i\mathrm{W}})$	$D(E) =$ $K_0 \sum d_i \delta(E - E_{i\mathrm{W}})^2$	0	3
Wire	$N(E) =$ $K_1 \sum d_i (E - E_{i\mathrm{W}})^{1/2}$	$D(E) =$ $\frac{1}{2} K_1 \sum d_i (E - E_{i\mathrm{W}})^{-1/2}$	1	2
Well	$N(E) = K_2 \sum d_i (E - E_{i\mathrm{W}})$	$D(E) = K_2 \sum d_i$	2	1
Bulk	$N(E) = K_3 (E)^{3/2}$	$D(E) = \frac{3}{2} K_3 (E)^{1/2}$	3	0

[a]The degeneracies d_i of the confined (square or parabolic well) energy levels depend on the particular level. The Heaviside step function $\Theta(x)$ is zero for $x < 0$ and one for $x > 0$; the delta function $\delta(x)$ is zero for $x \neq 0$, infinity for $x = 0$, and integrates to a unit area. The values of the constants K_1, K_2, and K_3 are given in Table A.3 of Appendix A.

dimensionality and of the confinement associated with a particular nanostructure have a pronounced effect on its properties. These considerations can be used to predict properties of nanostructures, and one can also identify types of nanostructures from their properties.

9.3.6. Properties Dependent on Density of States

We have discussed the density of states $D(E)$ of conduction electrons, and have shown that it is strongly affected by the dimensionality of a material. Phonons or quantized lattice vibrations also have a density of states $D_{\mathrm{PH}}(E)$ that depends on the dimensionality, and like its electronic counterpart, it influences some properties of solids, but our principal interest is in the density of states $D(E)$ of the electrons. In this section we mention some of the properties of solids that depend on the density of states, and we describe some experiments for measuring it.

The specific heat of a solid C is the amount of heat that must be added to it to raise its temperature by one degree Celsius (centigrade). The main contribution to this heat is the amount that excites lattice vibrations, and this depends on the phonon density of states $D_{\mathrm{PH}}(E)$. At low temperatures there is also a contribution to the specific heat C_{el} of a conductor arising from the conduction electrons, and this depends on the electronic density of states at the Fermi level: $C_{\mathrm{el}} = \pi^2 D(E_{\mathrm{F}}) k_{\mathrm{B}}^2 T/3$, where k_{B} is the Boltzmann constant.

The susceptibility $\chi = M/H$ of a magnetic material is a measure of the magnetization M or magnetic moment per unit volume that is induced in the material by the application of an applied magnetic field H. The component of the susceptibility arising from the conduction electrons, called the *Pauli susceptibility*, is given by the expression $\chi_{\mathrm{el}} = \mu_{\mathrm{B}}^2 D(E_{\mathrm{F}})$, where μ_{B} is the unit magnetic moment called the *Bohr magneton*, and is hence χ_{el} characterized by its proportionality to the

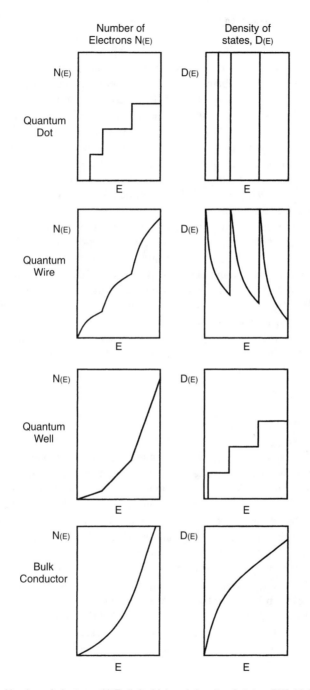

Figure 9.15. Number of electrons $N(E)$ (left side) and density of states $D(E)$ (right side) plotted against the energy for four quantum structures in the square well–Fermi gas approximations.

electronic density of states $D(E)$ at the Fermi level, and its lack of dependence on the temperature.

When a good conductor such as aluminum is bombarded by fast electrons with just enough energy to remove an electron from a particular Al inner-core energy level, the vacant level left behind constitutes a hole in the inner-core band. An electron from the conduction band of the aluminum can fall into the vacant inner-core level to occupy it, with the simultaneous emission of an X ray in the process. The intensity of the emitted X radiation is proportional to the density of states of the conduction electrons because the number of electrons with each particular energy that jumps down to fill the hole is proportional to $D(E)$. Therefore a plot of the emitted X-ray intensity versus the X-ray energy E has a shape very similar to a plot of $D(E)$ versus E. These emitted X rays for aluminum are in the energy range from 56 to 77 eV.

Some other properties and experiments that depend on the density of states and can provide information on it are photoemission spectroscopy, Seebeck effect (thermopower) measurements, the concentrations of electrons and holes in semiconductors, optical absorption determinations of the dielectric constant, the Fermi contact term in nuclear magnetic resonance (NMR), the de Haas–van Alphen effect, the superconducting energy gap, and Josephson junction tunneling in superconductors. It would take us too far afield to discuss any of these topics. Experimental measurements of these various properties permit us to determine the form of the density of states $D(E)$, both at the Fermi level E_F and over a broad range of temperature.

9.4. EXCITONS

Excitons, which were introduced in Section 2.3.3, are a common occurrence in semiconductors. When an atom at a lattice site loses an electron, the atom acquires a positive charge that is called a *hole*. If the hole remains localized at the lattice site, and the detached negative electron remains in its neighborhood, it will be attracted to the positively charged hole through the Coulomb interaction, and can become bound to form a hydrogen-type atom. Technically speaking, this is called a *Mott–Wannier* type of exciton. The Coulomb force of attraction between two charges $Q_e = -e$ and $Q_h = +e$ separated by a distance r is given by $F = -ke^2/\varepsilon r^2$, where e is the electronic charge, k is a universal constant, and ε is the dielectric constant of the medium. The exciton has a Rydberg series of energies E sketched in Fig. 2.20 and a radius given by Eq. (2.19): $a_{\text{eff}} = 0.0529(\varepsilon/\varepsilon_0)/(m^*/m_0)$, where $\varepsilon/\varepsilon_0$ is the ratio of the dielectric constant of the medium to that of free space, and m^*/m_0 is the ratio of the effective mass of the exciton to that of a free electron. Using the dielectric constant and electron effective mass values from Tables B.11 and B.8, respectively, we obtain for GaAs

$$E = 5.2 \, \text{meV} \qquad a_{\text{eff}} = 10.4 \, \text{nm} \qquad (9.10)$$

showing that the exciton has a radius comparable to the dimensions of a typical nanostructure.

The exciton radius can be taken as an index of the extent of confinement experienced by a nanoparticle. Two limiting regions of confinement can be identified on the basis of the ratio of the dimension d of the nanoparticle to the exciton radius a_{eff}, namely, the weak-confinement regime with $d > a_{\text{eff}}$ (but not $d \gg a_{\text{eff}}$) and the strong-confinement regime $d < a_{\text{eff}}$. The more extended limit $d \gg a_{\text{eff}}$ corresponds to no confinement. Under weak-confinement conditions the exciton can undergo unrestricted translational motion, just as in the bulk material, but for strong confinement this translation motion becomes restricted. There is an increase in the spatial overlap of the electron and hole wavefunctions with decreasing particle size, and this has the effect of enhancing the electron–hole interaction. As a result, the energy splitting becomes greater between the radiative and nonradiative exciton states. An optical index of the confinement is the blue shift (shift to higher energies) of the optical absorption edge and the exciton energy with decreasing nanoparticle size. Another result of the confinement is the appearance at room temperature of excitonic features in the absorption spectra that are observed only at low temperatures in the bulk material. Further details on exciton spectra are provided in Sections 2.3.3 and 8.2.1.

9.5. SINGLE-ELECTRON TUNNELING

We have been discussing quantum dots, wires, and wells in isolation, such as the ones depicted in Figs. 9.1–9.3. To make them useful, they need coupling to their surroundings, to each other, or to electrodes that can add or subtract electrons from them. Figure 9.16 shows an isolated quantum dot or island coupled through tunneling to two leads, a source lead that supplies electrons, and a drain lead that

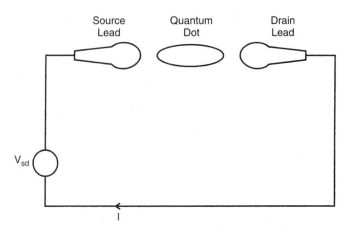

Figure 9.16. Quantum dot coupled to an external circuit through source and a drain leads.

removes electrons for use in the external circuit. The applied voltage V_{sd} causes direct current I to flow, with electrons tunneling into and out of the quantum dot. In accordance with Ohm's law $V = IR$, the current flow I through the circuit of Fig. 9.16 equals the applied source–drain voltage V_{sd} divided by the resistance R, and the main contribution to the value of R arises from the process of electron tunneling from source to quantum dot, and from quantum dot to drain. Figure 9.17 shows the addition to the circuit of a capacitor-coupled gate terminal. The applied gate voltage V_g provides a controlling electrode or gate that regulates the resistance R of the active region of the quantum dot, and consequently regulates the current flow I between the source and drain terminals. This device, as described, functions as a voltage-controlled or field-effect-controlled transistor, commonly referred to as an FET. For large or macroscopic dimensions the current flow is continuous, and the discreteness of the individual electrons passing through the device manifests itself by the presence of current fluctuations or shot noise. Our present interest is in the passage of electrons, one by one, through nanostructures based on circuitry of the type sketched in Fig. 9.17.

For an FET-type nanostructure the dimensions of the quantum dot are in the low nanometer range, and the attached electrodes can have cross sections comparable in size. For disk and spherical shaped dots of radius r the capacitance is given by

$$C = 8\varepsilon_0 \left(\frac{\varepsilon}{\varepsilon_0} \right) r \qquad \text{disk} \qquad (9.11)$$

$$C = 4\pi\varepsilon_0 \left(\frac{\varepsilon}{\varepsilon_0} \right) r \qquad \text{sphere} \qquad (9.12)$$

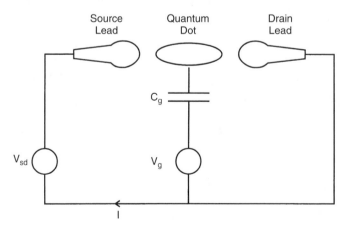

Figure 9.17. Quantum dot coupled through source and a drain leads to an external circuit containing an applied bias voltage V_{sd}, with an additional capacitor-coupled terminal through which the gate voltage V_g controls the resistance of the electrically active region.

where $\varepsilon/\varepsilon_0$ is the dimensionless dielectric constant of the semiconducting material that forms the dot, and $\varepsilon_0 = 8.8542 \times 10^{-12}$ F/m is the dielectric constant of free space. For the typical quantum dot material GaAs we have $\varepsilon/\varepsilon_0 = 13.2$, which gives the very small value $C = 1.47 \times 10^{-18} r$ farad for a spherical shape, where the radius r is in nanometers. The electrostatic energy E of a spherical capacitor of charge Q is changed by the amount $\Delta E \sim eQ/C$ when a single electron is added or subtracted, corresponding to the change in potential $\Delta V = \Delta E/Q$

$$\Delta V = e/C \cong 0.109/r \text{ volts} \tag{9.13}$$

where r is in nanometers. For a nanostructure of radius $r = 10$ nm, this gives a change in potential of 11 mV, which is easily measurable. It is large enough to impede the tunneling of the next electron.

Two quantum conditions must be satisfied for observation of the discrete nature of the single-electron charge transfer to a quantum dot. One is that the capacitor single-electron charging energy $e^2/2C$ must exceed the thermal energy $k_B T$ arising from the random vibrations of the atoms in the solid, and the other is that the Heisenberg uncertainty principle be satisfied by the product of the capacitor energy $e^2/2C$ and the time $\tau = R_T C$ required for charging the capacitor

$$\Delta E \, \Delta t = \left(\frac{e^2}{2C}\right)(R_T C) > h \tag{9.14}$$

where R_T is the tunneling resistance of the potential barrier. These two tunneling conditions correspond to

$$\frac{e^2}{2C} \gg k_B T \tag{9.15a}$$

$$R_T \gg \frac{h}{e^2} \tag{9.15b}$$

where $h/e^2 = 25.813$ kΩ is the quantum of resistance. When these conditions are met and the voltage across the quantum dot is scanned, then the current jumps in increments every time the voltage changes by the value of Eq. (9.13), as shown by the I-versus-V characteristic of Fig. 9.18. This is called a *Coulomb blockade* because the electrons are blocked from tunneling except at the discrete voltage change positions. The step structure observed on the I–V characteristic of Fig. 9.18 is called a *Coulomb staircase* because it involves the Coulomb charging energy $e^2/2C$ of Eq. (9.15a).

An example of single-electron tunneling is provided by a line of ligand-stabilized Au_{55} nanoparticles. These gold particles have what is referred to as a structural magic number of atoms arranged in a FCC close-packed cluster that approximates the shape of a sphere of diameter 1.4 nm, as was discussed in Section 2.1.3. The cluster of 55 gold atoms is encased in an insulating coating called a *ligand shell* that is adjustable in thickness, and has a typical value of 0.7 nm. Single-electron

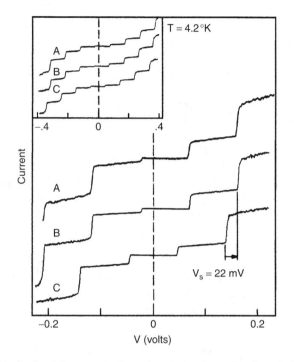

Figure 9.18. Coulomb staircase on the current *I*–voltage *V* characteristic plot from single-electron tunneling involving a 10-nm indium metal droplet. The experimental curve *A* was measured by a scanning tunneling microscope, and curves *B* and *C* are theoretical simulations. The peak-to-peak current is 1.8 nA. [After R. Wilkins, E. Ben-Jacob, and R. C. Jaklevic, *Phys. Rev. Lett.* **59**, 109 (1989).]

tunneling can take place between two of these Au_{55} ligand-stabilized clusters when they are in contact, with the shell acting as the barrier for the tunneling. Experiments were carried out with linear arrays of these Au_{55} clusters of the type illustrated in Fig. 9.19. An electron entering the chain at one end was found to tunnel its way through in a soliton-like manner. Estimates of the interparticle capacitance gave $C_{micro} \cong 10^{-18}$ F, and the estimated interparticle resistance was $R_T \cong 100$ MΩ [see Gasparian et al. (2000)]. Section 6.1.5 discusses electron tunneling along a linear chain of much larger (500-nm) gold nanoparticles connected by conjugated organic molecules.

9.6. APPLICATIONS

9.6.1. Infrared Detectors

Infrared transitions involving energy levels of quantum wells, such as the levels shown in Figs. 9.12 and 9.13, have been used for the operation of infrared

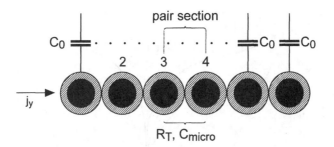

Figure 9.19. Linear array of Au$_{55}$ ligand-stabilized nanoparticles with interparticle resistance R_T, interparticle capacitance C_{micro}, and self-capacitance C_0. The single-electron current density j_y entering from the right, which tunnels from particle to particle along the line, is indicated. [From V. Gasparian et al., in Nalwa (2000), Vol. 2, Chapter 11, p. 550.]

photodetectors. Sketches of four types of these detectors are presented in Fig. 9.20. The conduction band is shown at or near the top of these figures, occupied and unoccupied bound-state energy levels are shown in the wells, and the infrared transitions are indicated by vertical arrows. Incoming infrared radiation raises electrons to the conduction band, and the resulting electric current flow is a measure of the incident radiation intensity. Figure 9.20a illustrates a transition from bound state to bound state that takes place within the quantum well, and Fig. 9.20b shows a transition from bound state to continuum. In Fig. 9.20c the continuum begins at the top of the well, so the transition is from bound state to quasi-bound state. Finally in Fig. 9.20d the continuum band lies below the top of the well, so the transition is from bound state to miniband.

The responsivity of the detector is the electric current (amperes, A) generated per watt (W) of incoming radiation. Figure 9.21 shows a plot of the dark-current density (before irradiation) versus bias voltage for a GaAs/AlGaAs bound state–continuum photodetector, and Fig. 9.22 shows the dependence of this detector's responsivity on

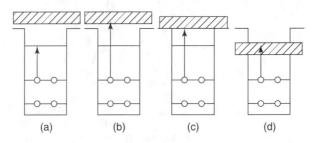

(a)   (b)   (c)   (d)

Figure 9.20. Schematic conduction band (shaded) and electron transition schemes (vertical arrows) of the following types: (a) bound state to bound state; (b) bound state to continuum; (c) bound state to quasi-bound; (d) bound state to miniband, for quantum-well infrared photodetectors. [Adapted from S. S. Li and M. Z. Tidrow, in Nalwa (2000), Vol. 4, Chapter 9, p. 563.]

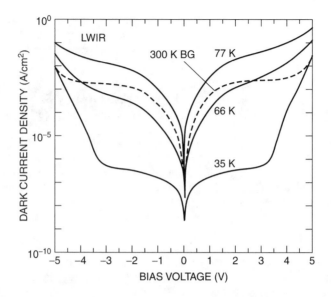

Figure 9.21. Dark-current density versus bias voltage characteristics for a GaAs/AlGaAs quantum-well long-wavelength infrared (LWIR) photodetector measured at three indicated temperatures. A 300-K background current plot (BG, dashed curve) is also shown. [From M. Z. Tidrow, J. C. Chiang, S. S. Li, and K. Bacher, *Appl. Phys. Lett.* **70**, 859 (1997).]

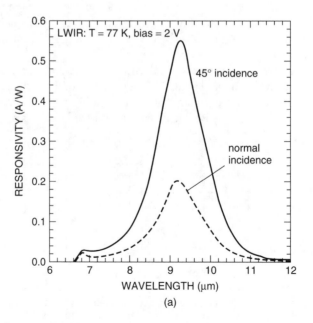

Figure 9.22. Peak responsivity versus wavelength at 77 K for a 2-V bias at normal and 45° angles of incidence. [From M. Z. Tidrow, J. C. Chiang, S. S. Li, and K. Bacher, *Appl. Phys. Lett.* **70**, 859 (1997).]

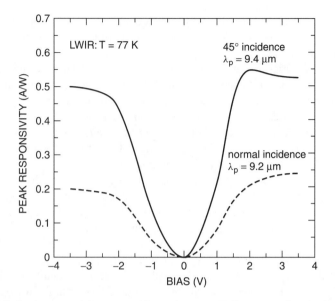

Figure 9.23. Peak responsivity versus bias voltage at 77 K at normal and 45° angles of incidence. The peak detection wavelengths λ_p are indicated. [From M. Z. Tidrow, J. C. Chiang, S. S. Li, and K. Bacher, *Appl. Phys. Lett.* **70**, 859 (1997).]

the wavelength for normal and 45° incidence. The responsivity reaches a peak at the wavelength $\lambda = 9.4$ μm, and Fig. 9.23 shows the dependence of this peak responsivity R_P on the bias voltage. The operating bias of 2 V was used to obtain the data of Fig. 9.22 because, as Fig. 9.23 shows, at that bias the responsivity has leveled off at a high value. This detector is sensitive for operation in the infrared wavelength range from 8.5 to 10 μm.

9.6.2. Quantum Dot Lasers

The infrared detectors described in the previous section depend on the presence of discrete energy levels in a quantum well between which transitions in the infrared spectral region can be induced. Laser operation also requires the presence of discrete energy levels, that is, levels between which laser emission transitions can be induced. The word *laser* is an acronym for *light amplification by stimulated emission of light*, and the light emitted by a laser is both monochromatic (single-wavelength) and coherent (in-phase). Quantum-well and quantum-wire lasers have been constructed that make use of these laser emission transitions. These devices have conduction electrons for which the confinement and localization in discrete energy levels takes place in one or two dimensions, respectively. Hybrid-type lasers have been constructed using "dots in a well", such as InAs quantum dots placed in a strained InGaAs quantum well. Another design employed what have been referred to as InAs quantum dashes, which are very short quantum wires, or from another point of view they are quantum dots elongated in one direction. The present section is devoted to a discussion of quantum-dot lasers, in which the confinement is in all three spatial directions.

Conventional laser operation requires the presence of a laser medium containing active atoms with discrete energy levels between which the laser emission transitions take place. It also requires a mechanism for population inversion whereby an upper energy level acquires a population of electrons exceeding that of the lower-lying ground-state level. In a helium–neon gas laser the active atoms are Ne mixed with He, and in a Nd-YAG solid-state laser the active atoms are neodymium ions substituted ($\sim 10^{19}$ cm^{-3}) in a yttruim aluminum garnet crystal. In the quantum-dot laser to be described here the quantum dots play the role of the active atoms.

Figure 9.24 provides a schematic illustration of a quantum-dot laser diode grown on an n-doped GaAs substrate (not shown). The top p-metal layer has a GaAs contact layer immediately below it. Between this contact layer above and the GaAs substrate (not shown) below the diagram, there are a pair of 2-μm-thick Al$_{0.85}$Ga$_{0.15}$As cladding or bounding layers that surround a 190-nm-thick waveguide made of Al$_{0.05}$Ga$_{0.95}$As. The waveguide plays the role of conducting the emitted light to the exit ports at the edges of the structure. Centered in the waveguide (dark

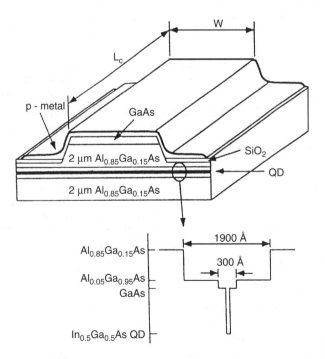

Figure 9.24. Schematic illustration of a quantum dot near-infrared laser. The inset at the bottom shows details of the 190-nm-wide (Al$_{0.85}$Ga$_{0.15}$As cladded) waveguide region that contains the 12 monolayers of In$_{0.5}$Ga$_{0.5}$As quantum dots (indicated by QD) that do the lasing. [From Park et al. (1999).]

horizontal stripe on the figure labeled QD) is a 30-nm-thick GaAs region, and centered in this region are 12 monolayers of $In_{0.5}Ga_{0.5}As$ quantum dots with a density of $1.5 \times 10^{10}/cm^2$. The inset at the bottom of the figure was drawn to represent the details of the waveguide region. The length L_C and the width W varied somewhat from sample to sample, with L_C = ranging from 1 to 5 mm, and W varying between 4 and 60 μm. The facets or faces of the laser were coated with $ZnSe/MgF_2$ high-reflectivity ($>95\%$) coatings that reflected the light back and forth inside to augment the stimulated emission. The laser light exited through the lateral edge of the laser.

The laser output power for continuous-wave (CW) operation at room temperature is plotted in Fig. 9.25 versus the current for the laser dimensions L_C = 1.02 mm and W = 9 μm. The near-infrared output signal at the wavelength of 1.32 μm for a current setting just above the 4.1-mA threshold value, labeled point a, is shown in the inset of the figure. The threshold current density increases sharply with the temperature above 200 K, and this is illustrated in Fig. 9.26 for pulsed operation.

9.7. SUPERCONDUCTIVITY

Superconductors exhibit some properties that are analogous to those of quantum dots, quantum wires, and quantum wells. This is the case, in part, because their characteristic length scales λ and ξ are in the nanorange, as the representative data in Table 9.6 indicate. The table also gives the transition temperature T_C below which each material superconducts, that is, has zero electrical resistance. The majority of the values of λ and ξ listed in the table are 200 nm or less, and several are below 6 nm. The penetration depth λ is a measure of the distance that an externally applied

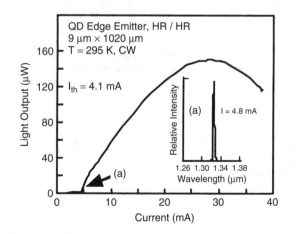

Figure 9.25. Dependence of the near-infrared light output power on the current for a continuous-wave, room-temperature, edge-emitting quantum dot laser of the type illustrated in Fig. 9.24. [From Park et al. (1999).]

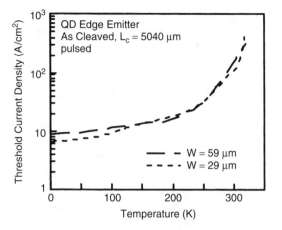

Figure 9.26. Dependence of the threshold current density on the temperature for a pulsed, edge-emitting quantum-dot laser with the design illustrated in Fig. 9.24. [From Park et al. (1999).]

magnetic field B_{app} can penetrate into a pure type I superconductor. Magnetic flux, that is, B fields, are excluded from the bulk of a type I superconductor, and applied magnetic fields that exceed the critical field B_C cause the superconductor to revert to being a normal metal. Superconductivity is present in a material when pairs of electrons condense into a bound state called a *Cooper pair* with dimensions of the order of the coherence length ξ. Thus Cooper pairs, the basic charge carriers of the supercurrent, can be looked on as nanoparticles.

A type II superconductor has both a lower and an upper critical magnetic field, $B_{C1} < B_{C2}$, and three regions of behavior in an applied magnetic field B_{app}. For low

Table 9.6. Transition temperatures T_C, coherence lengths ξ, and penetration depths λ of some typical superconductors

Material	Type	T_C (K)	ξ (nm)	λ (nm)
Cd	I	0.56	760	110
In	I	3.4	360	40
Pb	I	7.2	82	39
Pb–In alloy	II	7.0	30	150
Nb–N alloy	II	16	5	200
$PbMo_6S_8$	II	15	2	200
V_3Si	II	16	3	60
Nb_3Ge	II	23	3	90
K_3C_{60}	II	19	2.6	240

Source: Data are from C. P. Poole Jr, H. A. Farach, and R. J. Creswick, *Superconductivity*, Academic Press, San Diego, 1995, p. 271.

applied fields $B_{app} < B_{C1}$ the material acts like a type I superconductor and excludes magnetic flux, and at high applied fields $B_{app} > B_{C2}$ the material becomes normal. In the intermediate range, $B_{C1} < B_{app} < B_{C2}$, the magnetic field penetrates into the bulk in the form of tubes of magnetic flux, each of which contains one quantum of flux Φ_0, which has the value

$$\Phi_0 = \frac{h}{2e} = 2.0678 \times 10^{-15} \, \text{T m}^2 \tag{9.16}$$

Each vortex has a core of radius ξ within which the magnetic field is fairly constant, and an outside region of radius λ where the magnetic field decays with distance r from the core, a decay which has the exponential form $\exp(-r/\lambda)$ at large distances away. The length of a vortex is the thickness of the sample, which is typically in the centimeter range. The vortices viewed head-on form the two-dimensional hexagonal lattice shown sketched in Fig. 9.27, with the centers of the cores of the vortices a distances d apart that is approximately one penetration depth λ when the applied field equals the lower critical field, and approximately one coherence length ξ apart when the applied field equals the upper critical field. Vortices may be looked on as the magnetic analog of quantum wires in the sense that they confine one quantum unit of magnetic flux in the transverse direction, but set no limit longitudinally. The transverse dimensions of their core is in the nanometer range, but their length is ordinarily macroscopic.

A Josephson junction consists of two superconductors separated by a thin layer of insulating material. By the Josephson effect there can be a flow of DC current across the junction in the absence of applied electric or magnetic fields. An ultrasmall Josephson junction with an area of $0.01 \, \mu\text{m}^2$ and a thickness of $0.1 \, \text{nm}$ has a capacitance estimated from the expression $C = \varepsilon_0 A/d$ of about $10^{-15} \, \text{F}$, and the change in voltage arising from the tunneling of one electron across the barrier is given by $\Delta V = e/C = 0.16 \, \text{mV}$, which is an appreciable fraction of a typical junction voltage. This can be enough to impede the tunneling of the next electron, and the result is a Coulomb blockade. Figure 9.28 shows the observation of a Coulomb staircase on the I–V characteristic plot of a granular lead film Josephson junction. We see from the figure that the staircase features are much better resolved

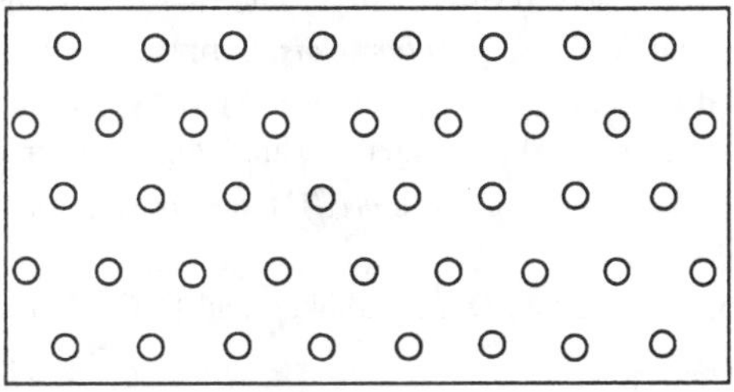

Figure 9.27. Two-dimensional hexagonal lattice of vortex cores. (From C. P. Poole, Jr., H. A. Farach, and R. J. Creswick, *Superconductivity*, Academic Press, Boston, 1995, p. 277.)

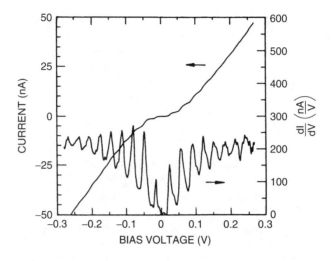

Figure 9.28. Coulomb staircase structure on plots of current I and derivative dI/dV versus the bias voltage V of a granular lead film Josephson junction. It is clear that the first-derivative response provides much better resolution. [From K. A. McGreer, J.-C. Wan, N. Anand, and A. M. Goldman, *Phys. Rev.* **B39**, 12260 (1989).]

by the differential plot of dI/dV versus V than they are on the initial I–V plot. This superconductor Coulomb staircase is similar to the much better resolved one presented in Fig. 9.18 for single-electron tunneling in quantum dots.

FURTHER READING

M. S. Feld and K. An, "Single Atom Laser", *Sci. Am.* 57 (July 1998).

D. K. Ferry and S. M. Goodnick, *Transport in Nanostructures*, Cambridge Univ. Press, Cambridge, UK, 1997.

V. Gasparian, M. Ortuño, G. Schön, and U. Simon, in Nalwa (2000), Vol. 2, Chapter 11.

D. Heitmann and J. P. Kotthaus, "The Spectroscopy of Quantum Dot Arrays", *Phys. Today* 56 (June 1993).

M. Henini, "Quantum Dot Nanostructures", *Materials Today*, 48 (June 2002).

L. Jacak, P. Hawrylak, and A. Wojs, *Quantum Dots*, Springer, Berlin, 1998.

L. P. Kouwenhoven and P. L. McEuen, "Single Electron Transport Through a Quantum Dot", in *Nanotechnology*, G. Timp, ed., Springer, Berlin, 1999, Chapter 13.

H. S. Nalwa, ed., *Handbook of Nanostructured Materials and Nanotechnology*, Vol. 1–5, Academic Press, Boston, 2000.

G. Park, O. B. Shchekin, S. Csutak, D. L. Huffaker, and D. G. Deppe, *Appl. Phys. Lett.* **75**, 3267 (1999).

10

SELF-ASSEMBLY AND CATALYSIS

10.1. SELF-ASSEMBLY

10.1.1. Process of Self-Assembly

Proteins are large molecules, with molecular weights in the tens of thousands, found in virtually all the cells and tissues of the body, and they are essential for life. They are formed by the successive addition of hundreds of amino acids, each brought to its site of attachment by a transfer RNA molecule, in an order prescribed by a messenger RNA molecule. Each amino acid readily bonds to the previous one when it arrives in place. Thus sequences of amino acids assemble into a polypeptide chain, which continually increases in length to eventually become a protein. This type of self-assembly process, which occurs naturally in all living systems, has its analog in nanoscience. Here we also encounter the spontaneous organization of small molecules into larger well-defined, stable, ordered molecular complexes or aggregates, and we encounter the spontaneous adsorption of atoms or molecules onto a substrate in a systematic, ordered manner. This process involves the use of weak, reversible interactions between parts of molecules without any central control, and the result is a configuration that is in equilibrium. The procedure is automatically

Introduction to Nanotechnology, by Charles P. Poole Jr. and Frank J. Owens.
ISBN 0-471-07935-9. Copyright © 2003 John Wiley & Sons, Inc.

error-checking, so faulty or improperly attached subunits can be replaced during the growth.

The traditional organic synthesis of very large molecules called *macromolecules* comprises a number of time-consuming steps that involve breaking and remaking strong covalent bonds, and these steps are carried out under kinetic control. The yields are small, and errors are not readily recognized or corrected. In contrast to this, the self-assembly variety of synthesis makes use of weak, noncovalent bonding interactions such as those involving hydrogen bonds and van der Waals forces, which permit the reactions to proceed under thermodynamic control, with the continual correction of errors. The initial individual molecules or subunits are usually small in size and number and easy to synthesize, and the final product is produced in a thermodynamic equilibrium state.

10.1.2. Semiconductor Islands

One type of self-assembly involves the preparation of semiconductor islands, and it can be carried out by a technique called *heteroepitaxy*, which involves the placement or deposition of the material that forms the island on a supporting substance called a *substrate* made of a different material with a closely matched interface between them. Heteroepitaxy has been widely used for research, as well as for the fabrication of many semiconductor devices, so it is a well-developed technique. It involves bringing atoms or molecules to the surface of the substrate where they do one of three things. They either are adsorbed and diffuse about on the surface until they join or nucleate with another adatom to form an island, attach themselves to or aggregate into an existing island, or desorb and thereby leave the surface. Small islands can continue to grow, migrate to other positions, or evaporate. There is a critical size at which they become stable, and no longer experience much evaporation. Thus there is an initial nucleation stage when the number of islands increases with the coverage. This is followed by an aggregation stage when the number of islands levels off and the existing ones grow in size. Finally there is the coalescence stage when the main events that take place involve the merger of existing islands with each other to form larger clusters.

The various stages can be described analytically or mathematically in terms of the rates of change dn_i/dt of the concentrations of individual adatoms n_1, pairs of adatoms n_2, clusters of size three n_3, and so on, and an example of a kinetic equation that is applicable at the initial or nucleation stage is the following expression for isolated atoms (Weinberg et al. 2000).

$$\frac{dn_1}{dt} = (R_{ads} + R_{det} + 2R_1) - (R_{evap} + R_{cap} + 2R_1')$$

(10.1)

where R_{ads} is the rate of adsorption, R_{det} is the rate of detachment of atoms from clusters larger than pairs, and R_1 is the rate of breakup of adatom pairs. The negative terms correspond to the rate of evaporation R_{evap}, the rate of capture of individual adatoms by clusters R_{cap}, and the rate of formation of pairs of adatoms $2R_1'$. The factor of 2 associated with R_1 and R_1' accounts for the participation of two atoms in each pair process. Analogous expressions can be written for the rate of change of the number of pairs dn_2/dt, for the rate of change of the number of triplet clusters dn_3/dt, and so forth. Some of the terms for the various rates R_i depend on the extent of the coverage of the surface, and the equation itself is applicable mainly during the nucleation stage.

At the second or aggregation stage the percentage of isolated adatoms becomes negligible, and a free-energy approach can provide some insight into the island formation process. Consider the Gibbs free-energy density $g_{\text{sur−vac}}$ between the bare surface and the vacuum outside, the free-energy density $g_{\text{sur−lay}}$ between the surface and the layers of adatoms, and the free-energy density $g_{\text{lay−vac}}$ between these layers and the vacuum. These are related to the overall Gibbs free-energy density g through the expression

$$g = g_{\text{sur−vac}}(1 - \varepsilon) + (g_{\text{sur−lay}} + g_{\text{lay−vac}})\varepsilon \tag{10.2}$$

where ε is the fraction of the surface covered. As the islands form and grow the relative contributions arising from these terms gradually change, and the growth process evolves to maintain the lowest thermodynamic free energy. These free energies can be used to define a spreading pressure $P_S = g_{\text{sur−vac}} - (g_{\text{sur−lay}} + g_{\text{lay−vac}})$ that involves the difference between the bare surface free energy $g_{\text{sur−vac}}$ and that of the layers ($g_{\text{sur−lay}} + g_{\text{lay−vac}}$), and it is associated with the spreading of adatoms over the surface. For the condition ($g_{\text{sur−lay}} + g_{\text{lay−vac}}) < g_{\text{sur−vac}}$, the addition of adatoms increases ε, and thereby causes the free energy to decrease. Thus the adatoms that adsorb will tend to remain directly on the bare surface leading to a horizontal growth of islands, and the eventual formation of a monolayer. The spreading pressure P_S is positive and contributes to the dispersal of the adatoms. This is referred to as the *Franck–van der Merwe growth mode*.

For the opposite condition ($g_{\text{sur−lay}} + g_{\text{lay−vac}}) > g_{\text{sur−vac}}$, the growth of the fractional surface coverage ε increases the free energy, so it is thermodynamically unfavorable for the adsorbed layer to be thin and flat. The newly added adatoms tend to keep the free energy low by aggregating on the top of existing islands, leading to a vertical rather than a horizontal growth of islands. This is called the *Volmer–Weber mode of growth*.

We mentioned above that heteroepitaxy involves islands or a film with a nearly matched interface with the substrate. The fraction f of mismatch between the islands and the surface is given by the expression (see p. 18)

$$f = \frac{|a_\text{f} - a_\text{s}|}{a_\text{s}} \tag{10.3}$$

where a_f is the lattice constant of the island or film and a_s is the lattice constant of the substrate. For small mismatches, less than 2%, very little strain develops at the growth of a film consisting of many successive layers on top of each other. If the mismatch exceeds 3%, then the first layer is appreciably strained, and the extent of the strain builds up as several additional layers are added. Eventually, beyond a transition region the strain subsides, and thick films are only strained in the transition region near the substrate. This mismatch complicates the free-energy discussion following Eq. (10.2), and favors the growth of three-dimensional islands to compensate for the strain and thereby minimize the free energy. This is called the *Stranski–Krastanov growth mode*. A common occurrence in this mode is the initial formation of a monolayer that accommodates the strain, and acquires a critical thickness. The next stage is the aggregation of three-dimensional islands on the two-dimensional monolayer. Another eventuality is the coverage of the surface with monolayer islands of a preferred size to better accommodate the lattice mismatch strain. This can be followed by adding further layers to these islands. A typical size of such a monolayer island is perhaps 5 nm, and it might contain 12 unit cells.

In addition to the three free energies included in Eq. (10.2), the formation and growth of monolayer islands involves the free energies of the strain due to the lattice mismatch, the free energy associated with the edges of a monolayer island, and the free energy for the growth of a three-dimensional island on a monolayer. A great deal of experimental work has been done studying the adsorption of atoms on substrates for gradually increasing coverage from zero to several monolayers thick. The scanning tunneling microscope images presented in Fig. 10.1 show the successive growth of islands of InAs on a GaAs (001) substrate for several fractional monolayer coverages. We see from the data in Table B.1 that the lattice constant $a = 0.606$ nm for InAs and $a = 0.565$ nm for GaAs, corresponding to a lattice mismatch $f = 7.0\%$ from Eq. (10.3), which is quite large.

10.1.3. Monolayers

A model system that well illustrates the principles and advantages of the self-assembly process is a self-assembled monolayer (Wilber and Whitesides 1999). The Langmuir–Blodgett technique, which historically preceded the self-assembled approach, had been widely used in the past for the preparation and study of optical coatings, biosensors, ligand-stabilized Au_{55} clusters, antibodies, and enzymes. It involves starting with clusters, forming them into a monolayer at an air–water interface, and then transferring the monolayer to a substrate in the form of what is called a *Langmuir–Blodgett film*. These films are difficult to prepare, however, and are not sufficiently rugged for most purposes. Self-assembled monolayers, on the other hand, are stronger, are easier to make, and make use of a wider variety of available starting materials.

Self-assembled monolayers and multilayers have been prepared on various metallic and inorganic substrates such as Ag, Au, Cu, Ge, Pt, Si, GaAs, SiO_2, and other materials. This has been done with the aid of bonding molecules or ligands such as alkanethiols RSH, sulfides RSR', disulfides RSSR', acids RCOOH, and siloxanes $RSiOR_3$, where the symbols R and R' designate organic molecule groups that bond to, for example, a thiol radical $-SH$ or an acid radical $-COOH$. The binding to the surface for the thiols, sulfides, and disulfides is via the sulfur atom;

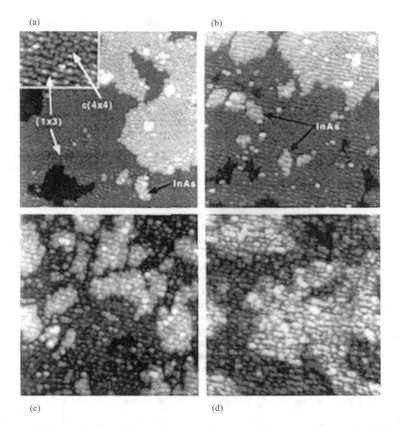

Figure 10.1. Transmission electron microscopy images of the successive growth of islands of InAs on a GaAs (001) substrate for fractional monolayer coverages of (a) 0.1, (b) 0.3, (c) 0.6, and (d) 1.0. The image sizes are 50 nm × 50 nm for (a) and 40 nm × 40 nm for (b) to (d). The inset for (a) is an enlarged view of the GaAs substrate. [From J. G. Belk, J. L. Sudijono, M. M. Holmes, C. F. McConville, T. S. Jones, and B. A. Joyce, *Surf. Sci.* **365**, 735 (1996).]

that is, the entity RS−Au is formed on a gold substrate, and the binding for the acid is $RCO_2-(MO)_n$, where MO denotes a metal oxide substrate ion, and the hydrogen atom H of the acid is released at the formation of the bond. The alkanethiols RSH are the most widely used ligands because of their greater solubility, their compatibility with many organic functional groups, and their speed of reaction. They spontaneously adsorb on the surface; hence the term *self-assemble* is applicable. In this section we consider the self-assembly of the thiol ligand $X(CH_2)_n$ SH, where the terminal group X is methyl (CH_3) and a typical value is $n = 9$ for decanethiol, corresponding to $C_{10}H_{21}$ for R.

To prepare a gold substrate as an extended site for the self-assembly, an electron beam or high-temperature heating element is used to evaporate a polycrystalline layer of gold that is 5–300 nm thick on to a polished support such as a glass slide, a

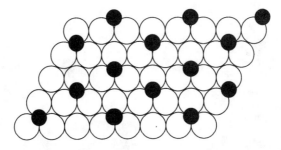

Figure 10.2. Hexagonal close-packed array of adsorbed *n*-alkanethiolate molecules (small solid circles) occupying one-sixth of the threefold sites on the lattice of close-packed gold atoms (large circles).

silicon wafer, or mica. The outermost gold layer, although polycrystalline, exposes local regions of a planar hexagonal close-packing atom arrangement, as indicated in Figs. 10.2 and 10.3. Various properties such as the conductivity, opacity, domain size, and surface roughness of the film depend on its thickness. The adsorption sites are in the hollow depressions between triplets of gold atoms on the surface. The number of depressions is the same as the number of gold atoms on the surface. Sometimes Cr or Ti is added to facilitate the adhesion. When molecules in the liquid (or vapor) phase come into contact with the substrate, they spontaneously adsorb on the clean gold surface in an ordered manner; that is, they self-assemble.

The mechanism or process whereby the adsorption takes place involves each particular alkanethiol molecule $CH_3(CH_2)_nSH$ losing the hydrogen of its sulfhydryl group HS−, acquiring a negative charge, and adhering to the surface as a thiolate compound by embedding its terminal S atom in a hollow between a triplet of Au

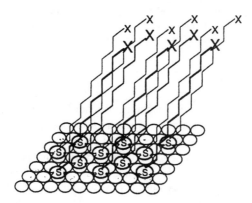

Figure 10.3. Illustration of the self-assembly of a monolayer of *n*-alkanethiolates on gold. The terminal sulphur resides in the hollow between three close-packed gold atoms, as shown in Fig. 10.2. The terminal groups labeled by X represent methyl. [From J. L. Wilber and G. M. Whitesides, in *Nanotechnology*, G. Timp, ed., Springer-Verlag, Berlin, 1999, Chapter 8, p. 336.]

atoms, as noted above and indicated in Fig. 10.3. The reaction at the surface might be written

$$CH_3(CH_2)_n SH + Au_m \Rightarrow CH_3(CH_2)_n S^-(Au_3^+) \cdot Au_{m-3} + \tfrac{1}{2}H_2 \qquad (10.4)$$

where Au_m denotes the outer layer of the gold film, which contains m atoms. The quantity (Au_3^+) is the positively charged triplet of gold atoms that forms the hollow depression in the surface where the terminal sulfur ion (S^-) forms a bond with the gold ion Au_3^+. Figure 10.3 gives a schematic view of this adsorption process. The sulfur–gold bond that holds the alkanethiol in place is a fairly strong one ($\sim$44 kcal/mol) so the adhesion is stable. The bonding to the surface takes place at one-sixth of the sites on the (111) close-packed layer, and these sites are occupied in a regular manner so they form a hexagonal close-packed layer with the lattice constant equal to $\sqrt{3}\, a_0 = 0.865$ nm, where $a_0 = 0.4995$ nm is the distance between Au atoms on the surface. Figures 10.2 and 10.3 indicate the location of the alkanethiol sites on the gold close-packed surface.

The alkanethiol molecules RS$-$, bound together sideways to each other by weak van der Waals forces with a strength of $\sim$1.75 kcal/mol, arrange themselves extending lengthwise up from the gold surface at an angle of $\sim$30° with the normal, as indicated in the sketch of Fig. 10.4. The alkyl chains R extend out $\sim$2.2 nm to form the layer of hexadecanethiol $CH_3(CH_2)_{10}S-$. A number of functional groups can be attached at the end of the alkyl chains in place of the methyl group such as acids, alcohols, amines, esters, fluorocarbons, and nitriles. Thin layers of alkanethiols with $n < 6$ tend to exhibit significant disorder, while thicker ones with $n > 6$ are more regular, but polycrystalline, with domains of differing alkane twist angle. Increasing the temperature can cause the domains to alter their size and shape.

For self-assembled monolayers to be useful in commercial microstructures, they can be arranged in structured regions or patterns on the surface. An alkanethiol "ink" can systematically form or write patterns on a gold surface with alkanethiolate. The monolayer-forming "ink" can be applied to the surface by a process called *microcontact printing*, which utilizes an elastomer, which is a material with rubberlike properties, as a "stamp" to transfer the pattern. The process can be employed to produce thin radiation-sensitive layers called *resists* for nanoscale lithography, as discussed in Section 9.2. The monolayers themselves can serve for a process called

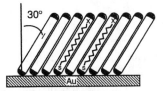

Figure 10.4. Sketch of *n*-alkanethiolate molecules adsorbed on a gold surface at an angle of 30° with respect to the normal. [From Wilber and Whitesides (1999), p. 336.]

passivation by protecting the underlying surface from corrosion. Alkanediols can assist in colloid preparation by controlling the size and properties of the colloids, and this application can be very helpful in improving the efficacy of catalysts.

10.2. CATALYSIS

10.2.1. Nature of Catalysis

Catalysis involves the modification of the rate of a chemical reaction, usually a speeding up or acceleration of the reaction rate, by the addition of a substance, called a *catalyst*, that is not consumed during the reaction. Ordinarily the catalyst participates in the reaction by combining with one or more of the reactants, and at the end of the process it is regenerated without change. In other words, the catalyst is being constantly recycled as the reaction progresses. When two or more chemical reactions are proceeding in sequence or in parallel, a catalyst can play the role of selectively accelerating one reaction relative to the others.

There are two main types of catalysts. *Homogeneous* catalysts are dispersed in the same phase as the reactants, the dispersal ordinarily being in a gas or a liquid solution. *Heterogeneous* catalysts are in a different phase than the reactants, separated from them by a phase boundary. Heterogeneous catalytic reactions usually take place on the surface of a solid catalyst, such as silica or alumina, which has a very high surface area that typically arises from their porous or spongelike structure. The surfaces of these catalysts are impregnated with acid sites or coated with a catalytically active material such as platinum, and the rate of the reaction tends to be proportional to the accessible area of a platinum-coated surface. Many reactions in biology are catalyzed by biological catalysts called *enzymes*. For example, particular enzymes can decompose large molecules into a groups of smaller ones, add functional groups to molecules, or bring about oxidation–reduction reactions. Enzymes are ordinarily specific for particular reactions.

Catalysis can play two principal roles in nanoscience: (1) catalysts can be involved in some methods for the preparation of quantum dots, nanotubes, and a variety of other nanostructures; (2) some nanostructures themselves can serve as catalysts for additional chemical reactions. See Moser (1996) for a discussion of the role of nanostructured materials in catalysis.

10.2.2. Surface Area of Nanoparticles

Nanoparticles have an appreciable fraction of their atoms at the surface, as the data in Tables 2.1, 9.1 demonstrate. A number of properties of materials composed of micrometer-sized grains, as well as those composed of nanometer-sized particles, depend strongly on the surface area. For example, the electrical resistivity of a granular material is expected to scale with the total area of the grain boundaries. The chemical activity of a conventional heterogeneous catalyst is proportional to the overall specific surface area per unit volume, so the high areas of nanoparticles provide them with the possibility of functioning as efficient catalysts. It does not follow, however, that catalytic activity will necessarily scale with the surface area in the nanoparticle range of sizes. Figure 4.14, which is a plot of the reaction rate of H_2 with Fe particles as a function of the particle size, does not show any trend in this direction, and neither does the dissociation rate plotted in Fig. 10.5 for atomic carbon formed on rhodium aggregates deposited on an alumina film. Figure 10.6 shows that the activity or turnover frequency (TOF) of the cyclohexene hydrogenation reaction (frequency of converting cyclohexene C_6H_4 to cyclohexane C_6H_6) normalized to the concentration of surface Rh metal atoms decreases with increasing particle size from 1.5 to 3.5 nm, and then begins to level off. The Rh particle size had been established by the particular alcohol $C_nH_{2n+1}OH$ (inset of Fig. 10.6) used in the catalyst preparation, where $n = 1$ for methanol, 2 for ethanol, 3 for 1-propanol, and 4 for 1-butanol.

The specific surface area of a catalyst is customarily reported in the units of square meters per gram, denoted by the symbol S, with typical values for commercial catalysts in the range from 100 to 400 m^2/g. The general expression for this specific surface area per gram S is

$$S = \frac{(area)}{\rho(volume)} = \frac{A}{\rho V} \tag{10.5}$$

where ρ is the density, which is expressed in the units g/cm^3. A sphere of diameter d has the area $A = \pi d^2$ and the volume $V = \pi d^3/6$, to give $A/V = 6/d$. A cylinder of

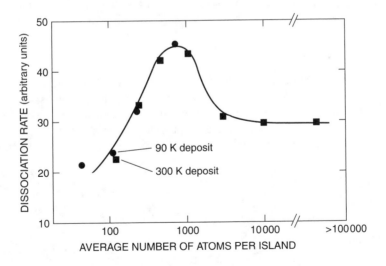

Figure 10.5. Effect of catalytic particle size on the dissociation rate of carbon monoxide. Rhodium aggregates of various sizes, characterized by the number of Rh atoms per aggregate, were deposited on alumina (Al$_2$O$_3$) films. The rhodium was given a saturation carbon monoxide (CO) coverage, then the material was heated from 90 to 500 K (circles), or from 300 to 500 K (squares), and the amount of atomic carbon formed on the rhodium provided a measure of the dissociation rate for each aggregate (island) size.

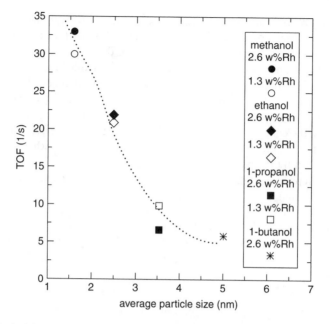

Figure 10.6. Activity of cyclohexene hydrogenation, measured by the turnover frequency (TOF) or rate of conversion of cyclohexene to cyclohexane, plotted as a function of the rhodium (Rh) metal particle size on the surface. The inset gives the alcohols (alkanols) used for the preparation of each particle size. [From G. W. Busser, J. G. van Ommen, and J. A. Lercher, "Preparation and Characterization of Polymer-Stabilized Rhodium Particles", in Moser (1996), Chapter 9, p. 225.]

diameter d and length L has the volume $V = \pi d^2 L/4$. The limit $L \ll d$ corresponds to the shape of a disk with the area $A \sim \pi d^2/2$, including both sides, to give $A/V \sim 2/d$. In like manner, a long cylinder or wire of diameter d and length $L \gg d$ has $A \sim 2\pi r L$, and $A/V \sim 4/d$. Figure 9.2 provides sketches of these figures. Using the units square meters per gram, m^2/g, for these various geometries we obtain the expressions

$$S(r) = \frac{6 \times 10^3}{\rho d} \qquad \text{sphere of diameter } d \qquad (10.6a)$$

$$S(r) = \frac{6 \times 10^3}{\rho a} \qquad \text{cube of side } a \qquad (10.6b)$$

$$S(r) \sim \frac{2 \times 10^3}{\rho L} \qquad \text{thin disk, } L \ll d \qquad (10.6c)$$

$$S(r) \sim \frac{4 \times 10^3}{\rho d} \qquad \text{long cylinder or wire } d \ll L \qquad (10.6d)$$

where the length parameters a, d, and L are expressed in nanometers, and the density ρ has the units g/cm^3. In Eq. (10.6c) the area of the side of the disk is neglected, and in Eq. (10.6d) the areas of two ends of the wire are disregarded. Similar expressions can be written for distortions of the cube [Eq. (10.6b)] into the quantum-well and quantum-wire configurations of Fig. 9.1.

The densities of types III–V and II–VI semiconductors, from Table B.5, are in the range from 2.42 to 8.25 g/cm^3, with GaAs having the typical value $\rho = 5.32$ g/cm^3. Using this density we calculated the specific surface areas of the nanostructures represented by Eqs. (10.6a), (10.6c), and (10.6d), for various values of the size parameters d and L, and the results are presented in Table 10.1. The specific surface areas for the smallest structures listed in the table correspond to quantum dots (column 2, sphere), quantum wires (column 3, cylinder), and quantum wells (column 4, disk), as discussed in Chapter 9. Their specific surface areas are within the range typical of commercial catalysts.

The data tabulated in Table 10.1 represent minimum specific surface areas in the sense that for a particular mass, or for a particular volume, a spherical shape has the lowest possible area, and for a particular linear mass density, or mass per unit length, a wire of circular cross section has the minimum possible area. It is of interest to examine how the specific surface area depends on the shape. Consider a cube of side a with the same volume as a sphere of radius r

$$\frac{4\pi r^3}{3} = a^3 \tag{10.7}$$

so $a = (4\pi/3)^{1/3} r$. With the aid of Eqs. (10.6a) and (10.6b) we obtain for this case $S_{\text{cub}} = 1.24 S_{\text{sph}}$, so a cube has 24% more specific surface than a sphere with the same volume.

To obtain a more general expression for the shape dependence of the area : volume ratio, we consider a cylinder of diameter D and length L with the

Table 10.1. Specific surface areas of GaAs spheres, long cylinders (wires) and thin disks as a function of their size[a]

Size (nm)	Surface Area (m^2/g)		
	Sphere	Wire	Disk
4	281	187	94
6	187	125	62
10	112	76	37
20	57	38	19
30	38	26	13
40	29	19	10
60	19	13	6
100	11	8	4
200	6	4	2

same volume as a sphere of radius r, specifically, $4\pi r^3/3 = \pi D^2 L/4$, which gives $r = \frac{1}{2}[3D^2L/2]^{1/3}$. It is easy to show that the specific surface area $S(L/D)$ from Eq. (10.5) is given by

$$S\left(\frac{L}{D}\right) = \frac{\pi DL + \frac{1}{2}\pi D^2}{\rho\pi D^2 L/4} = 0.382 S_{\text{sph}}\left[2\left(\frac{L}{D}\right)^{1/3} + \left(\frac{D}{L}\right)^{2/3}\right] \quad (10.8)$$

This expression $S(L/D)$, which has a minimum $S_{\text{min}} = 1.146 S_{\text{sph}}$ for the ratio $(L/D) = 1$, is plotted in Fig. 10.7, normalized relative to S_{sph}. The normalization factor S_{sph} was chosen because a sphere has the smallest surface area of any object with a particular volume. Figure 10.7 shows how the surface area increases when a sphere is distorted into the shape of a disk with a particular L/D ratio, without changing in its volume. This figure demonstrates that nanostructures of a particular mass or of a particular volume have much higher surface areas S when they are flat or elongated in shape, and further distortions from a regular shape will increase the area even more.

10.2.3. Porous Materials

In the previous section we saw that an efficient way to increase the surface area of a material is to decrease its grain size or its particle size. Another way to increase the surface area is to fill the material with voids or empty spaces. Some substances such as zeolites, which are discussed in Sections 6.2.3 and 8.4, crystallize in structures in which there are regularly spaced cavities where atoms or small molecules can lodge, or they can move in and out during changes in the environmental conditions. A molecular sieve, which is a material suitable for filtering out molecules of particular sizes, ordinarily has a controlled narrow range of pore diameters. There

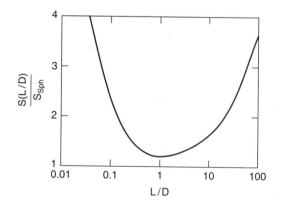

Figure 10.7. Dependence of the surface area $S(L/D)$ of a cylinder on its length:diameter ratio L/D. The surface area is normalized relative to that of a sphere $S_{\text{sph}} = 3/\rho r$ with the same volume.

are also other materials such as silicas and aluminas which can be prepared so that they have a porous structure of a more or less random type; that is, they serve as sponges on a mesoscopic or micrometer scale. It is quite common for these materials to have pores with diameters in the nanometer range. Pore surface areas are sometimes determined by the Brunauer–Emmett–Teller (BET) adsorption isotherm method in which measurements are made of the uptake of a gas such as nitrogen (N_2) by the pores.

Most commercial heterogeneous catalysts have a very porous structure, with surface areas of several hundred square meters per gram. Ordinarily an heterogeneous catalyst consists of a high-surface-area material that serves as a catalyst support or substrate, and the surface linings of its pores contain a dispersed active component, such as acid sites or platinum atoms, which bring about or accelerate the catalytic reaction. Examples of substrates are the oxides silica (SiO_2), gamma-alumina (γ-Al_2O_3), titania (TiO_2 in its tetragonal anatase form), and zirconia (ZrO_2). Mixed oxides are also in common use, such as high-surface-area silica-alumina. A porous material ordinarily has a range of pore sizes, and this is illustrated by the upper right spectrum in Fig. 10.8 for the organosilicate molecular sieve MCM-41, which has a mean pore diameter of 3.94 nm (39.4 Å). The introduction of relatively large trimethylsilyl groups ($CH_3)_3Si$ to replace protons of silanols SiH_3OH in the pores occludes the pore volume, and shifts the distribution of pores to a smaller range of sizes, as shown in the lower left spectrum of the figure. The detection of the nuclear magnetic resonance (NMR) signal from the ^{29}Si isotope of the trimethylsilyl groups in these molecular sieves, with its $+12$ ppm chemical shift shown in Fig. 10.9, confirmed its presence in the pores after the trimethylsilation treatment.

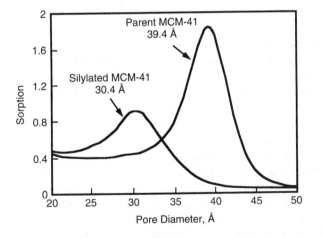

Figure 10.8 Distribution of pore diameters in two molecular sieves with mean pore diameters of 3.04 and 3.94 nm, determined by the physisorption of argon gas. [From J. S. Beck, J. C. Vartuli, W. J. Roth, M. E. Leonowicz, C. T. Kresge, K. D. Schmitt, C. T.-W. Chu, D. H. Olson, E. W. Sheppard, S. B. McCullen, J. B. Higgins, and J. L. Schenkler, *J. Am. Chem. Soc.* **114**, 10834 (1962).]

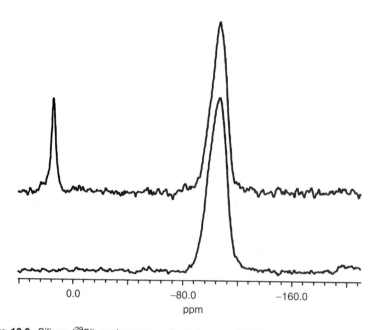

Figure 10.9. Silicon (^{29}Si) nuclear magnetic resonance (NMR) spectrum of an organosilicate molecular sieve before (lower spectrum) and after (upper spectrum) the introduction of the large trimethylsilyl groups (CH$_3$)$_3$Si to replace the protons of the silanols SiH$_3$OH in the pores. The signal on the left with a chemical shift of +12 ppm (parts per million) arises from ^{29}Si of trimethylsilyl, and the strong signal on the right at −124 ppm is due to ^{29}Si in silanol SiH$_3$OH. [From J. C. Vartuli et al., in Moser (1996), Chapter 1, p. 13.]

The active component of an heterogeneous catalyst can be a transition ion, and traditionally over the years the most important active component has been platinum dispersed on the surface. Examples of some metal oxides that serve as catalysts, either by themselves or distributed on a supporting material, are NiO, Cr$_2$O$_3$, Fe$_2$O$_3$, Fe$_3$O$_4$, Co$_3$O$_4$, and β-Bi$_2$Mo$_2$O$_9$. Preparing oxides and other catalytic materials for use ordinarily involves calcination, which is a heat treatment at several hundred degrees Celsius. This treatment can change the structure of the bulk and the surface, and Fig. 10.10 illustrates this for the catalytically active material β-Bi$_2$Mo$_2$O$_9$. We deduce from the figure that for calcination in air the grain sizes grow rapidly between 300 and 350°C, reaching $\cong$ 20 nm, with very little additional change up to 500°C. Sometimes a heat treatment induces a phase change of catalytic importance, as in the case of hydrous zirconia (ZrO$_2$), which transforms from a high-surface-area amorphous state to a low-area tetragonal phase at 450°C, as shown in Fig. 10.11. The change is exothermic, that is, one accompanied by the emission of heat, as shown by the exotherm peak at 450°C in the differential thermal analysis (DTA) curve of Fig. 10.12.

For some reactions the catalytic activity arises from the presence of acid sites on the surface. These sites can correspond to either Brønsted acids, which are proton

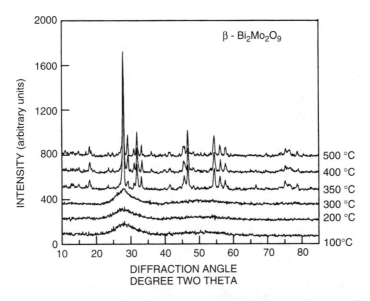

Figure 10.10. X-ray diffraction patterns of the catalyst ß-$Bi_2Mo_2O_9$ taken at 100°C intervals during calcination in air over the range 100–500°C. [From W. R. Moser, J. E. Sunstrom, and B. Marshik-Guerts, in Moser (1996), Chapter 12, p. 296.]

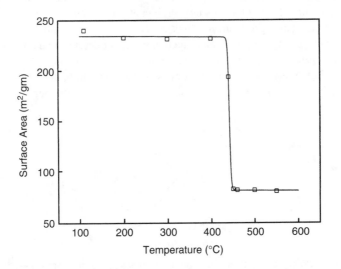

Figure 10.11. Temperature dependence of the surface area of hydrous zirconia, showing the phase transition at 450°C [From D. R. Milburn et al., in Moser (1996), Chapter 6A, p. 138.]

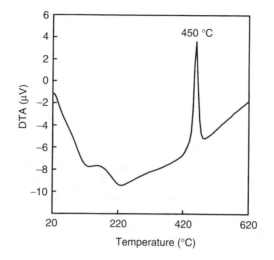

Figure 10.12. Differential thermal analysis curve of a hydrous zirconia catalyst. The heat emission peak at 450°C arises from the exothermic phase transition shown in Fig. 10.11. [From D. R. Milburn et al., in Moser (1996), Chapter 6A, p. 137.]

donors, or Lewis acids, which are electron pair acceptors. Figure 10.13 sketches the structures of a Lewis acid site on the left and a Brønsted acid site on the right located on the surface of a zirconia sulfate catalyst, perhaps containing platinum (Pt/ ZrO_2-SO_4). The figure shows a surface sulfate group SO_4 bonded to adjacent Zr atoms in a way that creates the acid sites. In mixed-oxide catalysts the surface acidity depends on the mixing ratio. For example, the addition of up to 20 wt% ZrO_2 to gamma-alumina (γ-Al_2O_3) does not significantly alter the activity of the Brønsted acid sites, but it has a pronounced effect in reducing the strength of the Lewis acid sites from very strong for pure γ-alumina to a rather moderate value after the addition of the 20 wt% ZrO_2. The infrared stretching frequency of the $-CN$ vibration of adsorbed CD_3CN is a measure of the Lewis acid site strength, and Fig. 10.14 shows the progressive decrease in this frequency with increasing incorporation of ZrO_2 in the γ-Al_2O_3.

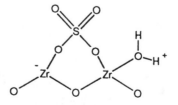

Figure 10.13. Configuration of sulphate group on the surface of a zirconia–sulfate catalyst, showing the Lewis acid site Zr^- at the left, and the Brønsted acid site H^+ at the right. [From G. Strukul et al., in Moser (1996), Chapter 6B, p. 147.]

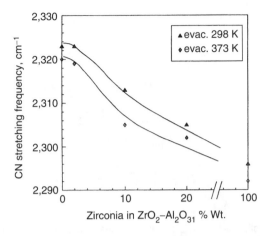

Figure 10.14. Cyanide group (−CN) infrared stretching frequency of CD$_3$CN adsorbed on zirconia alumina catalysts as a function of zirconia (ZrO$_2$) content after evacuation at 25°C (upper curve) and 100°C (lower curve). The stretching frequency is a measure of the Lewis acid site strength. [From G. Centi et al., in Moser (1996), Chapter 4, p. 81.]

10.2.4. Pillared Clays

Clays are layered minerals with spaces between the layers where they can adsorb water molecules or positive and negative ions, that is, cations and anions, and undergo exchange interactions of these ions with the outside. The clays that we will discuss (Clearfield 1996, Lerf 2000) can swell and thereby increase the space between their layers to accommodate the adsorbed water and ionic species. Figure 10.15b shows positive ions or cations between the layers, and Fig. 10.16 provides a more detailed sketch of an idealized clay structure with the prototype talc formula Mg$_3$Si$_4$O$_{10}$(OH)$_2$. The latter figure shows one and part of a second so-called silicate layer, and indicates the presence between them of exchangeable hydrated cations M^{q+}(H$_2$O)$_x$, such as alkali or transition metal ions, with charge $+q$ and n waters of hydration. In this idealized structure the four oxygen planes of the silicate

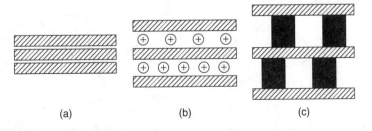

Figure 10.15. Saponite clay layers shown (a) before adsorption, (b) after adsorption of cations, and (c) after addition of pillars.

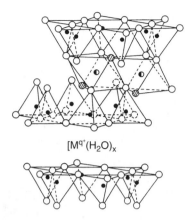

$$[M^{q+}(H_2O)_x$$

Figure 10.16. Smectite clay structure with the ideal talc formula $Mg_3Si_4O_{10}(OH)_2$. (From R. Grim, *Clay Mineralogy*, McGraw-Hill, New York, 1968.)

layer are perfectly flat, with two-thirds of the oxygens in the two centrally located planes replaced by hydroxyl (OH) groups. All the tetrahedral sites are occupied by Si, and all the octahedral sites are occupied by Mg, forming the coordination groups SiO_4 and MgO_6 with the structures sketched in Figs. 10.17a and 10.17b, respectively.

The clays to be discussed here are of the montmorillonite class, and the discussion will center around the particular clay called *saponite*, in which some aluminum (Al^{3+}) replaces silicon in tetrahedral sites, and some divalent iron (Fe^{2+}) replaces magnesium in octahedral sites, corresponding to the typical formula $[Na_{0.3}Ca_{0.015}][Mg_{2.9}Fe^{2+}_{0.1}][Si_{3.6}Al_{0.4}]O_{10}(OH)_2 \cdot 4H_2O$. Each layer is 0.94 nm thick,

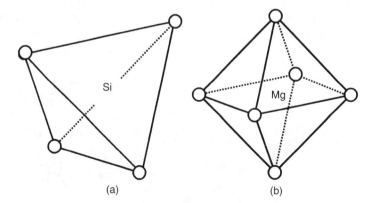

(a) (b)

Figure 10.17. Structures of (a) SiO_4 showing the silicon atom centered in a regular tetrahedron of oxygens and (b) MgO_6 showing the magnesium atom centered in a regular octahedron of oxygens.

with a top and bottom surface area of $660\,m^2/g$. It has been found possible to intercalate or place between the layers bulky ions that form pillars that hold the layers apart, as shown in Fig. 10.15c, and thereby provide a system of spaces or pores where various small molecules can reside. The dimensions of the pores produced by this layering process are in the low-nanometer range. These materials are called pillared inorganic layered compounds (PILCs).

The pillaring is often carried out with the aid of the positively charged Al_{13} Keggin ion, which has the formula $[Al_{13}O_4(OH)_{24}(H_2O)_{12}]^{7+}$, or alternately $\{AlO_4[Al(OH)_2H_2O]_{12}\}^{7+}$. The aluminum atom groups depicted by circles (i.e., spheres) are arranged in the close-packed structure sketched in Fig. 10.18, which depicts a midplane containing six visible and one centrally located but obscured atom group, plus three such atom groups above them. The three below are not shown. In this ion the centrally located aluminum is tetrahedrally coordinated, and the surrounding 12 aluminums are octahedrally coordinated, as delineated in Fig. 10.19. The Keggin ion is sometimes prepared in $AlCl_3$ solutions by transforming pairs of the trivalent hexaaquo ions $Al(H_2O)_6^{3+}$ to dimers $Al_2(OH)_2(H_2O)_8^{4+}$, which are subsequently coalesced to form the Keggin ion. The presence of this ion can be unambiguously established by an ^{27}Al nuclear magnetic resonance (NMR) measurement because its central tetrahedrally coordinated aluminum ion produces a narrow NMR line that is chemically shifted by 62.8 ppm relative to hexaaquo aluminum $Al(H_2O)_6^{3+}$. The twelve octahedrally coordinated aluminums of the Keggin ion do not appear on the NMR spectrum because they are rapidly quadrupolar relaxed by the aluminum nucleus, which has nuclear spin $I = \frac{5}{2}$, and hence their NMR spectral lines are broadened beyond detection.

One way to carry out the pillaring process is to suspend the clay in a solution containing Keggin ions at a controlled pH (i.e., acidity) and a controlled OH : Al ratio. Some of the initial charge of $+7$ on the ion is compensated for when they bond with the silicate layers to form the pillars. The pillars themselves may be considered as cylinders with a diameter of 1.1 nm. There are about 6.5 saponite unit cells with the oxygen content $O_{20}(OH)_4$ per pillar. Taking the value $A = 3.15\,nm^2$ for the area

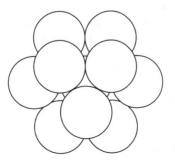

Figure 10.18. A close-packed Al_{13} cluster showing one atom (almost obscured) in the center, six atoms in the plane around it, and three atoms at close-packed sites above it. Three more atoms, not shown, are at close-packed sites in the plane below.

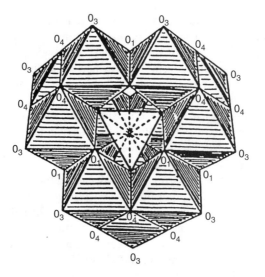

Figure 10.19. Sketch of the structure of the $\{AlO_4[Al(OH)_2H_2O]_{12}\}^{7+}$ Keggin ion with one tetrahedral group AlO_4 in the center position of Fig. 10.18, surrounded by 12 AlO_6 octahedra at the remaining sites of Fig. 10.18, where the oxygens O_i of the octahedra belong to hydroxyl groups OH, or to water molecules H_2O. [From A. Clearfield, in Moser (1996), Chapter 14, p. 348.]

of the silicate layers in 6.5 unit cells, the distance between nearest-neighbor pillars can be estimated. If the pillars are assumed to form a square lattice, then the spacing between the center points of nearest-neighbor pillars is $(A)^{1/2} = 1.77$ nm, and if they are arranged on a regular triangular or hexagonal lattice, then their separation is $(2A/\sqrt{3})^{1/2} = 1.90$ nm. Taking an average of these two values, one obtains a free space of about 0.74 nm between pillars. Experimental measurements indicate that the resulting pillared clay has a basal spacing of $\cong 1.85$ nm, a surface area of $\cong 250 \, \text{m}^2/\text{g}$, and a pore volume of $\cong 0.2 \, \text{cm}^3/\text{g}$.

One important aspect of pillared clays that contributes to their catalytic properties is the presence of acid sites, which can be of the Lewis or Brønsted types, as explained in the previous section. When a pillared clay is heated, the water and hydroxyl groups split out protons to balance the negative charges of the layers as the pillars approach electrical neutrality, and this generates considerable Brønsted acidity. Lewis acid sites are generated on the layers by defect formation, and on the pillars by dehydroxylation, or the removal of OH groups. To confirm the presence of these sites on the PILC surface, the heterocyclic ring compound pyridine (C_5H_5N) was adsorbed and an infrared spectrum was recorded. This spectrum exhibited a strong IR band at 1453 cm^{-1} arising from Lewis acid sites, and a weaker band at 1550 cm^{-1} produced by Brønsted sites.

The discussion until now has centered around the clay saponite pillared by the Keggin ion. Other types of montmorillonite clay materials have been used, and other metal oxide polymers have served as pillars. Examples of nonalumina pillars are based on the proposed zirconium ion $Zr_{18}O_4(OH)_{36}(SO_4)_{14}$, the trivalent titanium ion $[Ti(CH_3COO)_{6.4}(OH)_{0.4}Cl_{1.2}]Cl \cdot 11H_2O$, hexavalent chromium octahedra forming the ions $[Cr_4(OH)_6(H_2O)_{11}]^{6+}$ and $[Cr_4O(OH)_6(H_2O)_{10}]^{5+}$, an alumina–silica Al_2O_3–SiO_2 combination, and silica SiO_2 supplemented by some titania TiO_2. The availability of these nonaluminum pillars, which differ in their dimensions, can provide catalysts with a wide range of pore sizes. These catalysts have been studied for their capability in carrying out various chemical reactions, such as cracking, in which hydrocarbons or other molecules are broken up and their fragments are recombined into desirable product molecules. An example is crude oil and gas cracking to produce gasoline. One of the liabilities of these pillared catalysts is their tendency toward coke formation whereby the surface becomes coated with carbon, and acid sites become deactivated or unable to function.

10.2.5. Colloids

Nanosized particles of metals are ordinarily insoluble in inorganic or organic solvents, but if they can be prepared in colloidal form, they can function more readily as catalysts. A colloid is a suspension of particles in the range from 1 nm to 1 µm (i.e., 1000 nm) in size, larger than most ordinary molecules, but still too small to be seen by the naked eye. Many colloidal particles can, however, be detected by the way they scatter light, such as dust particles in air. These particles are in a state of constant random movement called *Brownian motion* arising from collisions with solvent molecules, which themselves are in motion. Particles are kept in suspension by repulsive electrostatic forces between them. The addition of salt to a colloid can weaken these forces and cause the suspended particles to gather into aggregates, and eventually they collect as a sediment at the bottom of the solvent. This process of the settling out of a colloid is called *flocculation*. Some of the colloidal systems to be discussed are colloidal dispersions of insoluble materials (e.g., nanoparticles) in organic liquids, and these are called *organosols*. Analogous colloidal dispersions in water are called *hydrosols*.

In Section 2.1.3 we discussed the formation of face-centered cubic nanoparticles such as Au_{55} with structural magic numbers of atoms, in this case 55. This nanoparticle has been ligand-stabilized in the form $Au_{55}(PPh_3)_{12}Cl_6$ to make it less reactive, and hence more stable. This sturdiness is brought about by adding atomic or organic groups between the atoms of the cluster, or on their surfaces. These FCC metallic nanoparticles can be stabilized as colloids by the use of surfactants, which can operate, for example, by lowering the surface tension. The ring compounds tetrahydrofuran (THF) and tetrahydrothiophene, with structures sketched in Fig. 10.20, have been used to stabilize metallic nanoparticles as colloids. Figure 10.21 shows a Ti_{13} nanocluster coordinated with the oxygen atoms of six THF molecules in an octahedral configuration. In this cluster the Ti–Ti distance (0.2804 nm) is slightly less than that (0.289 nm) in the bulk metal.

A way to obtain colloidal dispersions in organic liquids is to stabilize a metallic core using a lipophilic surfactant tetraalkylammonium halide NR_4X, where X is a halogen such as chlorine (Cl) or bromine (Br), and R represents the alkyl group

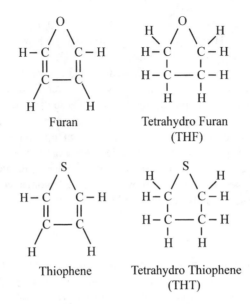

Furan

Tetrahydro Furan
(THF)

Thiophene

Tetrahydro Thiophene
(THT)

Figure 10.20. Sketches of the structures of the heterocyclic ring compound furan, its hydrated analogue tetrahydrofuran (THF), and the corresponding sulfur compounds thiophene and tetrahydrothiophene (THT).

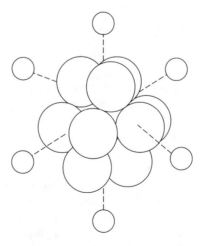

Figure 10.21. Titanium neutral metal cluster Ti_{13} bonded to the oxygen atoms of octahedrally coordinated tetrahydrofuran (C_4H_8O) molecules. Only the oxygen atoms of the THF are shown. [From H. Bönnemann and W. Brijoux, in Moser (1996), Chapter 7, p. 169.]

C_nH_{2n+1}, which is a hydrocarbon radical obtained by removing a hydrogen atom from the terminal carbon atom of a straight-chain or normal alkane molecule, with n ordinarily in the range of 6–20. For example, the hexyl radical with $n = 6$

$$\begin{array}{cccccc} H & H & H & H & H & H \\ | & | & | & | & | & | \\ H-C-&C-&C-&C-&C-&C\cdot \\ | & | & | & | & | & | \\ H & H & H & H & H & H \end{array}$$

is derived from hexane

$$\begin{array}{cccccc} H & H & H & H & H & H \\ | & | & | & | & | & | \\ H-C-&C-&C-&C-&C-&C-H \\ | & | & | & | & | & | \\ H & H & H & H & H & H \end{array}$$

by removing the hydrogen atom on the extreme right, where the dot "·" on the right terminal carbon of hexyl indicates the presence of an unpaired electron on the carbon. To stabilize the metal nanoparticle of the colloid, the NR_4X molecules dissociate into their cation NR_4^+ and anion X^- parts at the surface of the core, as illustrated in Fig. 10.22. Metal cores of various diameters from 1.3 to 10 nm can be obtained by varying the transition metal of the core: Ru (1.3 nm), Ir (1.5 nm), Rh (2.1 nm), Pd (2.5 nm), Co, Ni, Pt (2.8 nm), Fe (3.0 nm), Cu (8.3 nm), and Au (10 nm). The hydrocarbon chains pointing outward from the core, as shown in Fig. 10.22, are lipophilic; hence they attract organic solvent molecules, forming stable dispersions in organic liquids. An analogous colloidal dispersion in water, or a hydrosol, can be made by attaching an SO_3^- group to the end of one of the hydrocarbon chains of each NR_4^+ ion to make it hydrophilic or water-attracting, as shown in Fig. 10.23. This particular metal-stabilizing hydrophilic compound is called a *sulfobetaine*.

Figure 10.22. Metallic colloidal particle stabilized by tetraalkylammonium halide NR_4X molecules. [From H. Bönnemann and W. Brijoux, in Moser (1996), Chapter 7, p. 173.]

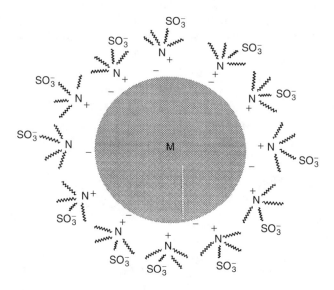

Figure 10.23. Metallic colloidal particle stabilized by sulfobetaine compounds related to NR_4X, but with an SO_3^- group at the end of one of the alkyl chains R. [From H. Bönnemann and W. Brijoux, in Moser (1996), Chapter 7, p. 174.]

FURTHER READING

A. Clearfield, "Preparation of Pillared Clays and Their Catalytic Properties," in Moser (1996), Chapter 14, p. 345.

A. Lerf, *Intercalation Compounds in Layered Host Lattices: Supramolecular Chemistry in Nanodimensions*, in Nalwa (2000), Vol. 5, Chapter 1, p. 1.

W. R. Moser, ed., *Advanced Catalysts and Nanostructured Materials*, Academic Press, San Diego, 1996.

H. S. Nalwa, ed., *Handbook of Nanostructured Materials and Nanotechnology*, Vols. 1–5, Academic Press, Boston, 2000.

S. Sugano and H. Koizumi, *Microcluster Physics*, Springer, Berlin, 1998.

W. H. Weinberg, C. M. Reaves, B. Z. Nosho, R. I. Pelzel, and S. P. DenBaars, *Strained-Layer Heteroepitaxy to Fabricate Self-Assembled Semiconductor Islands*, in Nalwa (2000), Vol. 1, Chapter 6, p. 300.

J. L. Wilber and G. M. Whitesides, in *Nanotechnology*, G. Timp, ed., Springer-Verlag, Berlin, 1999, Chapter 8.

11

ORGANIC COMPOUNDS AND POLYMERS

11.1. INTRODUCTION

Nanoparticles have been prepared from various types of large organic molecules, as well as from polymers formed from organic subunits, and in this chapter some of these nanostructures are described. Organic compounds are those compounds that contain the atom carbon (C), with a few exceptions such as carbon monoxide (CO), carbon dioxide (CO_2), and carbonates (e.g, $CaCO_3$), which are classified as inorganic compounds Almost all organic compounds also contain hydrogen atoms (H), and those that contain only carbon and hydrogen are called *hydrocarbons*.

Before proceeding to discuss organic nanoparticles it will be helpful to present a little background material on some of the chemical concepts that will be used. Carbon has a valence of 4 and hydrogen has a valence of 1, so they can be written

$$-\overset{|}{\underset{|}{C}}- \qquad H- \qquad (11.1)$$

Introduction to Nanotechnology, by Charles P. Poole Jr. and Frank J. Owens.
ISBN 0-471-07935-9. Copyright © 2003 John Wiley & Sons, Inc.

to display their chemical bonds indicated by horizontal ($-$) and vertical ($|$) dashes. The compound methane, CH_4, which has a structural formula that may be written in one of two ways

$$
\begin{array}{ccc}
\begin{array}{c}
H \\
| \\
H-C-H \\
| \\
H
\end{array}
& \text{or} &
\begin{array}{c}
H \\
HCH \\
H
\end{array}
\end{array}
\qquad (11.2)
$$

is one of the simplest organic compounds, and it is a gas at room temperature. Figure 11.1 shows structural formulas of some examples of other hydrocarbons. The linear pentane molecule C_5H_{12} has all single bonds, and the π-conjugated compound butadiene C_4H_6 has alternating single and double bonds. The simplest aromatic compound, or π-conjugated ring compound, that is, one with alternating single and double bonds, is benzene C_6H_6, and the figure shows two ways to represent it. It can also be expressed in the form $H\phi$, where the phenyl group ϕ is a benzene ring that is missing a hydrogen atom, corresponding to $-C_6H_5$, where the dash ($-$) denotes an incomplete chemical bond. Naphthalene ($C_{10}H_8$) is the simplest condensed ring (fused ring) aromatic compound. The bottom panel of Fig. 11.1 shows the triple-bonded compounds acetylene and diacetylene. Aromatic compounds can contain other atoms besides carbon and hydrogen, such as chlorine (Cl), nitrogen (N), oxygen (O), and sulphur (S), as well as atomic groups or radicals such as amino ($-NH_2$) nitro ($-NO_2$), and the acid group $-COOH$.

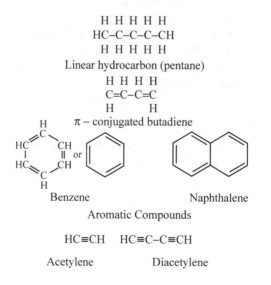

Figure 11.1. Examples of organic molecules. Acetylene and diacetylene serve as monomers for the formation of polymers.

The present chapter emphasizes experimental aspects of polymeric nanoparticles. Insights can also be obtained about these nanoparticles by computer simulation and modeling, and the article by Sumpter et al. (2000) can be consulted for some details on this approach.

11.2. FORMING AND CHARACTERIZING POLYMERS

11.2.1. Polymerization

A polymer is a compound of high molecular weight that is formed from repeating subunits called *monomers*. The monomer generally has a parent compound with a double chemical bond that opens up to form a single bond during the polymerization reaction that forms the polymer. For example, consider the chemical compound styrene $\phi CH{=}CH_2$, which has the structural formula

$$
\begin{array}{cc}
\text{H} & \text{H} \\
\text{C} & = & \text{C} \\
\phi & \text{H}
\end{array}
\tag{11.3}
$$

Opening up the double bond forms the monomer

$$
\begin{array}{cc}
\text{H} & \text{H} \\
-\text{C}-\text{C}- \\
\phi & \text{H}
\end{array}
\tag{11.4}
$$

and a sequence of many of these monomers forms the linear polymer

$$
\begin{array}{c}
\text{H H H H H H H H H H H H H H H H} \\
\text{R}-\text{C}-\text{C}-\text{C}-\text{C}-\text{C}-\text{C}-\text{C}-\text{C}-\text{C}-\text{C}-\text{C}-\text{C}-\text{C}-\text{C}-\text{C}-\text{C}-\text{R}' \\
\phi\ \text{H}\ \phi\ \text{H}\ \phi\ \text{H}\ \phi\ \text{H}\ \phi\ \text{H}\ \phi\ \text{H}\ \phi\ \text{H}\ \phi\ \text{H}
\end{array}
\tag{11.5}
$$

which can be written in the abbreviated form

$$
\text{R}\left[\begin{array}{c} \text{H H} \\ -\text{C}-\text{C}- \\ \phi\ \text{H} \end{array}\right]_n \text{R}'
\tag{11.6}
$$

where in the structural formula (11.5) the index $n = 8$. The groups R and R' are added to the ends to satisfy the chemical bonds of the terminal carbons. This particular compound, (11.6), is called *polystyrene*, and it is referred to as a *linear polymer*, although in reality it is staggered since the chemical bonds C—C—C in the carbon chain subtend the tetrahedral angle $\theta \cong 109°28'$ between them.

The more complex monomer

$$
\begin{array}{cc}
\text{H} & \text{A} \\
-\text{C}-\text{C}- \\
\text{H} & \text{B}
\end{array}
\qquad (11.7)
$$

forms the polymer

$$
R\left[\begin{array}{cc}
\text{H} & \text{A} \\
-\text{C}-\text{C}- \\
\text{H} & \text{B}
\end{array}\right]_n R'
\qquad (11.8)
$$

When A is the methyl radical $-CH_3$ and B is the radical $-COOCH_3$, then the polymer is polymethyl methacrylate, which was mentioned in Chapters 9 and 12. It generally has a molecular weight between 10^5 and 10^6 Daltons (Da, g/mol), and since the molecular weight of the monomer $CH_2=C(CH_3)CO_2CH_3$ is 100 Da, there are between 1000 and 10,000 monomers in this polymer. Polymers are also formed from a number of other radicals, such as allyl $(CH_2=CHCH_2-)$ and vinyl $(CH_2=CH-)$. Natural rubber is a polymer based on the isoprene molecule:

$$
\begin{array}{ccc}
\text{H} & \text{H} & \text{CH}_3 \\
| & | & | \\
\text{H}-\text{C}=\text{C}-\text{C}=\text{CH}_2
\end{array}
\qquad (11.9)
$$

which has a pair of double bonds. The second double bond is used to form crosslinkages between the linear polymer chains. The methyl group $-CH_3$ at the top of the molecule is an example of a side chain on an otherwise linear molecule.

11.2.2. Sizes of Polymers

Polymers are generally classified by their molecular weight, and to discuss them from the nanoparticle aspect, we need a convenient way to convert molecular weight to a measure of the polymer size d. The volume V in the units cubic nanometers (nm^3) of a substance of molecular weight M_w and density ρ is given by

$$
V = 0.001661 \, \frac{M_w}{\rho}
\qquad (11.10)
$$

where M_w has the unit dalton or g/mol (grams per mole) and ρ has the conventional units g/cm^3. If the shape of the nanoparticle is fairly uniform, with very little stretching or flattening in any direction, then a rough measure of its size is the cube root of the volume (11.10), which we call the size parameter d:

$$
d = 0.1184 \left(\frac{M_w}{\rho}\right)^{1/3} \, nm
\qquad (11.11)
$$

This expression is exact for the shape of a cube, but it can be used to estimate average diameters of polymers of various shapes. If the molecule is a sphere of diameter D_0, then we know from solid geometry that its volume is given by $V = \pi D_0^3/6$, and inserting this in Eq. (11.10) provides the expression $d_{SPH} = D_0 = 0.1469 \, (M_w/\rho)^{1/3}$ nm for a spherical molecule. For a molecule shaped like a cylinder of diameter D and length (or height) L with the same volume as a sphere of diameter D_0, we have the expression $\pi D_0^3/6 = \pi D^2 L/4$, which gives

$$D_0 = \left(\frac{3}{2}\right)^{1/3} (D^2 L)^{1/3} = \left(\frac{3}{2}\right)^{1/3} D \left(\frac{L}{D}\right)^{1/3} = \left(\frac{3}{2}\right)^{1/3} L \left(\frac{D}{L}\right)^{2/3} \qquad (11.12)$$

These equivalent relationships permit us to write expressions for the diameter and the length of the cylinder in terms of its length : diameter ratio, and the molecular weight of the molecule

$$D = 0.128 \left(\frac{M_w}{\rho}\right)^{1/3} \left(\frac{D}{L}\right)^{1/3} \qquad (11.13)$$

$$L = 0.128 \left(\frac{M_w}{\rho}\right)^{1/3} \left(\frac{L}{D}\right)^{2/3} \qquad (11.14)$$

where D and L have the units of nanometers. These expressions are plotted in Figs. 11.2 and 11.3 for $D > L$ and $L > D$, respectively. The figures can be employed to estimate the size parameter for an axially shaped, flat or elongated, polymer if its molecular weight, density, and length : diameter ratio are known. The curves in these figures were drawn for the density $\rho = 1 \, g/cm^3$, but the correction for the density is easily made since most polymer densities are close to 1. Typical polymers have molecular weights in the range from 10^4 to 10^7 Da.

11.3. NANOCRYSTALS

11.3.1. Condensed Ring Types

A great deal of work has been done in the preparation, testing, and applications of inorganic nanocrystals, especially those of the semiconductor type, such as CdS, CdSe, and GaAs, as well as Ag- and Au-doped glasses, but much less effort has been expended in the study of organic nanocrystals. Examples of some organic compounds that have been used by Kasai et al. (2000) to prepare these nanocrystals are given in Fig. 11.4. The compounds listed at the top part of Table 11.1 were readily prepared by a reprecipitation method that involves pouring a rich solution into a poor solvent, which is generally water, during vigorous agitation such as sonication. After the pouring there are initially widely scattered droplets that gather

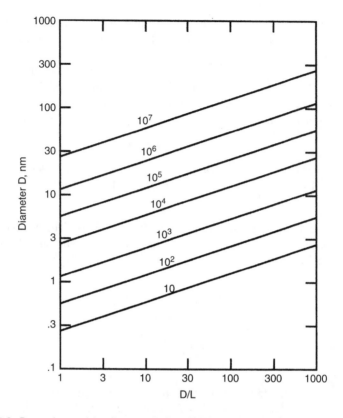

Figure 11.2. Dependence of the diameter D of a cylindrical polymer on its diameter : length ratio D/L for molecular weights from 10 to 10^7 Da, as indicated on the curves. A density $\rho = 1\,g/cm^3$ was assumed in Eq. (11.13) for plotting these curves.

into dispersed clusters that undergo a process of nucleation and growth, until they finally produce the nanocrystals.

As these crystallites form, they scatter light, and the intensity of the scattered light $I_s(t)$ relative to the incident light intensity I_0, after a time t has elapsed, can be used to monitor the rate at which the growth takes place. Figure 11.5 plots the normalized scattered light intensity $I_s(t)/I_0$ versus the time, and establishes that the growth is much faster at higher temperatures. This time dependence of $I_s(t)/I_0$ follows the expression $[1 - \exp(\alpha_{app}t)]^2$, where the growth rate constant α_{app} depends on the temperature in the manner shown in Fig. 11.6. The linearity of this latter plot, which is called an *Arrhenius plot*, provides the activation energy for the crystal growth process, and for perylene nanocrystals this is 68 kJ/mol. The activation energy is the minimum amount of energy that must be supplied for the nanocrystals to form. The size of the crystallite can be regulated by varying the concentration, temperature, and mixing procedure, and also by the use of surfactants that modify the surface of the particles, or reduce the surface tension of the solution. In the particular case of

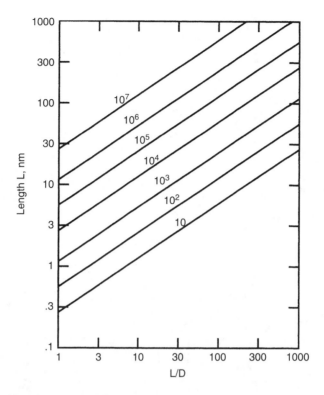

Figure 11.3. Dependence of the length L of a cylindrical polymer on its length : diameter ratio L/D for molecular weights from 10 to 10^7 Da, as indicated on the curves. A density $\rho = 1\,\text{g/cm}^3$ was assumed in Eq. (11.14) for plotting these curves.

perylene the size of the resulting nanocrystals is almost the same for growth at various temperatures.

Many π-conjugated organics in single-crystal and thin-film configurations exhibit third-order nonlinear optical properties that makes them useful for converting visible light frequencies to higher frequencies in the ultraviolet region of the spectrum. They also have an ultrafast response time, so they can be used for switching light beams on and off by the use of optical Kerr shutters in which a strong applied electric field activates their birefringence (double refraction). Single crystals containing individual conjugated polymer chains that stretch across the entire length of the crystallite should constitute favorable material for molecular device design. Exciton spectroscopy indicates the presence, in nanocrystals of perylene, pyrene, and anthracene, of quantum confinement effects of the type discussed in Chapter 9. In bulk perylene crystals it is the self-trapped exciton that provides the luminescence at $\lambda = 560\,\text{nm}$ because the free exciton state is so much less stable. However, in nanocrystals free exciton luminescence is observed with a shift in wavelength from $\lambda = 470$ to $482\,\text{nm}$ for a size change from 50 to 200 nm.

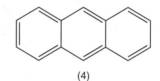

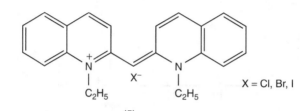

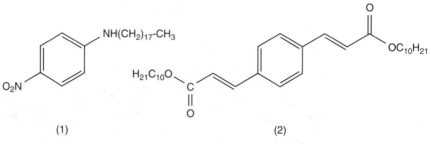

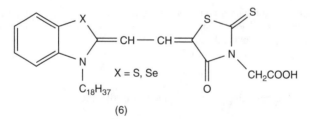

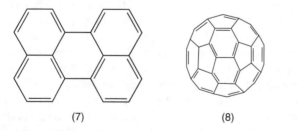

Figure 11.4. Chemical structures of π-conjugated organic compounds used to prepare organic nanocrystals: (1) *N*-octadecyl-4-nitroaniline, (2) didecyl-1,4-phenylenediacrylate, (3) naphthalene, (4) anthracene, (5) pseudocyanine rhodanine dye (PIC), (6) merocyanine rhodanine dye, (7) perylene, and (8) fullerene C_{60}. [From H. Kasai et al., in Nalwa (2000), Vol. 5, Chapter 8, p. 440.]

Table 11.1. Nanocrystals prepared by recrystallization of four π-conjugated compounds listed in Fig. 11.4, and by recrystallization of four diacetylene polymers with the side chains listed in Fig. 11.7

Material	Type	Crystallite size
Anthracene $C_{14}H_{10}$	π-Conjugated compound	150 nm–1 µm
PIC	π-Conjugated compound	200–300 nm
Perylene $C_{20}H_{12}$	π-Conjugated compound	50–200 nm
Fullerene C_{60}	π-Conjugated compound	200 nm
4-BCMU	Polydiacetylene	200 nm–1 µm
DCHD	Polydiacetylene	15 nm–1 µm
DCHD	Polydiacetylene	1-µm-long, 60-nm-diameter microfiber
14-8ADA	Polydiacetylene	15–200 nm

Source: Data were obtained from H. Kasai et al., in Nalwa (2000), Vol. 5, Chapter 8, p. 441.

11.3.2. Polydiacetylene Types

The nanocrystals listed toward the bottom of Table 11.1 were prepared by recrystallization from a polydiacetylene polymer with a backbone formed from the diacetylene molecule, which has the structure sketched at the lower part of Fig. 11.1. The starting compound is $RC{\equiv}C{-}C{\equiv}CR'$, where examples of the R and

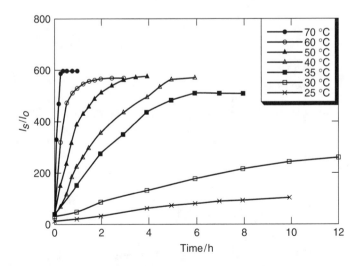

Figure 11.5. Growth of the scattered light intensity $I_s(t)/I_0$ from perylene particles dispersed in water at various temperatures during the reprecipitation process for the formation of nano-crystals, where I_0 is the incident light intensity [From H. Kasai et al., in Nalwa (2000), Vol. 5, Chapter 8, p. 450.]

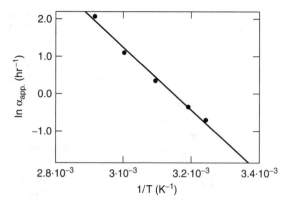

Figure 11.6. Arrhenius plot of the dependence on the temperature of the logarithm of the growth rate constant α_{app} for the growth of perylene nanocrystals. The slope of this plot provides the activation energy for the crystal growth reprecipitation process. [From Kasai et al. (2000), p. 452.]

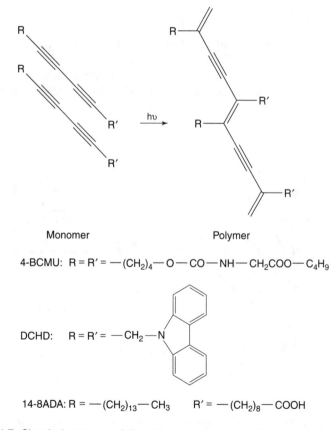

Figure 11.7. Chemical structures of diacetylene monomers used to form polyacetylene nanocrystals. [From Kasai et al. (2000), p. 441.]

R′ groups are given in Fig. 11.7. The monomer for the polymer is formed by interchanging triple and single bonds in the following manner

$$
\begin{matrix} \text{R} & & \text{R}' \\ \text{C} \equiv \text{C} - \text{C} \equiv \text{C} \end{matrix} \xrightarrow{h\nu} \begin{matrix} \text{R} & & \text{R}' \\ = \text{C} - \text{C} \equiv \text{C} - \text{C} = \end{matrix} \qquad (11.15)
$$

where the resulting open bonds (=) at the ends of the structure are used to attach successive monomers to each other during the polymerization. The polymerization process in the solid state is brought about by the application of heat, ultraviolet light, or γ irradiation (via γ-rays), as indicated by the photon notation $h\nu$ above the arrow in Eq. (11.15). The polymer backbone is conjugated by its system of alternating single, double and triple bonds. Both the solid forms and the solutions of diacetylene polymers exhibit many bright colors: red, yellow, green, blue, and gold.

Diacetylene polymers have the capability of forming perfect crystals in the solid state, with every polymer chain reaching from one end of the crystal to the other. This occurs when the crystal size is less than the usual length of the polymer in a bulk material, and it causes the polymer molecular weight to depend on the nanocrystal size. A typical molecular weight is 10^6 Da. Figure 11.8 shows a 130-nm rectangular microcrystal of the polymer 4-BCMU, which has the chemical structure given in Fig. 11.7. The compound DCHD was found to form both nanocrystals like the one illustrated in Fig. 11.8, as well as nanofibers about 7 µm long, with diameters of ~ 60 nm.

These materials have a number of important applications, such as in nonlinear optics. Small nanocrystals of the polydiacetylene compound DCHD exhibit quantum size effects, with the excitonic absorption peak of 70-, 100-, and 150-nm crystals

Figure 11.8. Scanning electron microscope picture of a poly(4-BCMU) single nanocrystal about 130 nm in size. [From Kasai et al. (2000), p. 443.]

appearing at wavelengths of 640, 656, and 652 nm, respectively, which is the expected shift toward lower energies (i.e., longer wavelengths) with increasing particle size.

11.4. POLYMERS

11.4.1. Conductive Polymers

Many nanoparticles are metals, such as the structural magic number particle Au_{55}. The metals under consideration in their bulk form are good conductors of electricity. There are also polymers, called *conductive polymers* or *organic metals*, which are good conductors of electricity, and polyacetylene is an example. Many polyaniline-based polymers are close to silver in the galvanic series, which lists metals in the order of their potential or ease of oxidation.

Acetylene HC≡CH has the monomer

$$\begin{array}{cc} H & H \\ -C=C- \end{array} \tag{11.16}$$

corresponding to the repeat unit $[-CH=CH-]_n$. Other examples of compounds that produce conducting polymers are the benzene derivative aniline $C_6H_5NH_2$, which may also be written ϕNH_2, and the two 5-membered ring compounds pyrrole (C_4H_4NH) and thiophene (C_4H_4S). The structure of thiophene is sketched in Fig. 10.20, and pyrrole has the same structure with the sulfur atom S replaced by a nitrogen atom N bonded to a hydrogen H. These molecules all have alternating double–single chemical bonds, and hence they form polymers that are π-conjugated. The π conjugation of the carbon bonds along the oriented polymer chains provides pathways for the flow of conduction electrons, and hence it is responsible for the good electrical conduction along individual polymer nanoparticles. Polarons, or electrons surrounded by clouds of phonons, may also contribute to this intrinsic conductivity. The overall conductivity, however, is less than this intrinsic conductivity, and must take into account the particular nature of the polymer.

Wessling (2000) has proposed an explanation of the high electrical conductivity of conductive polymers such as polyacetylene and polyaniline on the basis of their nanostructure involving primary particles with a metallic core of diameter $\cong 8$ nm surrounded by an amorphous nonconducting layer 0.08 nm thick of the same $[C_2H_2]_n$ composition. Figure 11.9 presents a sketch of the model proposed by Wessling based on scanning electron microscope pictures of conductive polymers. The individual nanoparticles are seen joined together in networks comprising 30–50 particles, with branching every 10 or so particles. Several of the nanoparticles are pictured with their top halves removed to display the inner metallic core, and their surrounding amorphous coating. The electrical conductivity mechanism is purely metallic within each particle, and involves thermally activated tunneling of the

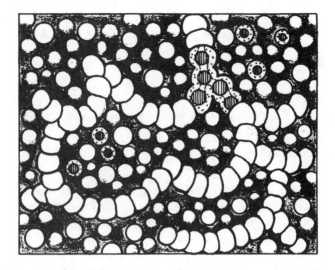

Figure 11.9. Sketch of the nanostructure of a polyacetylene conductive polymer showing the $\cong 9.6$-nm-diameter nanoparticles. The top halves of several of these nanoparticles have been removed to display the $\cong 8$-nm-diameter metallic core and the $\cong 0.8$-nm-thick surrounding amorphous coating. This illustration was reconstructed from scanning electron microscope pictures. [From B. Wessling, in Nalwa (2000), Vol. 5, Chapter 10, p. 512.]

electrons responsible for the passage of electrical current through the outer amorphous layer from one particle to the next. Thus bulk conductive polymers are truly nanomaterials because of their $\cong 10$-nm microstructure. In many cases it is easier to prepare conductive polymers in the nanoparticle range of dimensions than it is to prepare conventional metal particles in this size range.

Polyaniline and its analogs change color with the application of particular voltages and suitable chemicals; that is, they are electrochromic and chemochromic. This makes them appropriate candidates for use in light-emitting diodes (LEDs). Other applications are the surface finish of printed-circuit boards, corrosion protection for metal surfaces, semitransparent antistatic coatings for electronic products, polymeric batteries, and electromagnetic shielding.

11.4.2. Block Copolymers

We have seen that a polymer is a very large molecule composed of a chain of individual basic units called *monomers* joined together in sequence. A copolymer is a macromolecule containing two or more types of monomers, and a block copolymer has these basic units or monomer types joined together in long individual sequences called blocks (Liu 2000). Of particular interest is a diblock polymer $(A)_m(B)_n$, which contains a linear sequence of m monomers of type A joined through a transition section to a linear sequence of n monomers of type B. An example of a diblock polymer is *polyacetylene–transition section–polystyrene* with the following structure

[End group]—[polyacteylene]—[transition member]—[polystyrene]—[end group]

$$(11.17)$$

which for a particular case may be written in a more detailed manner as

$$C_2H_5-[-CH=CH-]_m-CH_2-C\phi_2-[-C\phi H-CH_2-]_n-CH_3 \quad (11.18)$$

where the end groups have been chosen as the ethyl $-C_2H_5$ and methyl $-CH_3$ radicals, and the transition member that joins the two polymer sequences is the chemical group $-CH_2-C\phi_2-$. Of greater practical importance are more complex copolymers that contain several or many monomer sequences of the types $(A)_m$ and $(B)_n$.

If the conditions are right, then individual polymers are able to self-assemble to produce copolymers. In many cases one polymer component is water-soluble and the other is not. Some examples of nanostructures fabricated from copolymers are hairy nanospheres, star polymers, and polymer brushes, which are illustrated in Fig. 11.10. The nanosphere can be constructed from one long polymer $(A)_m$ that coils up and develops crosslinks between adjacent lengths of strands to give rigidity to the sphere. The projections from the nanosphere surface are sets of the other copolymer element $(B)_n$ attached to the spherical surface formed from polymer $(A)_m$. If the lengths of the projections formed by $(B)_n$ are short compared to the sphere diameter, the

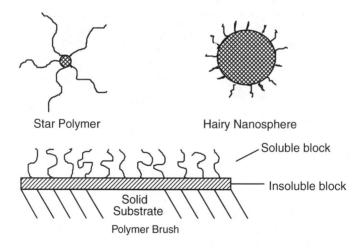

Figure 11.10. Sketch of a star polymer (top left), a hairy nanosphere (top right), and a polymer brush (bottom). [From G. Liu, in Nalwa (2000), Vol. 5, Chapter 9, pp. 479, 488.]

nanostructure is called a "hairy nanosphere," whereas if the sphere is small and the projections are long, it is called a "star polymer". If the nanosphere is hollow, then fibers of the polymers $(B)_n$ can also project inward from the inner surface of the spherical shell. Copolymers can be used to construct structures that resemble the micelles discussed in Section 12.4.2.

Star polymers are used in industry to improve the melt strength, that is, the mechanical properties of molten plastic materials. Hairy nanospheres have been employed for the removal of organic compounds from water, both in a dispersed form and as solid microparticles. Polymer brushes are effective for dispersing latex and pigment particles in paint. Nanostructures consisting of block copolymers function as catalysts, are utilized in the production of nanosized electronic devices, and find applications for water reclamation. Otsuka et al. (2001) pointed out that block copolymers adsorbed on surfaces in brush or micelle forms, or self-assembled into micelles, provide a powerful tool for manipulating the characteristics of surfaces and interfaces. An example from this article is described in Section 11.5.4. Block copolymers are expected to have novel applications, especially of the biomedical type.

11.5. SUPRAMOLECULAR STRUCTURES

11.5.1. Transition-Metal-Mediated Types

Supramolecular structures are large molecules formed by grouping or bonding together several smaller molecules. In this section we will follow Stang and Olenyuk (2000) and discuss the assembly of supramolecular structures containing transition metals in the form of molecular squares with a high degree of symmetry. Analogous patterns have been synthesized in the shapes of equilateral triangles, pentagons, hexagons, and even three-dimensional octahedra. These configurations can often be constructed by the process of self-assembly. The use of self-assembly procedures in industry could lead to lower manufacturing costs for chemical products.

A square supramolecular structure can be fabricated by starting with an angular subunit and combining it with either a linear subunit, or another angular subunit, in the manner sketched in Fig. 11.11. The former process, outlined in Fig. 11.12, was used to produce the assembly shown in Fig. 11.13 in which either palladium (Pd) or platinum (Pt) is the transition metal. Eight nearby singly charged counterions $^-OSO_2CF_3$ compensate for the +2 charges on each of the four metal ions M^{2+}. The latter process for forming an approximately square molecule, outlined in Fig. 11.14, produced the molecular square with the structure sketched in Fig. 11.15. This figure lists the bond lengths and bond angles, and indicates that the Pd—Pd and Pt—Pt separation distances are 1.4 nm and 1.3 nm, respectively. Once again, the +2 charge on each Pd ion is balanced by four nearby $^-OSO_2CF_3$ counterions. The overall geometry of the center square is nearly flat, with only minor deviations from a perfect plane. The stacking diagram of Fig. 11.16 clarifies how adjacent molecules fit together in space.

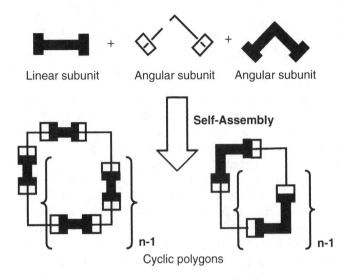

Figure 11.11. Cyclic polygons (bottom) constructed from building blocks (top) by the use of linear and angular subunits (lower left), and by the use of the two types of angular subunits (lower right). [From P. J. Stang and B. Olenyuk, in Nalwa (2000), Vol. 5, Chapter 2, p. 169.]

11.5.2. Dendritic Molecules

We are all familiar with the branching construction of a tree, how one trunk forms several large branches, each large branch forms additional smaller branches, and so on. The roots of the tree exhibit the same branched mode of growth. This type of architecture is a fractal one, associated with a space whose dimensions are not an integer such as 2 or 3, but rather the dimensionality is a fraction. There are molecules

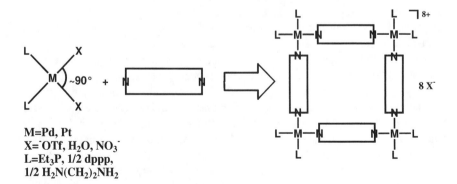

Figure 11.12. Programmed self-assembly of four linear and four 90° angular subunits to form a molecular square. [From Stang and Olenyuk, Nalwa (2000), p. 171.]

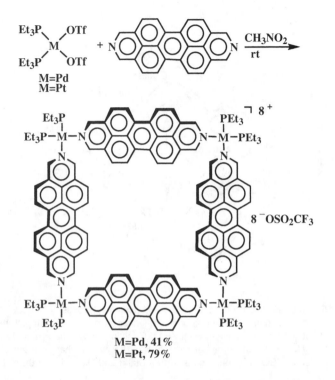

Figure 11.13. Molecular square formed by the self-assembly process of Fig. 11.11 from the subunits acyclic bisphosphane on the upper left and 2,9-diazabenzo[*cd,lm*]perylene on the upper right, where Et denotes an ethyl group —C_2H_5, and M can be either a Pd or a Pt atom. The overall +8 charge of the structure is balanced by eight —OSO_2CF_3 counterions, as indicated. [From Stang and Olenyuk, Nalwa (2000), p. 173.]

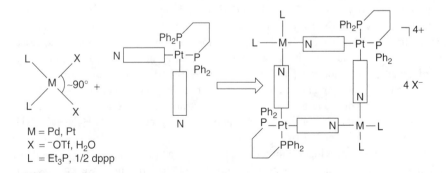

Figure 11.14. Programmed self-assembly of two types of angular subunits to form a molecular square [From Stang and Olenyuk (2000), p. 180.]

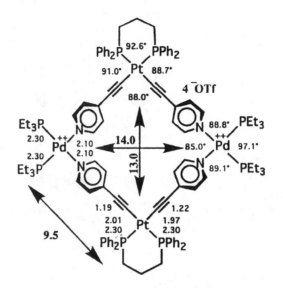

Figure 11.15. Molecular square formed by the self-assembly process of Fig. 11.14 showing bond angles, bond lengths, and interatomic distances expressed in angstrom (Å) units (10 Å = 1 nm), obtained from a single-crystal X-ray diffraction study. The symbol Et denotes an ethyl group $-C_2H_5$ and Ph, a phenyl group $-C_6H_5$. The overall charge of the structure +4 is balanced by four $-OSO_2CF_3$ counterions. [From Stang and Olenyuk (2000), p. 181.]

called *dendrimers* or *cascade molecules*, which form by this type of cascade process. See Archut and Vögtle (2000) for more details.

The first dendrimer preparation scheme, outlined in Fig. 11.17, started with three different diamine compounds: diamine itself (H_2N-NH_2), *m*-phenylenediamine, and 2,6-diaminopyridine. Initially the starting compound, $H_2N-X-NH_2$, was reacted with vinylcyanide ($H_2C=CH-CN$) to replace the hydrogens of the amino groups (NH_2) with cyano groups CN, and form the double cyanide derivative $(CN)_2N-X-N(CN)_2$. This derivative compound was then reacted with sodium borohydride ($NaBH_4$) using a Co^{2+} catalyst to form the second-generation compound $(H_2N)_2N-X-N(NH_2)_2$ of the dendrimer iteration sequence. Figure 11.17 outlines the iteration sequence leading up to the formation of the third-generation compound $[(H_2N)_2N]_2N-X-N[N(NH_2)_2]_2$ of the sequence (not shown in the figure). The notation is to call the compound obtained by two iterations the third-generation compound. The chemical structures for X in the *m*-phenylene and the 2,6-pyridine varieties of these diamine starting compounds $H_2N-X-NH_2$ are shown in the lower left of Fig. 11.17.

Figure 11.18 provides another example of a polyaminoamine dendrimer (PAMAM). To estimate the size of a dendrimer such as this, we can use the crystallographic data of Wyckoff (1966) for the three normal (i.e., unbranched) aliphatic compounds C_nH_{2n+2} with $n = 8, 18, 36$. The average transverse unit-cell

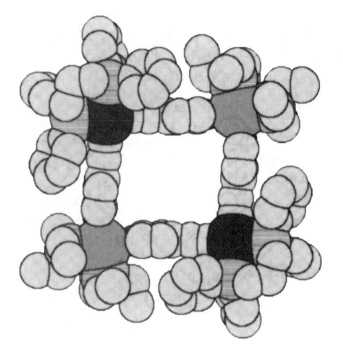

Figure 11.16. Space-filling model of the molecular square of Fig. 11.15, constructed from X-ray diffraction data. [From Stang and Olenyuk (2000), p. 182.]

dimensions are $a_0 = 0.415$ nm and $b_0 = 4.96$ nm corresponding to the cross-sectional area $a_0 b_0 = 0.206$ nm^2. The length of the unit cell c_0 is proportional to the number of carbon atoms n, and has the average value 0.137 nm/carbon. This is consistent with the ~ 2.2-nm extension of the hexadecanethiol [$CH_3(CH_2)_{10}S-$] compounds in the self-assembled monolayer discussed in Section 10.1.3. Each PAMAM monomer has five carbons and two nitrogens corresponding to $n = 7$, which gives a length of 0.96 nm. Taking into account bending at the splitting points or bifurcations, the radius increases by perhaps 1.3 nm per generation, which gives a total of 13 nm for 10 generations. Thus dendrimers of this type have sizes typical of nanoparticles.

The dendrimers discussed so far are ones in which the number of terminal groups doubles at each branching point. The polyamine of Fig. 11.17 grows by the series 2,4,8,16,..., and the polyamidoamine of Fig. 11.18 grows in accordance with the sequence 3,6,12,24,.... This continuous doubling is referred to as *divergent growth*, and the process is termed *divergent synthesis*. Each main branching complex emanating from the core is called a *wedge*, which means that the polyamine dendrimer has two wedges, and the polyamidoamine has three wedges. Thus a typical dendrimer consists of a central core plus two, three or more wedges, each of which ends with an outer region or periphery consisting of terminal groups.

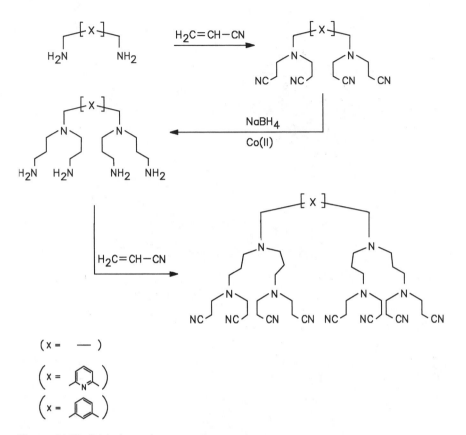

Figure 11.17. Original step-by-step self-assembly of a polyamine dendrimer by alternating between steps that replace the hydrogens of amino groups by cyanide groups [$-NH_2 \Rightarrow -N(CN)_2$], and then add hydrogens to the cyanide nitrogens [$-CN \Rightarrow -CNH_2$]. [From E. Buhleier, W. Wehner, and F. Vögtle, *Synthesis* **155** (1978).]

Since the steps of a dendrimer synthesis repeat, as occurs in the diamine case with the alternating vinylcyanide/sodium–borohydride reactions shown in Fig. 11.17, the growth process leading to the final array of terminal groups is related to, or perhaps analogous to, the self-assembly processes discussed in Chapter 10. To obtain a functionally useful molecule, the branching or bifurcation process can be terminated after several generations, with the ending groups serving as receptors for the attachment of functional groups such as catalysts, molecular switches, or light-sensitive chromophors. A *chromophor* is a compound which becomes colored when it is exposed to visible or ultraviolet light.

Dendrimers with terminal groups containing catalytic sites are sometimes referred to as *dendralysts*. An example is the Si core dendrimer, which has a carbosilane skeleton constructed from silane molecules SiH_4 by replacing the hydrogen atoms H of SiH_4 by carbon atoms C of hydrocarbon molecules C_nH_{2n+2}. These long-chain

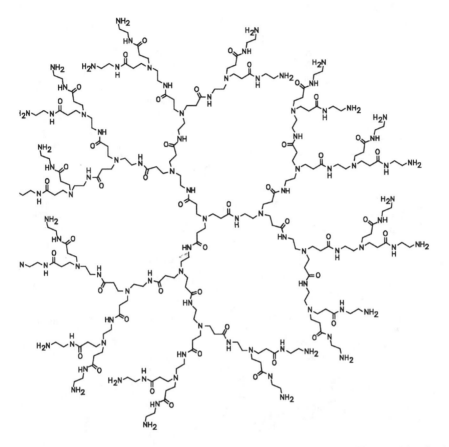

Figure 11.18. Fifth-generation polyaminoamine (PAMAM) dendrimer. [Prepared by D. A. Tomalia, H. Baker, J. R. Dewald, M. Hall, G. Kallos, S. Martin, J. Roeck, J. Ryder, and P. Smith, *Polym. J.* **17**, 117 (1985).]

hydrocarbon molecules are represented by zigzag lines in Fig. 11.19. This dendrimer has aryl–nickel complexes (aromatic compounds containing Ni) bound at the periphery of the carbosilane skeleton. It is capable, for example, of catalyzing the particular chemical reaction called the Kharash addition of tetrachloromethane (CCl_4) to the polymer precursor material methyl methacrylate ($CH_2{=}C(CH_3)CO_2CH_3$). Because of their large size, dendralysts like the one illustrated in Fig. 11.19 are easily separated from the reaction mixture after the end of the reaction, which would not be the case if the catalytic sites were situated on much smaller molecules or solid polymer supports. Thus dendralysts, which are really heterogeneous catalysts, can function in solution as homogeneous catalysts (see Section 10.2.1). Functional groups can also be associated with the core of the dendrimer, with the surrounding bifurcation regions serving as a shield for the active group of the core region.

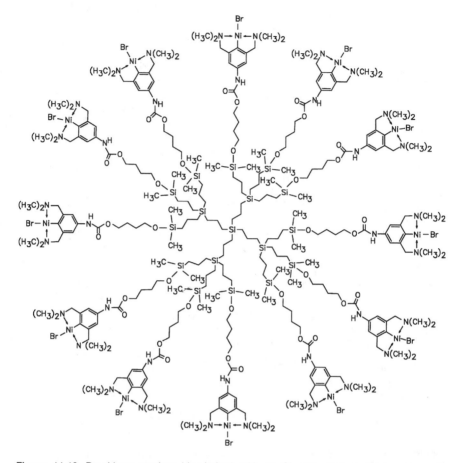

Figure 11.19. Dendrimer catalyst (dendralyst) with an Si core, and terminal group aryl–nickel complexes as the catalytically active functional groups. [From J. W. J. Knapen, A. W. van der Made, J. C. de Wilde, P. W. N. M. van Leeuwenn, P. Wijkens, D. M. Grove, and G. van Koten, *Nature* **372**, 659 (1994).]

11.5.3. Supramolecular Dendrimers

Individual dendritic compounds can be linked together to form larger structures called *supramolecular dendrimers*, and this linking procedure can be carried out by a process called *supramolecular self-assembly*. As an example consider the dendrimer in the upper left corner of Fig. 11.20, in which the core consists of a central benzene ring associated with a complicated aromatic, polycyclic complex attached above it, plus two branched wedges shown bifurcated below it. To simplify the notation, this dendrimer is presented in abbreviated form in the upper right corner with a boldfaced triangle representing the complicated complex attached to the central benzene ring. The lower part of the figure shows a supramolecular dendritic complex

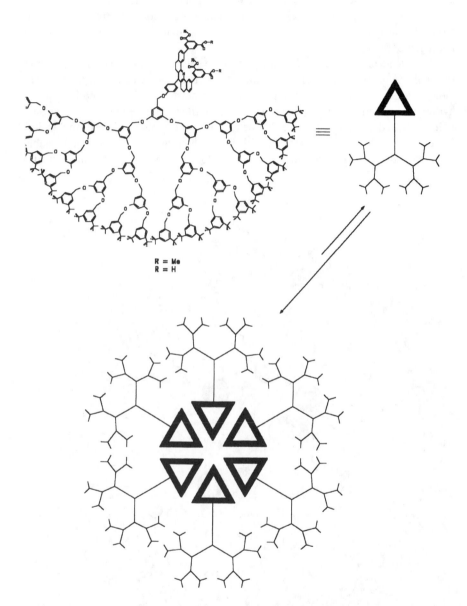

R = Me
R = H

Figure 11.20. Sketch of sixfold (hexamer) supramolecular dendritic complex showing the structural formula of an individual component dendrimer at the upper left, the same component in a compact notation at the upper right, and the final self-assembled configuration of six components at the bottom. [Adapted from A. Archut and F. Vögtle, in Nalwa (2000), Vol. 5, Chapter 5, p. 367.]

called a hexamer self-assembled via hydrogen bonding, with six of the original dendrimers forming its wedges. Supramolecular dendritic structures with high molecular weights have also been prepared by utilizing metal coordination and hydrogen bonding.

Many dendrimers tenaciously hold back some solvent, and some of them can trap molecules such as radicals, charged moieties (parts of molecules), and dyes. When molecules of different sizes are trapped, they can be selectively released by gradual hydrolysis (reaction with water) of the dendrimer outer and middle layers. Figure 11.21 depicts a dendrimer that has trapped two sizes of "guest" molecules. Dendrimers of this type can be useful for prolonging the lifetime of unstable chemical molecules. The torroidal shaped ß-cyclodextrin molecule with the structure

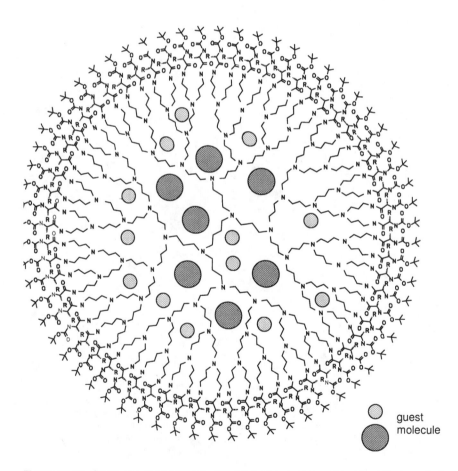

guest
molecule

Figure 11.21. Small and large "guest" molecules shown enclosed in 'dendritic boxes' of a polyamine dendrimer. [From Archut and Vögtle (2000), p. 364.]

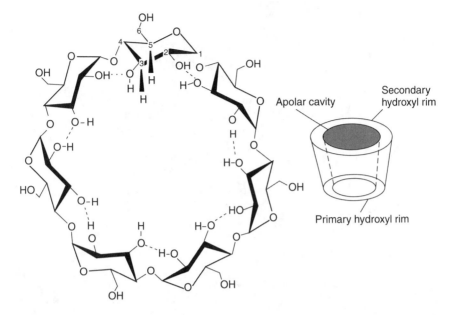

Figure 11.22. Chemical structure (left) and sketch of torroidal shaped cavity (right) of the B-cyclohydextrin polymer. [From F. J. Owens and S. Bulusu (to be published).]

sketched in Fig. 11.22 has a hydrophobic central cavity that has a range of inner radii from 0.5 to 0.8 nm, depending on the number of D-glycosyl units in the cyclic polymer chain. The number can vary from six to eight or more. This molecule is able to trap energetic molecules such as trinitroazedine and remove them from water effluents so that they will not contaminate the environment.

11.5.4. Micelles

Micelles, described in Section 12.4.2, are globular configurations or arrays of molecules containing hydrophobic (water-avoiding) tails that form a cluster on the inside, and hydrophilic (water-seeking) heads that point outward toward the surrounding water solvent. Dendrimers have been made that are equivalent to unimolecular micelles with an inner structure of (mostly) hydrophobic hydrocarbon chains, and an outer region or periphery with hydrophilic terminal groups. Figure 11.23 sketches the structure of a *micellanoic acid dendrimer*, which is a synthetic micelle containing terminal hydrophilic acid groups (—COOH). Inverse dendrimer micelles have also been synthesized that have an interior structure that is water-seeking, and an exterior or peripheral structure that is water-avoiding.

We shall conclude this chapter with an example of a block polymer that is formed into a micelle that modifies a surface. The block copolymer is formed from two

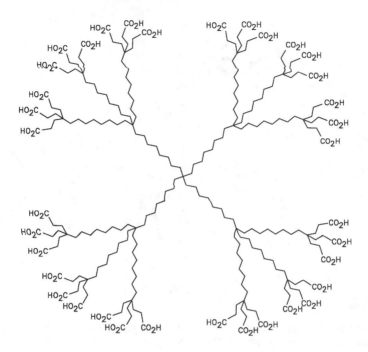

Figure 11.23. Dendritic micelle with an internal structure of hydrophobic hydrocarbon chains depicted as wavy lines, and hydrophilic acid groups —COOH attached around the outer perimeter. [From Archut and Vögtle (2000), p. 368.]

segments linked to each other. One is a polylactide (PLA) polymer with a terminal polymerizable double bond, and the other is a polyethylene glycol (PEG) polymer with an acetal $[CH_3CH(OC_2H_5)_2]$ functional group at its outer or distal end, as shown in the center part of Fig. 11.24. The upper part of the figure shows how the polylactide components form into a substrate matrix with a layer of polyethylene glycol segments comprising a dense polymer block above it of the type illustrated at the bottom of Fig. 11.10. The functional groups that terminate the polyethyene glycol segments are shown reacting with specific cells and proteins. The lower part of the figure shows how dialysis, or membrane separation in water, is used to form spherical micelles from the block copolymer with acetal or aldehyde (—CHO) functional groups on the surface. The micelles are used to modify a surface, as shown.

Electron paramagnetic resonance (EPR) was employed to study the aldehyde (—CHO) groups on a copolymer surface with the aid of the 2,2,6,6-tetramethyl-1-piperidineloxyl (TEMPO) derivative spin-label compound that has the abbreviated formula sketched in Fig. 11.25. The three-line EPR spectrum characteristic of the spin label that is shown in part (a) of the figure indicated the conjugation or attachment of the spin label TEMPO to the aldehyde group at the end of a

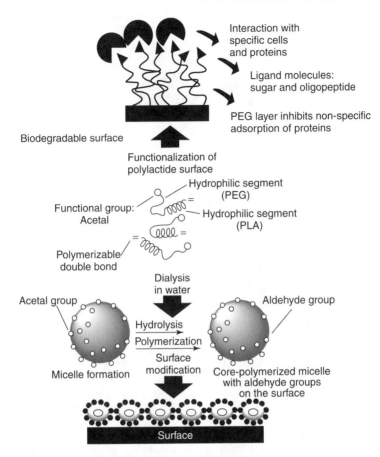

Figure 11.24. Schematic representation of an aggregation of polylactide (PLA)/polyethylene-glycol (PEG) block copolymers (center of figure). The upper illustration shows how these copolymers have formed a PEG layer that inhibits the nonspecific adsorption of proteins. The lower illustration shows how these copolymers form micelles that coat a surface. [From Otsuka et al. (2001).]

polyethylene glycol (PEG) segment on the surface. When the acetal surface was treated with TEMPO prior to replacing acetal groups with aldehydes, Fig. 11.25b indicates that only a very weak EPR triplet spectrum was observed, probably due to the direct physical adsorption of TEMPO molecules on the surface. Figure 11.25c shows that no EPR spin-label signal appears when the aldehyde surface was treated with a variety of TEMPO that lacked an amino ($-NH_2$) group. The upper part of the figure shows how the TEMPO molecule bonds to the aldehyde ($-CHO$) group located at the end of the ethylene glycol copolymer segment.

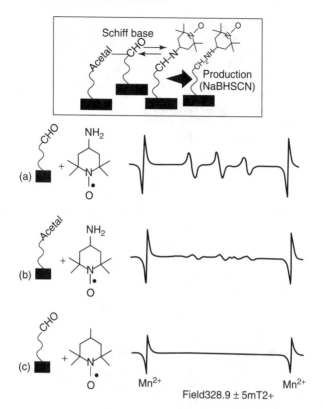

Figure 11.25. Electron paramagnetic resonance spectra (lower right) of a PEG-PLA surface containing acetal or aldehyde groups after interaction with the spin label TEMPO, which produces an EPR triplet spectrum. Spectra are illustrated for (a) an aldehyde surface interacting with 4-amino TEMPO, (b) an acetal surface interacting with 4-amino TEMPO, and (c) an aldehyde surface interacting with TEMPO, which lacks an amino group —NH$_2$. The strong outer calibration lines with a 90 mT spacing on the EPR spectra are due to the presence of Mn^{2+} ions. [From Otsuka et al. (2001).]

FURTHER READING

A. Archut and F. Vögtle, "Dendritic Molecules—Historic Development and Future Applications," in Nalwa (2000), Vol. 5, Chapter 5, p. 333.

H. Kasai, H. S. Nalwa, S. Okada, H. Oikawa, and H. Nakanishi, "Fabrication and Spectrographic Characterization of Organic Nanocrystals," in Nalwa (2000), Vol. 5, Chapter 8, p. 334.

G. Liu, "Polymeric Nanostructures," in Nalwa (2000), Vol. 5, Chapter 9, p. 475.

H. S. Nalwa, ed., *Handbook of Nanostructured Materials and Nanotechnology* Vol. 5, *Organics, Polymers and Biological Compounds*, Academic Press, Boston, 2000.

H. Otsuka, Y. Nagasaki, and K. Kataoka, *Materials Today* (3), 30, (2001).

F. J. Owens and S. Bulusu, "Nanoporous Cyclodextrin Polymers for Removal of Contaminant Molecules from Effluents," to be published.

P. J. Stang and B. Olenyuk, "Transition-Metal-Mediated Self-Assembly of Discrete Nanoscopic Species with Well-Defined Shapes and Geometries," in Nalwa (2000), Vol. 5, Chapter 2, p. 167.

B. G. Sumpter, K. Fukui, M. D. Barnes, and D. W. Noid, *Materials Today* (2), 323, (2000).

B. Wessling, "Conductive Polymers as Organic Nanometals," in Nalwa (2000), Vol. 5, Chapter 10, p. 501.

R. W. G. Wyckoff, *Crystal Structures*, Vol. 5, Wiley, New York, 1996.

12

BIOLOGICAL MATERIALS

12.1. INTRODUCTION

It is customary to define nanoparticles or nanostructures as entities in the range of sizes from 1 to 100 nm, so many biological materials are classified as nanoparticles. Bacteria, which range in size between 1 and 10 μm, are in the mesoscopic range, while viruses with dimensions from 10 to 200 nm are at the upper part of the nanoparticle range. Proteins, which ordinarily come in sizes between 4 and 50 nm, are in the low nanometer range. The building blocks of proteins are 20 amino acids, each about 0.6 nm in size, which is slightly below the official lower limit of a nanoparticle. More than 100 amino acids occur naturally, but only 20 are involved in protein synthesis. To construct a protein, combinations of these latter amino acids are tied together one after the other by strong peptide chemical bonds and form long chains called *polypeptides* containing hundreds, and in some cases thousands, of amino acids; hence they correspond to nanowires. The polypeptide nanowires undergo twistings and turnings to compact themselves into a relatively small volume corresponding to a polypeptide nanoparticle with a diameter that is typically in the range of 4–50 nm. Thus a protein is a nanoparticle consisting of a compacted polypeptide nanowire. The genetic material desoxyribonucleic acid (DNA) also has the structure of a compacted nanowire. Its building blocks are four nucleotide molecules that bind together in a long double-helix nanowire to form chromosomes, which in humans contain about 140×10^6 nucleotides in sequence. Thus the DNA

Introduction to Nanotechnology, by Charles P. Poole Jr. and Frank J. Owens.
ISBN 0-471-07935-9. Copyright © 2003 John Wiley & Sons, Inc.

molecule is a double nanowire, two nucleotide nanowires twisted around each with a repeat unit every 3.4 nm, and a diameter of 2 nm. This long double-stranded nanowire also undergoes systematic twistings and turnings to become compacted into a chromosome about 6 μm long and 1.4 μm wide. The chromosome itself is not small enough to be a nanoparticle; rather, it is in the mesoscopic range of size.

To gain some additional perspective about the overall scope of nanometer-range sizes involved in the buildup of biological structures, let us consider the human tendon as a typical structure (Tirrell 1994). The function of a tendon is to attach a muscle to a bone. From the viewpoint of biology, the fundamental building block of a tendon is the assemblage of amino acids (0.6 nm) that form the gelatinlike protein called *collagen* (1 nm), which coils into a triple helix (2 nm). There follows a threefold sequence of fiberlike or fibrillar nanostructures: a microfibril (3.5 nm), a subfibril (10–20 nm), and a fibril itself (50–500 nm). The final two steps in the buildup, specifically, the cluster of fibers called a *fascicle* (50–300 μm) and the tendon itself (10–50 cm), are far beyond the nanometer range of sizes. The fascicle is considered mesoscopic and the tendon, macroscopic in size. Since the smallest amino acid, glycine, is ~0.42 nm in size, and some viruses reach 200 nm, it seems appropriate to define a biological nanostructure as being in the nominal range from 0.5 to 200 nm. With this in mind, the present chapter focuses on nanometer-size constituents of biological materials. In addition, we also comment on some special cases in which artificially constructed nanostructures are of importance in biology. See Gross (1999) for some additional discussions of biological nanostructures.

12.2. BIOLOGICAL BUILDING BLOCKS

12.2.1. Sizes of Building Blocks and Nanostructures

There are a number of ways to determine or estimate the size parameters d of the fundamental biological building blocks, which are amino acids for proteins and nucleotides for DNA. If the crystal structure is known for the building-block molecule, and there are n molecules in the crystallographic unit cell, then one can divide the unit cell volume V_U by n and take the cube root of the result to obtain an average size or average dimension:

$$d = \left(\frac{V_U}{n}\right)^{1/3} \tag{12.1}$$

If the crystal structure is orthorhombic (see Section 3.3), then the unit cell is a rectangular box of length a, width b, and height c with the volume $V_U = a \times b \times c$, to give for the average size of the molecule $d = (a \times b \times c/n)^{1/3}$, where n is the number of molecules in the cell. In a typical case n is 2 or 4. For the higher-symmetry tetragonal case we set $a = b$ in this expression, and the cubic case for $n = 1$ has the special result $a = b = c = d$. One can also deduce the size by reconstructing the molecule from knowledge of its atomic constitution, taking into account the lengths and angles of the chemical bonds between its atoms.

Another common way to determine the size of a biological molecule is to observe it using an electron microscope, which can provide images called *electron micrographs* taken from various molecular orientations. This approach is especially useful for larger nanosized objects such as proteins or viruses. Figure 12.1 presents a micrograph of the poliomyelitis virus enlarged 74,000 times, and Fig. 12.2 shows a bacteriophage attacking a bacterium at an enlargement of 41,600×. A bacteriophage is a type of virus that attacks and infects bacteria. The poliomyelitis virus has a diameter of 30 nm, and the bacteriophage shown in Fig. 12.2 has a 40-nm head attached to a tail that is 100 nm long and 13 nm wide. The bacterium in the illustration is 1600 nm (i.e., 1.6 μm) long and 360 nm wide. These figures confirm our earlier observation that viruses are nanoparticles, and that bacteria are mesoscopic, beyond the nanoparticle size range. Figure 12.3 provides sketches of the shapes of four well-known proteins with dimensions ranging from 4 to 76 nm.

Many biological macromolecules such as proteins are characterized by their molecular weight M_W, and it was shown in Section 11.2.2 that the size d of a molecule or nanoparticle is related to its molecular weight M_W and its density ρ through the expression

$$d = 0.1184 \left(\frac{M_W}{\rho} \right)^{1/3} \text{nm} \tag{12.2}$$

where M_W is expressed in daltons (Da, g/mol) and ρ in the usual units g/cm^3. This formula assumes that the nanoparticle is fairly uniform in shape, with very little stretching or compression in any direction. If the molecule is flat, or perhaps elongated like γ-globulin or fibrinogen, which have the shapes depicted in Fig. 12.3, then either Fig. 11.2 or Fig. 11.3 can be used to deduce the size.

Crystallographic data can be used to calculate the density of amino acids, and the *Handbook of Chemistry and Physics* reports the densities 1.43, 1.607, and

Figure 12.1. Micrograph of poliomyelitis virus, with an amplification of 74,000× [From R. C. Williams, in A. Nason, *Textbook of Modern Biology*, Wiley, New York, 1965, p. 81.]

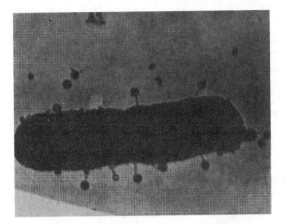

Figure 12.2. Micrograph of bacteriophages attacking a bacterium, with an amplification of 41,580×. [From T. F. Anderson, in A. Nason, *Textbook of Modern Biology*, Wiley, New York, 1965, p. 82.]

1.316 g/cm³ for the amino acids alanine, glycine, and valine, respectively and, of course, $\rho = 1$ g/cm³ for water. Proteins have a less compact structure and hence lower values of ρ than their constituent amino acids. On the basis of these considerations, we arrive at the following approximate expression

$$d = 0.12(M_W)^{1/3} \text{ nm} \tag{12.3}$$

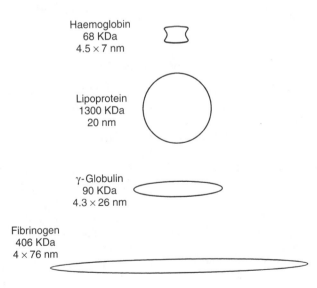

Figure 12.3. Approximate sizes and molecular weights of four proteins.

which we will use to estimate the sizes of biological macromolecules. For example, the protein haemoglobin, which has a molecular weight $M_W = 68,000$ Da, has the length parameter $d = 4.8$ nm, well within the nanoparticle range.

The twistings and turnings of polypeptide nanowires to form a compact protein structure held together by weak hydrogen and disulfide ($-S-S-$) bonds can be somewhat loose, with spaces present between the polypeptide nanowire sections, so the density of the protein is less than that of its constituent amino acids in the crystalline state. This can cause Eq. (12.3) to underestimate the size parameter of a protein. If the molecular weight of a protein is known and the volume is determined from an electron micrograph such as those pictured in Figs. 12.1 and 12.2, then the density ρ (in g/cm^3) can be calculated by an inversion of Eq. (11.10)

$$\rho = 0.001661 \frac{M_W}{V} \tag{12.4}$$

where the molecular weight M_W is in daltons, and the volume V is in cubic nanometers.

Table 12.1 lists the molecular weights M_W and various length parameters d for a number of biological nanoparticles, Fig. 12.3 provides estimated molecular weights and dimensions of four proteins, and Table 12.2 lists sizes for biological structures and quantities that are larger than nanoparticles, in the micrometer region. All the amino acids have the common structure sketched in Fig. 12.4, with the acid or carboxyl group $-COOH$ at one end, and an adjacent carbon atom that is bonded to a hydrogen atom, an amino group NH$_2$, and a group R that characterizes the particular amino acid. Figure 12.5 presents the structures of six of the amino acids, including the smallest acid, glycine, for which the R group is simply a hydrogen atom H, and the largest tryptophan in which R is a conjugated double-ring system. The structures of the nucleotide building blocks of DNA and RNA are presented in Section 12.3.1.

12.2.2. Polypeptide Nanowire and Protein Nanoparticle

Figure 12.6 illustrates the manner in which amino acids combine together in chains through the formation of a peptide bond. To form this bond, the hydroxy ($-OH$) of the carboxyl group of one amino acid combines with the hydrogen atom H of the amino group of the next amino acid, with the establishment of a C$-$N peptide bond accompanied by the release of water (H$_2$O), as displayed in the figure. The figure shows the formation of a tripeptide molecule, and a typical protein is composed of one or more very long polypeptide molecules. Small peptides are called *oligopeptides*, and amino acids incorporated into polypeptide chains are often referred to as *amino acid residues* to distinguish them from free or unbound amino acids. The protein haemoglobin, for example, contains four polypeptides, each with about 300 amino acid residues.

The stretched-out polypeptide chain, of the type shown at the top of Fig. 12.7, is called the *primary structure*. To become more compact locally, the chains either coil up in a what is called an *alpha helix* (α helix), or they combine in sheets called *beta*

Table 12.1. Typical sizes of various biological substances in the nanometer range

Class	Material	M_w (Da)	Size d (nm)
Amino acids	Glycine (smallest amino acid)	75	0.42
	Tryptophan (largest amino acid)	246	0.67
Nucleotides	Cytosine monophosphate (smallest DNA nucleotide)	309	0.81
	Guanine monophosphate (largest DNA nucleotide)	361	0.86
	Adenosine triphosphate (ATP, energy source)	499	0.95
Other molecules	Steric acid $C_{17}H_{35}CO_2H$	284	0.87
	Chlorophyll, in plants	720	1.1
Proteins	Insulin, polypeptide hormone	6,000	2.2
	Hemoglobin, carries oxygen	68,000	7.0
	Albumin, in white of egg	69,000	9.0
	Elastin, cell-supporting material	72,000	5.0
	Fibrinogen, for blood clotting	400,000	50
	Lipoprotein, carrier of cholesterol (globular shape)	1,300,000	20
	Ribosome (where protein synthesis occurs)		30
	Glycogen granules of liver		150
Viruses	Influenza		60
	Tobacco mosaic, length		120
	Bacteriophage T_2		140

sheets (β sheets) held together by hydrogen bonds, as shown in Fig. 12.7b. The sheets might also be called *nanofilms*. These two configurations constitute what is referred to as the secondary structure. An overall compactness is achieved by a tertiary structure consisting of a series of twistings and turnings, held in place by disulfide bonds, as shown in Fig. 12.7c. If there is more than one polypeptide present, they position themselves relative to each other in a quaternary structure, as shown in Fig. 12.7d. This type of quaternary structure packing is a characteristic of globe-shaped proteins. Some proteins are globular, and others are elongated, as

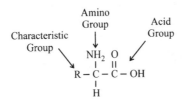

Figure 12.4. Chemical structure of an amino acid.

Table 12.2. Typical sizes in micrometers of various biological substances in the mesoscopic range

Class	Material	Size d (μm)
Organelles (structures in cells outside nucleus)	Mitochondrion, where aerobic respiration produces ATP molecules	$0.5 \times 0.9 \times 3$
	Chloroplast, site of photosynthesis, length	4
	Lysosome (vesicle with enzymes for digesting macromolecules)	0.7
	Vacuole of amoeba	10
Cells	*Escherichia coli* (*E. coli*) bacterium, length	8
	Human blood platelet	3
	Leukocytes (white blood cells), globular shape	8–15
	Erythrocytes (red blood cells), disk shape	1.5×8
Miscellaneous	Human chromosome	9
	Fascicle in tendon	50–300

illustrated in Fig. 12.3. It is clear from the two bottom sketches of Fig. 12.7 that the tertiary and quaternary structures are not very closely packed, so the density is lower than that of the amino acids in the crystalline state, as was mentioned above. In practice, some of the space within a protein molecule residing in the cytoplasm of a cell will contain water of hydration between the twistings and turnings. We conclude from these considerations that the structure of protein nanoparticles is often complex.

12.3. NUCLEIC ACIDS

12.3.1. DNA Double Nanowire

The basic building block of DNA, which is a nucleotide with the chemical structure sketched in Fig. 12.8, is more complex than an amino acid. It contains a five-membered desoxyribose sugar ring in the center with a phosphate group (PO_4H_2) attached at one end, and a nucleic acid base R attached at the other end. The figure also indicates by arrows on the left side the attachment points to other nucleotides to form the sugar–phosphate backbone of a DNA strand. Figure 12.9 presents the structures of the four nucleotide bases that can attach to the sugar on the upper right of Fig. 12.8. It is clear from a comparison of Figs. 12.5 and 12.9 that the nucleic acid base molecules are about the same sizes as the amino acid molecules. The

NAME	SYMBOL	M_w, Da size d, nm	RNA WORDS (CODONS)	STRUCTURE
Glycine	Gly	75.07 Da 0.42 nm	GGU GGA GGC GGG	
Alanine	Ala	89.09 Da 0.47 nm*	GCU GCA GCC GCG	
Valine	Val	117.12 Da 0.54 nm	GUU GUA GUC GUG	
Threonine	Thr	119.12 Da 0.54 nm	ACU ACA ACC ACG	
Glutamate (Glutamic acid)	GluN	147.13 Da 0.70 nm	GAA GAG	
Tryptophan	Try	246.27 Da 0.90 nm	UGG	

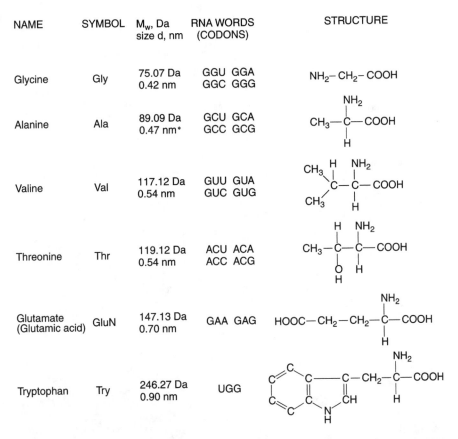

Figure 12.5. Names, symbols, molecular weights M_w, size parameters d, three-letter RNA genetic codewords (codons), and structures of 6 of the 20 amino acids, with the smallest amino acid, glycine, at the top, and the largest, tryptophan, at the bottom.

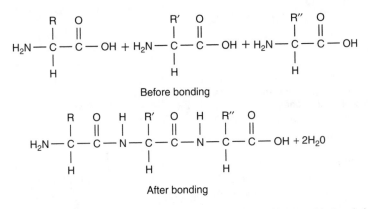

Before bonding

After bonding

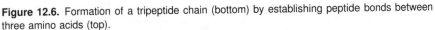

Figure 12.6. Formation of a tripeptide chain (bottom) by establishing peptide bonds between three amino acids (top).

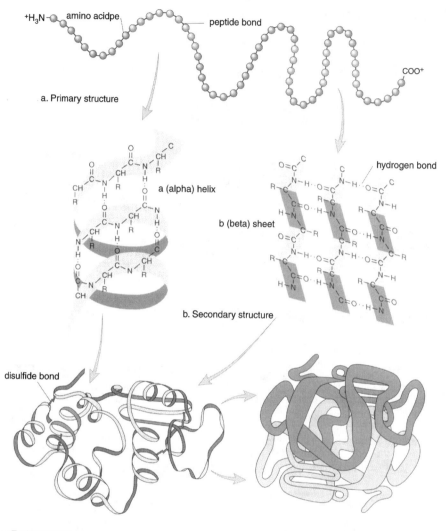

Figure 12.7. The four levels of structure of a protein: (a) primary structure of a stretched-out polypeptide, (b) secondary structures of an α helix (left) and a β sheet (right), (c) tertiary structure of a polypeptide held together by disulfide bonds (—S—S—), and quaternary structure formed by two polypeptides. [From S. S. Mader, *Biology*, McGraw-Hill, Boston, 2001, p. 47.]

attachment points where the bases bond to the desoxyribose sugar are indicated by vertical arrows in Fig. 12.9, and the bases pair off with each other in the double-stranded DNA in accordance with the horizontal arrows in the figure; that is, cytosine (C) pairs with guanine (G), and thymine (T) pairs with adenine (A), in the manner shown in Fig. 12.10. We see from this figure that the acid–phosphate group

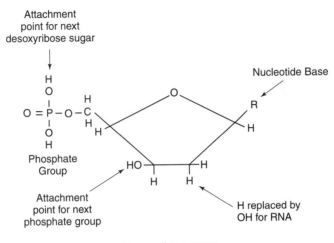

Figure 12.8. Structure of a nucleotide molecule, showing the points of attachment for the next ribose sugar (upper left) and the next phosphate group (lower left), the location of the nucleotide base (upper right), and the hydrogen atom H to be replaced by an hydroxyl group —OH to convert the desoxyribose sugar to ribose.

and the sugar group parts of adjacent nucleotides bond together to form the sugar–phosphate backbone of a DNA strand, resulting in a macroscopically long double-stranded molecule. The complementary base pairs C-G and T-A are held together between the two strands by hydrogen bonds, as shown. Weak hydrogen bonds are used to accomplish this, so the double helix can easily unwind for the purposes of transcription (forming RNA) or replication (duplicating itself). The individual strand is 0.34 nm thick, the double helix has a diameter of 2 nm, and the repeat unit containing 10 nucleotide pairs is 3.4 nm long, as indicated in the upper left of Fig. 12.10. The 0.84-nm size of a nucleotide listed in Table 12.1 is greater than the 0.34 distance between base pairs because, in accordance with Fig. 12.8, the distance between the two attachment points on the nucleotide is much less than the overall length of the molecule. It is also clear from Fig. 12.10 that the pairs of nucleotides stretch lengthwise between the sugar–phosphate backbones of the two DNA strands, resulting in a 2-nm separation between them. To accomplish this coupling together of the two nanostrands in an efficient manner, a small single-ring pyrimidine base always pairs off with a larger two-ring purine base, namely, cytosine with guanine, and thymine with adenine, as indicated in Fig. 12.10.

The 2-nm-wide strands are many orders of magnitude too long to fit lengthwise in the nucleus of a 6-μm-diameter human cell, so they undergo several stages of coiling, depicted in Fig. 12.11. Figure 12.11a shows the double-stranded DNA that we have been describing. The next coiling stage consists of an ∼140-base-pair length of DNA winding around a group of proteins called *histones* to form what is sometimes called a "bead", which has a diameter of 11 nm, as shown in Fig. 12.11b.

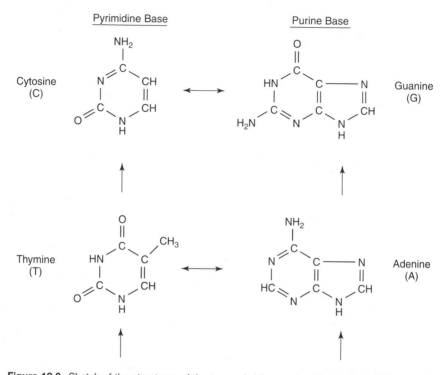

Figure 12.9. Sketch of the structures of the two pyrimidine nucleotide bases, cytosine (C) and thymine (T), and the two purine bases, guanine (G) and adenine (A). The points of attachment on the desoxyribose sugar of Fig. 12.8, entailing the loss of a hydrogen atom H, are indicated by vertical arrows. The horizontal arrows designate the complementary amino acid pairs.

The histone beads are joined together by lengths of double-stranded DNA in the "linker region" between the beads, shown in Fig. 12.11b. The DNA associated with the histones is called *chromatin*, and the histone bead with encircling DNA strands is called a *nucleosome*. The linker regions provide the nucleosome sequence with the great flexibility that is required for subsequent stages of folding.

In the next stage of compaction the nucleosomes stack one above the other, alternating between two coiled columns and connected by the linker strands, as shown laid out lengthwise in Fig. 12.11c. They now form a structure of 1-mm-long chromatin fibers 30 nm in diameter, a configuration called the "packing of the nucleosomes". The chromatin fibers then undergo the next higher order of folding shown in Fig. 12.11d, and this becomes condensed into 700-nm-wide hyperfoldings of the 300-nm-wide foldings, in the manner shown in Fig. 12.11e. These various stages of compaction are held in place largely by relatively weak hydrogen bonds, which makes it feasible for the overall structure to unfold, partially or completely, for replication during cell division, or for transcription during the formation of ribonucleic acid (RNA) molecules, which bring about or direct the synthesis

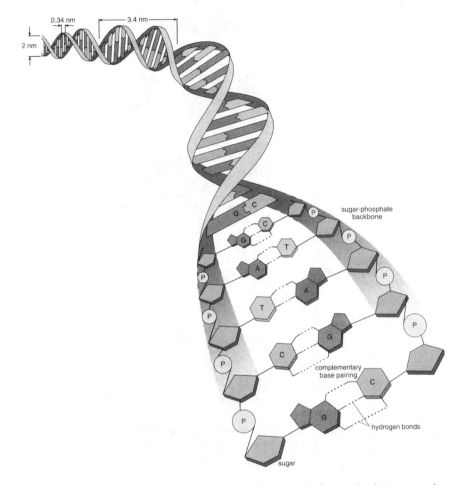

Figure 12.10. Model of the DNA double helix. The width (2 nm), the spacing between nucleo-tides (0.34 nm), and the length of the repeat unit or pitch (3.4 nm) of the helix are indicated in the upper left. The sugar–phosphate backbones, the arrangements of the nucleotides (C, T, G, A), and the four cases of nucleotide base pairing are sketched in the lower right. [From Mader (2001), p. 225.]

of proteins. The final condensed structure of the *metaphase chromosome*, sketched in Fig. 12.11f, is small enough to fit inside the nucleus of a cell.

The genome is the complete set of hereditary units called *genes*, each of which is responsible for a particular structure or function in the body, such as determining the color of one's eyes. In human beings the genome consists of 46 chromosomes (two haploid sets of 23 chromosomes), with an average of about 1600 genes arranged lengthwise along each chromosome. The human haploid genome contains about 3.4×10^9 base pairs, an average of about 150×10^6 base pairs per chromosome.

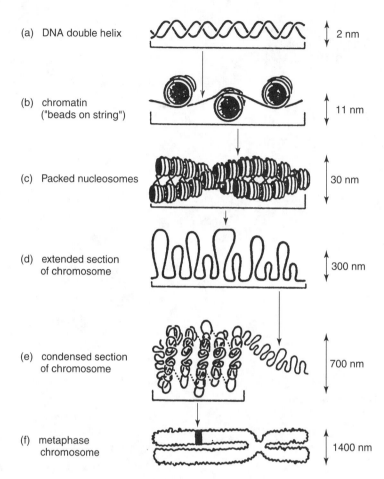

(a) DNA double helix — 2 nm

(b) chromatin
     ("beads on string") — 11 nm

(c) Packed nucleosomes — 30 nm

(d) extended section
     of chromosome — 300 nm

(e) condensed section
     of chromosome — 700 nm

(f) metaphase
     chromosome — 1400 nm

Figure 12.11. Successive twistings and foldings during the packing of DNA into mammalian chromosomes, with the sizes at successive stages given in nanometers. [From R. J. Nossal and H. Lecar, *Molecular and Cell Biophysics*, Addison-Wesley, Boston, 1991, Fig. 4.9 (p. 118).]

12.3.2. Genetic Code and Protein Synthesis

The DNA molecule that we have just described contains the information for directing the synthesis of proteins by the assignment of codewords called *codons* to the individual amino acids, which bond together to form the proteins. One of the DNA strands, called the *coding strand*, stores this information, while the other strand of the pair, called the *complementary strand*, does not have useful data. The information for the synthesis is contained in a sequence of three-letter words using a four-letter alphabet A, C, G, and T based on the four nucleic acid bases adenine (A), cytosine (C), guanine (G), and thymine (T) whose structures are sketched in Fig. 12.9. Since there are three letters in a word and four letters in the alphabet there are $4^3 = 64$ possible words, and 61 of these are used as codewords for amino acids.

DNA is the carrier of heredity in the human body. This double-stranded molecule has a companion single-stranded molecule, *ribonucleic acid* (RNA), which is involved in the synthesis of proteins using the information transcribed or passed on to it from DNA. To form RNA, the DNA uncoils, and sections of the RNA strand are synthesized one nucleotide at a time, in sequence, a process called *transcription*. The RNA strand structure differs from the DNA strand through the replacement of one hydrogen atom (H) of the sugar molecule by an hydroxyl group (OH), thereby forming the sugar ribose (instead of desoxyribose), as indicated at the lower right of Fig. 12.8. RNA also utilizes the nucleotide base uracil with the structure shown in Fig. 12.12 in place of the base thymine. Both of these nucleic acid macromolecules—DNA and RNA—can be classified as nanowires because their diameters are so small and their stretched-out lengths are so much greater than their diameters.

To carry out the synthesis of a particular protein, a segment of the DNA molecule uncoils, and the region of the double helix that stores the codewords for that particular protein serves as a template for the synthesis of a single-stranded messenger RNA molecule (mRNA) containing these codewords. In transcribing the code, each nucleotide base of DNA is replaced by its complementary base on the RNA, with the base uracil substituting for thymine in the RNA. Thus from Fig. 12.10 the transcription takes place by rewriting the codewords in accordance with the scheme

$$
\begin{aligned}
A &\to U \\
C &\to G \\
G &\to C \\
T &\to A
\end{aligned}
\tag{12.5}
$$

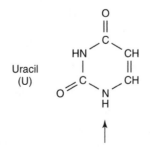

Figure 12.12. Structure of the uracil pyrimidine base nucleic acid that replaces thymine in the RNA molecule. The point of attachment on the ribose sugar of Fig. 12.8, entailing the loss of a hydrogen atom H, is indicated by a vertical arrow.

so each word or codon on the mRNA is a composed of three letters from the set A, C, G, U. Each codon corresponds to an amino acid, and the sequence of codons of the mRNA provides the sequence in which the corresponding amino acids are incorporated into the protein being formed. Since DNA inventories the codewords for thousands of proteins, and mRNA contains the codewords for one protein, the mRNA molecule is very short compared to DNA; it is a short nanowire. The mRNA brings the message of the amino acid sequence from the nucleus of the cell where the transcription from DNA takes place to nanoparticles called *ribosomes* in the cytoplasm region outside the nucleus where the proteins are synthesized. A number of additional protein nanoparticles called *enzymes* function as catalysts for the processes of protein synthesis.

Two of the twenty amino acids have a single word or codon allocated to them, and this word on RNA is UGG for tryptophan, as noted in Fig. 12.5. Other amino acids have two to six words assigned to them, and some examples are given in Fig. 12.5. In the terminology of linguistics, all the words assigned to a particular amino acid are synonyms, and in the language of science the code is called *degenerate* because there are synonyms. Three of the words, namely, UAA, UAG, and UGA, are stop codons used to indicate the end of a specification of amino acids in a part of a polypeptide. For example, the sequence of codons or words ACU   GCA   GGC   UAG corresponds to the triple-amino-acid polypeptide *Thr–Ala–Gly*, where the symbols for these amino acids are given in Fig. 12.5, and the final codon UAG indicates the end of the tripeptide.

12.4. BIOLOGICAL NANOSTRUCTURES

12.4.1. Examples of Proteins

There are many varieties of proteins, and they play a large number of roles in animals and plants. For example, biological catalysts, called enzymes, are ordinarily proteins. In this section we describe several representative proteins.

The protein hemoglobin, with molecular weight 68,000 Da, consists of four polypeptides, each of which has a sequence of about 300 amino acids. Each of these polypeptides contains a heme molecule ($C_{34}H_{32}O_4N_4Fe$) with an iron (Fe) atom that serves as a site for the attachment of oxygen molecules O_2 to the hemoglobin for transport to the tissues of the body. Every red blood cell or erythrocyte contains about 250 million hemoglobin molecules, so each cell is capable of carrying approximately one billion oxygen molecules.

Collagen constitutes 25–50% of all the proteins in mammals. It is the main component of connective tissue, and hence the major load-bearing constituent of soft tissue. It is found in cartilages, bones, tendons, ligaments, skin, and the cornea of the eye. The fibers of collagen are formed by triple helices of proteins containing repetitions of tripeptide sequences *–GlyProXxx–*, where the amino acid denoted by *Xxx* is usually proline (*Pro*) or hydroxyproline (*Hpro*). Collagen differs from other proteins in its high percentage of the amino acids proline (12%) and hydroxyproline

(10%), and it also has a high percentage of glycine (34%) and alanine (10%). The proline and hydroxyproline residues have rigid five-membered rings containing nitrogen, so collagen cannot form an α helix, but these residues do interject bends into the polypeptide chains, facilitating the establishment of conformations that strengthen the triplet helical structure. The smaller glycine residues make it easy to establish close packing of the strands held together by hydrogen bonds. Artificial genes in the size range from 40 to 70 kDa have been prepared that encode tripeptide sequences such as *–GlyProPro–* that resemble those in natural collagen.

The protein elastin possesses many properties similar to those of collagen. It provides elasticity to skin, lungs, tendons, and arteries in mammals. Elastin has a precursor protein called *tropoelastin*, which has a molecular weight of 72 kDa. Its structure involves sequences of oligopeptides, short polypeptides that are only four to nine amino acid residues in length. An analog of such a chain has been synthesized that contains the pentapeptide sequence *–ValProGlyValGly–*, which exhibits a reverse turn around a *ProGly* dipeptide. Additional amino acid residues can occupy the space between turns, and make use of the high flexibility of glycine as a spacer. Many turns joined together adapt a springlike structure that can be stretched to over 3 times its resting length, and then return to rest without experiencing any residual deformation.

The silk caterpillar *Bombyx mori* produces silk formed from hydrogen-bonded β sheets of the protein fibroin. The fibers of the sheets are closely packed and highly oriented, which gives them a large tensile strength. The amino acid residues are 46% glycine, 26% alanine, and 12% serine (*Ser*), and the principal repeat sequence is the hexapeptide *–GlyAlaGlyAlaGlySer–*. Artificial fibers with the properties of silk have been prepared according to this hexapeptide sequence, and proteins of this type with molecular weights between 40 and 100 kDa have been expressed (i.e., formed) in the bacterium *Escherichia coli*. The polypeptide sequences *–ArgGlyAspSer–* from the protein fibronectin and *–ValProGlyValGly–* from elastin and have been incorporated into artificial silk polymers. The former serves as a substrate for cell culture, and the latter renders the polymer more soluble and easier to process. The larvae of midge spiders (*Chironomus tentans*) produce a silklike fiber with a molecular weight of ∼1 MDa.

Deming et al. (1999) have undertaken the de novo design and synthesis of well-defined polypeptides to assess the feasibility of creating novel useful proteins, and to determine the extent to which chain folding and large-scale or supramolecular organization can be controlled at the molecular level. They have incorporated unnatural amino acids into artificial proteins by using intact cellular protein synthesis techniques. "Unnatural" amino acids are those not included in the set of 20 that have DNA codons. These artificial proteins formed folded-chain, layered, or lamellar crystals of controlled surface configuration and thickness. One such crystal was prepared via sequence-controlled crystallization by employing a gene that incorporated the genetic codewords for 36 repeats of the octapeptide *–(GlyAla)₃GlyGlu–* that could be expressed in the bacterium *E. coli*.

12.4.2. Micelles and Vesicles

A *surfactant* (surface-active agent) is an amphiphilic chemical compound, so named because it contains a hydrophilic or water-seeking head group at one end, and a hydrophobic or water-avoiding (i.e., lipophilic or oil-seeking) tail group at the other end, as shown in Fig. 12.13. The hydrophilic part is polar, with a charge that renders it either anionic ($-$), cationic ($+$), zwitterionic ($\pm$), or nonionic in nature, and the lipophilic portion consists of one or perhaps two nonpolar hydrocarbon chains. Surfactants readily adsorb at an oil–water or air–water interface, and decrease the surface tension there. The hydrocarbon chain might consist of a monomer that can take part in a polymerization reaction.

A surfactant molecule is characterized by a dimensionless packing parameter p defined by

$$p = \frac{V_T}{A_H L_T} \tag{12.6}$$

where A_H is the area of the polar head and V_T and L_T are, respectively, the volume and length of the hydrocarbon tail (Nakache et al. 2000). If the average cross-sectional area of the tail (V_T/L_T) is appreciably less than that of the head (e.g., $p < \frac{1}{3}$), then the tails will pack conveniently inside the surface of a sphere enclosing oil with a radius $r > L_T$ suspended in an aqueous medium, as indicated in Fig. 12.14a. Such a structure is called a *micelle*. The elongated or cylinder-shaped micelle of Fig. 12.14b appears over the range $\frac{1}{3} < p < \frac{1}{2}$. Larger fractional values of the packing parameter, $\frac{1}{2} < p < 1$, lead to the formation of vesicles, which have a double-layer surface structure, as shown in Fig. 12.14c. For example, sodium di-2-ethylhexyl-phosphate can form nanosized vesicles with $V_T \sim 0.5\,\text{nm}^3$, $L_T \sim 0.9\,\text{nm}$, and $A_H \sim 0.7\,\text{nm}^2$, corresponding to $p \sim 0.8$, which is in the vesicle range. If the packing

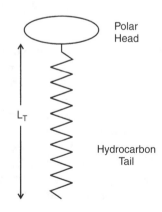

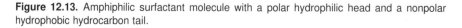

Figure 12.13. Amphiphilic surfactant molecule with a polar hydrophilic head and a nonpolar hydrophobic hydrocarbon tail.

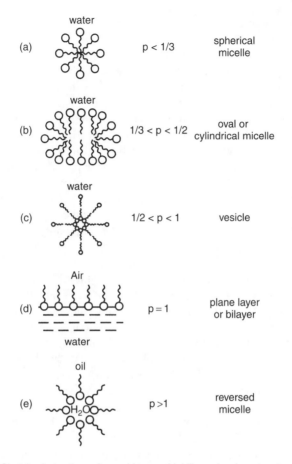

Figure 12.14. Sketch of structures formed by amphiphilic molecules at water–oil or water–air interfaces for various values of the packing parameter *p* of Eq. (12.6). [Adapted from E. Nakache et al., in Nalwa (2000), Vol. 5, Chapter 11, p. 580.]

parameter is unity, $p = 1$, then the average transverse cross-sectional area of the tail V_T/L_T will be the same as the area A_H of the head, and the tails will pack easily at a planar interface, as shown in Fig. 12.14d, or they can form a bilayer. The inverse of a spherical micelle, called an *inverse micelle*, occurs with $p > 1$ for surfactants at the surface of a spherical drop of water in oil, as illustrated in Fig. 12.14e.

An *emulsion* is a cloudy colloidal system of micron size droplets of one immiscible (i.e., nonmixing) liquid dispersed in another, such as oil in water. It is formed by vigorous stirring, and is thermodynamically unstable because the sizes of the droplets tend to grow with time. If surfactants are present, then nanosized particles, ~100 nm, can spontaneously form as a thermodynamically stable, transparent microemulsion that persists for a long time. Surfactant molecules in a solvent can organize in various ways, depending on their concentration. For low concentrations they can adsorb at an air–water interface. Above a certain surfactant concentration, called the *critical micellar concentration*, distributions of micelles in the size range from 2 to 10 nm can form in equilibrium with free surfactant molecules, being continuously constituted and disassembled with lifetimes measured in microseconds or seconds. Synthetic surfactants with more bulky hydrophobic groups, meaning larger packing parameters, produce extended bilayers that can close in on themselves to form vesicles that are generally spherical. These structures form above a critical vesicular concentration. Vesicles typically have lifetimes measured in weeks or months, so they are much more stable than micelles.

If the vesicles are formed from natural or synthetic phospholipids, they are called *unilaminar* or *single-layer liposomes*, that is, liposomes containing only one bilayer. A *phospholipid* is a lipid (fatty or fatlike) substance containing phosphorus in the form of phosphoric acid, which functions as a structural component of a membrane. The main lipid part is hydrophobic, and the phosphoryl or phosphate part is hydrophilic. The hydration or uptake of water by phospholipids causes them to spontaneously self-assemble into unilaminar liposomes. Mechanical agitation of these unilaminar liposomes can convert them to multilayer liposomes that consist of concentric bilayers. Unilaminar liposomes have diameters from the nanometer to the micrometer ranges, with bilayers that are 5–10 nm thick. Proteins can be incorporated into unilaminar liposomes to study their function in an environment resembling that of their state in phospholipid bilayers of a living cell.

If polymerizable surfactants are employed, such as those containing acrylate, acrylamido, allyl ($CH_2{=}CHCH_2{-}$), diallyl, methacrylate, or vinyl ($CH_2{=}CH{-}$) groups, then polymerization interactions can be carried out. When vesicles are involved, the characteristic time for the polymerization is generally shorter than the vesicle lifetime, so the final polymer is one that would be expected from the monomers or precursors associated with the surfactant. However, when micelles are involved, their lifetimes are generally short compared to the characteristic times of the polymerization processes, and as a result the final product may differ considerably from the starting materials. Surfactants with highly reactive polymerizable groups such as acrylamide or styryl have been found to produce polymers with molecular weights in excess of a million daltons. Those with polymerizable groups of low reactivity such as allyl produce much smaller products, namely, products with degrees of polymerization that can bring them close to the micelle size range prior to polymerization.

Micelles and bilayers or liposomes have a number of applications in chemistry and biology. Micelles can assist soap solutions to disperse insoluble organic compounds, and permit them to be cleaned from surfaces. Micelles play a similar role in digestion by permitting components of fat such as fatty acids, phospholipids, cholesterol, and several vitamins (A, D, E, and K) to become soluble in water, and thereby more easily processed by the digestive system. Liposomes can enclose enzymes, and at the appropriate time they can break open and release the enzyme so that it can perform its function, such as catalyzing digestive processes.

Many biological membranes such as the plasma membrane of a red blood cell or erythrocyte are composed of proteins and lipids, where a lipid is a fat or fatlike

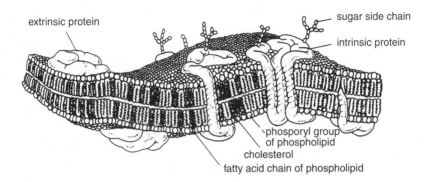

Figure 12.15. Sketch of the structure of a plasma membrane. The phospholipids arrange themselves into a lipid bilayer with their hydrophilic phosphoryl groups at the outside and their fatty acid lipophilic chains on the inside. There are also cholesterol chains in parallel with them. In addition, the membrane contains tightly bound intrinsic proteins that pass through the lipid bilayer, and loosely bound extrinsic proteins on the outside. Some intrinsic proteins have sugar side chains at their surface. (From C. de Duve, *A Guided Tour of the Living Cell*, Scientific American Books, 1984.)

substance. Much of the lipid material is present as phospholipids and cholesterol that form bilayers, typically with the packing parameter $p \sim 1$. The overall structure of a plasma membrane is a lipid bilayer of the type illustrated in Fig. 12.15, which is formed by the self-assembly of lipid molecules. On the outside of the bilayer are found the hydrophilic phosphoryl groups PO_3^{2-} of the phospholipids, which are in contact with the blood plasma aqueous solution that surrounds the red blood cell, and in which it resides. The lipophilic groups are inside pointing toward the suspension of hemoglobin molecules that fills the interior of the red blood cell. We see from the figure that most of the membrane structure consists of fatty acid chains of the phospholipids, with the remainder of the bilayer formed from cholesterol chains. The figure also shows that the plasma membrane contains extrinsic proteins on the surface, intrinsic proteins that penetrate through the bilayer, and sugar side chains attached to the outside.

12.4.3. Multilayer Films

Biomimetrics involves the study of synthetic structures that mimic or imitate structures found in biological systems [see Cooper (2000)]. It makes use of large-scale or supramolecular self-assembly to build up hierarchical structures similar to those found in nature. This approach has been applied to the development of techniques to construct films in a manner that imitates the way nature sequentially adsorbs materials to bring about the biomineralization of surfaces, as discussed below. The process of biomineralization involves the incorporation of inorganic compounds such as those containing calcium into soft living tissue to convert it to a hardened form. Bone contains, for example, many rod-shaped inorganic mineral crystals with typical 5 nm diameters, and lengths ranging from 20 to 200 nm.

The kinetics for the self-assembly of many of these films involved in biomineralization can be approximately modeled as the initial joining together or dimerization of two monomers

$$R + R \Rightarrow R_2 \tag{12.7}$$

with a low equilibrium constant $K_D = C_D/(C_F)^2$, followed by the stepwise or sequential addition of more monomers

$$R_n + R \Rightarrow R_{n+1} \tag{12.8}$$

with a much larger equilibrium constant $K = C_{n+1}/C_F C_n$. These two constants exercise control over the rate at which the reaction proceeds. For the case under consideration, $K_D < K$, the concentration of free or unbound monomers C_F always remains below a critical concentration $C_0 = 1/K$, specifically, $C_F < C_0$. When the total concentration of free and clustered (i.e., bound) monomers C_T satisfies the condition $C_T < C_0$, then the free monomer concentration C_F increases with increases in C_T. When a high enough concentration is provided so that C_T becomes larger than the critical value (i.e., $C_T > C_0$), then the aggregate forms and grows for further increases in C_T. In analogy with this model, self-assembly kinetics often involves a slow dimer formation step followed by faster propagation steps, with $K_D \ll K$.

There are many cases of multilayer thin films in biology, such as structural colors in insects that change when the films are subjected to pressure, shrinking, or swelling. For example, scale cells from some butterflies can produce iridescent multicoloring affects due to optical interference of thin-film layers or lamellae formed from the secretion of networks of filaments that condense on cell boundaries.

The biomineralization of mollusc cells begins by laying down a sheet of organic material so that calcium carbonate can be deposited on its surface and in its pores, and $CaCO_3$ layers can build up. Proteins from the mollusc shell containing high concentrations of particular amino acid residues control the form of the calcium carbonate layering, and these proteins can be altered to vary the layering morphology. Multiple layers either grow in sequences that are organic in nature or contain the rhombohedral calcite form, or the orthorhombic aragonite variety of $CaCO_3$. It is possible to imitate some aspects of these natural biomineralization processes for the preparation of synthetic multilayer thin films, although the resulting films themselves do not closely resemble those in molluscs. For example, consider a positively charged substrate placed in a solution with a negative electrolyte, that is, a solution containing negative ions that can carry electric current. The positive substrate attracts the negative electrolyte, and the latter can adsorb on its surface, forming a structure called a *polyion sheet*, as shown in Fig. 12.16. This sheet is rinsed and dried, and then placed into another electrolyte solution from which it adsorbs a second positive layer. The sequential adsorption process can be repeated, as indicated in Fig. 12.16, to form a multilayer of alternating positively and negatively

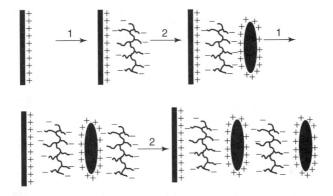

Figure 12.16. Sketch of the sequential adsorption process for the formation of a polyion film. The upper figure shows a positively charged substrate (left) that has adsorbed a negatively charged polyelectrolyte by being dipped into a negative electrolyte solution (center), and then adsorbed a positively charged layer from a positive electrolyte solution (right). The lower figure shows two additional steps in the sequential adsorption process. [From T. M. Cooper, in Nalwa (2000), Vol. 5, Chapter 13, p. 720.]

charged polyion sheets. The nature of the layering can be controlled by varying the types of electrolytes in the successive solutions.

FURTHER READING

T. M. Cooper, "Biometric Thin-Films", in Nalwa (2000), Vol. 5, Chapter 13, p. 711.

T. J. Deming, V. P. Conticello, and D. A. Tirrell, in *Nanotechnology*, G. Timp, ed., Springer-Verlag, Berlin, 1999, Chapter 9.

M. Gross, *Travels to the Nanoworld*, Plenum, New York, 1999.

S. S. Mader, *Biology*, McGraw-Hill, Boston, 2001.

E. Nakache, N. Poulain, F. Candau, A. M. Orecchioni, and J. M. Irache, "Biopolymer and Polymer Nanoparticles and their Biomedical Applications", in Nalwa (2000), Vol. 5, Chapter 11, p. 577.

H. S. Nalwa, ed., *Handbook of Nanostructured Materials and Nanotechnology*, Vol. 5, *Organics, Polymers and Biological Compounds*, Academic Press, Boston, 2000.

D. A. Tirrell, ed., *Hierarchical Structures in Biology as a Guide for New Materials Technology*, National Academy Press, Washington DC, 1994.

13

NANOMACHINES AND NANODEVICES

In previous chapters some potential applications of nanotechnology have been described. In this final chapter we discuss research aimed at developing other applications such as tiny machines, devices having nanosized components, and nanosized molecules as the writing and reading elements of faster computers. We also discuss some of the issues involved in the fabrication of such small devices.

13.1. MICROELECTROMECHANICAL SYSTEMS (MEMSs)

Although microelectromechanical systems do not technically fall under the subject of nanotechnology, it is useful to briefly discuss them at the beginning of the chapter because they represent a more mature technology, and many of the differences in behavior observed in the micromechanical world could well apply to the nano-regime, thereby providing a basis for the design of nanomachines.

The extensive fabrication infrastructure developed for the manufacture of silicon integrated circuits has made possible the development of machines and devices having components of micrometer dimensions. Lithographic techniques, described in previous chapters, combined with metal deposition processes, are used to make MEMS devices. Microelectromechanical systems involve a mechanical response to

Introduction to Nanotechnology, by Charles P. Poole Jr. and Frank J. Owens.
ISBN 0-471-07935-9. Copyright © 2003 John Wiley & Sons, Inc.

an applied electrical signal, or an electrical response resulting from a mechanical deformation.

The major advantages of MEMS devices are miniaturization, multiplicity, and the ability to directly integrate the devices into microelectronics. *Multiplicity* refers to the large number of devices and designs that can be rapidly manufactured, lowering the price per unit item. For example, *miniaturization* has enabled the development of micrometer-sized accelerometers for activating airbags in cars. Previously an electromechanical device the size of a soda can, weighing several pounds and costing about $15, triggered airbags. Presently used accelerometers based on MEMS devices are the size of a dime, and cost only a few dollars. The size of MEMS devices, which is comparable to electronic chips, allows their integration directly on the chip. In the following paragraphs we present a few examples of MEMS devices and describe how they work. But before we do this, let us examine what has been learned about the difference between the mechanical behavior of machines in the macro- and microworlds.

In the microworld the ratio of the surface area to the volume of a component is much larger than in conventional-sized devices. This makes friction more important than inertia. In the macroworld a pool ball continues to roll after being struck because friction between the ball and the table is less important than the inertia of its forward motion. In the microregime the surface area : volume ratio is so large that surface effects are very important. In the microworld mechanical behavior can be altered by a thin coating of a material on the surface of a component. We shall describe MEMS sensors that take advantage of this property. Another characteristic of the microworld is that molecular attractions between microscale objects can exceed mechanical restoring forces. Thus the elements of a microscale device, such as an array of cantilevers, microsized boards fixed at one end, could become stuck together when deflected. To prevent this, the elements of micromachines may have to be coated with special nonstick coatings. In the case of large motors and machines electromagnetic forces are utilized, and electrostatic forces have little impact. In contrast to this, electromagnetic forces become too small when the elements of the motors have micrometer-range dimensions, while electrostatic forces become large. Electrostatic actuation is often used in micromachines, which means that the elements are charged, and the repulsive electrostatic force between the elements causes them to move. We will describe below an actuator, which uses the electrostatic interaction between charged carbon nanotubes. Many of these differences between micromachines, and macromachines become more pronounced in the nanoregime. There are many devices and machines that have micrometer-sized elements. Since this book is concerned primarily with nanotechnology, we give only a few examples of the microscale analogs.

Figure 13.1 illustrates the principle behind a MEMS accelerometer used to activate airbags in automobiles. Figure 13.1a shows the device, which consists of a horizontal bar of silicon a few micrometers in length attached to two vertical hollow bars, having flexible inner surfaces. The automobile is moving from left to right in the figure. When the car suddenly comes to a halt because of impact, the horizontal bar is accelerated to the right in the figure, which causes a change in the separation

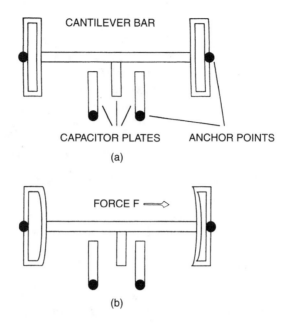

CANTILEVER BAR

CAPACITOR PLATES   ANCHOR POINTS

(a)

FORCE F ⟹

(b)

Figure 13.1. Illustration of a MEMS device (a) used to sense impact and initiate expansion of airbags in cars. The automobile is moving from left to right. On impact (b) the horizontal cantilever bar is accelerated to the right and changes the separation of the capacitor plates, thereby triggering a pulse of electric current that activates the bag expansion mechanism. (Adapted from M. Gross, *Travels to the Nanoworld*, Plenum, New York, 1999, p. 169.)

between the plates of the capacitor, as shown in Fig. 13.1b. This changes the value of the electrical capacitance of the capacitor, which in turn electronically triggers a pulse of current through a heating coil embedded in sodium azide, NaN_3. The instantaneous heating causes a rapid decomposition of the azide material, thereby producing nitrogen gas N_2 through the reaction $2NaN_3 \rightarrow 2Na + 3N_2$, which inflates the airbag.

Coated cantilever beams are the basis of a number of sensing devices employing MEMS. A cantilever is a small supported beam. The simplest of such devices consist of arrays of singly supported polysilicon cantilevers having various length to width ratios in the micrometer range. The beams can be made to vibrate by electrical or thermal stimuli. Optical reflection techniques are used to measure the vibrational frequency. As shown in Fig. 13.2, the vibrational frequency is very sensitive to the length of the beam. Thermal sensors have been developed using these supported micrometer-sized cantilevers by depositing on the beams a layer of a material that has a coefficient of thermal expansion different from that of the polysilicon cantilever itself. When the beam is heated, it bends because of the different coefficients of expansion of the coating and the silicon, and the resonant frequency of the beam changes. The sensitivity of the device is in the micro degree range, and it can be used as an infrared (IR) sensor. A similar design can be used to make a sensitive detector of DC magnetic fields. In this case the beam is coated with a material that displays *magnetorestrictive effects*, meaning that the material changes

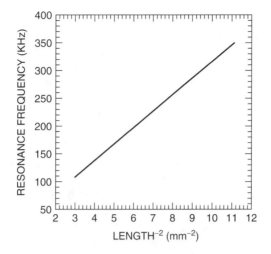

Figure 13.2. Plot of the resonance frequency of a MEMS cantilever versus the square of the reciprocal of the length of the beam. [Adapted from R. C. Benson, *Johns Hopkins Tech. Digest* **16**, 311 (1995).]

its dimensions when a DC magnetic field is applied. This causes the beam to bend and change its resonance frequency. These devices can detect magnetic fields as small as 10^{-5} G (gauss) [10^{-9} T (tesla)].

13.2. NANOELECTROMECHANICAL SYSTEMS (NEMSs)

13.2.1. Fabrication

Nanomechanical machines and devices are in the early stages of development, and many are still in conceptual stages. Numerous computer simulations of possibilities and ideas have been proposed. It turns out that nature is far ahead of us in its ability to produce nanosized machines. Nanomotors exist in biological systems such as the flagellar motor of bacteria. Flagellae are long, thin, blade-like structures that extend from the bacteria. The motion of these flagellae propel the bacteria through water. These whip-like structures are made to move by a biological nanomotor consisting of a highly structured conglomerate of protein molecules anchored in the membrane of the bacterium. The motor has a shaft and a structure about the shaft resembling an armature. However, the motor is not driven by electromagnetic forces, but rather by the breakdown of adenosine triphosphate (ATP) energy-rich molecules, which causes a change in the shape of the molecules. Applying the energy gained from ATP to a molecular ratchet enables the protein shaft to rotate. Perhaps the study of biological nanomachines will provide insights that will enable us to improve the design of mechanical nanomachines.

Optical lithography is an important manufacturing tool in the semiconductor industry. However, to fabricate semiconductor devices smaller than 100 nm, ultraviolet light of short wavelengths (193 nm) is required, but this will not work because the materials are not transparent at these wavelengths. Electron-beam and X-ray lithography, discussed in earlier chapters, can be used to make nanostructures, but these processes are not amenable to the high rate of production that is necessary for large-scale manufacturing. Electron-beam lithography uses a finely focused beam of electrons, which is scanned in a specific pattern over the surface of a material. It can produce a patterned structure on a surface having 10-nm resolution. Because it requires the beam to hit the surface point by point in a serial manner, it cannot produce structures at sufficiently high rates to be used in assembly-line manufacturing processes. X-ray lithography can produce patterns on surfaces having 20-nm resolution, but its mask technology and exposure systems are complex and expensive for practical applications.

More recently, a technique called *nanoimprint lithography* has been developed that may provide a low-cost, high-production rate manufacturing technology. Nanoimprint lithography patterns a resist by physically deforming the resist shape with a mold having a nanostructure pattern on it, rather than by modifying the resist surface by radiation, as in conventional lithography. A resist is a coating material that is sufficiently soft that an impression can be made on it by a harder material. A schematic of the process is illustrated in Fig. 13.3. A mold having a nanoscale structured pattern on it is pressed into a thin resist coating on a substrate (Fig. 13.3a), creating a contrast pattern in the resist. After the mold is lifted off (Fig. 13.3b), an etching process is used to remove the remaining resist material in the compressed regions (Fig. 13.3c). The resist is a thermoplastic polymer, which is a material that softens on heating. It is heated during the molding process to soften the polymer relative to the mold. The polymer is generally heated above its glass transition temperature, thereby allowing it to flow and conform to the mold pattern. The mold can be a metal, insulator, or semiconductor fabricated by conventional lithographic methods. Nanoimprint lithography can produce patterns on a surface having 10-nm resolution at low cost and high rates because it does not require the use of a sophisticated radiation beam generating patterns for the production of each structure.

The scanning tunneling microscope (STM), described in detail in Chapter 3, uses a narrow tip to scan across the surface of the material about a nanometer above it. When a voltage is applied to the tip, electrons tunnel from the surface of the material and a current can be detected. If the tip is kept at a constant distance above the surface, then the current will vary as the tip scans the surface. The amount of detected current depends on the electron density at the surface of the material, and this will be higher were the atoms are located. Thus, mapping the current by scanning the tip over the surface produces an image of the atomic or molecular structure of the surface.

An alternate mode of operation of the STM is to keep the current constant, and monitor the deflection of the cantilever on which the tip is held. In this mode the recorded cantilever deflections provide a map the atomic structure of the surface.

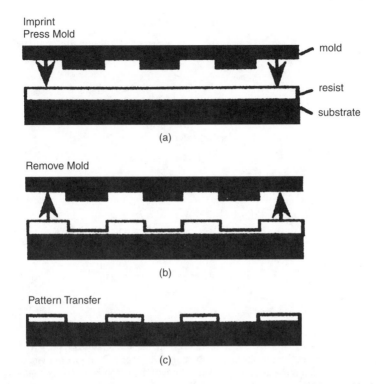

Figure 13.3. Schematics of steps in a nanoimprint lithography process: (a) a mold of a hard material made by electron beam lithography is pressed into a resist, a softer material, to make an imprint; (b) the mold is then lifted off; (c) the remaining soft material is removed by etching from the bottom of the groves. [Adapted from S. Y. Chou et al., *Science* **272**, 85 (1996).]

The scanning tunneling microscope has been used to build nanosized structures atom by atom on the surface of materials. An adsorbed atom is held on the surface by chemical bonds with the atoms of the surface. When such an atom is imaged in an STM, the tip has a trajectory of the type shown in Fig. 13.4a. The separation between the tip and the adsorbed atom is such that any forces between them are small compared to the forces binding the atom to the surface, and the adsorbed atom will not be disturbed by the passage of the tip over it. If the tip is moved closer to the adsorbed atom (Fig. 13.4b) such that the interaction of the tip and the atom is greater than that between the atom and the surface, the atom can be dragged along by the tip. At any point in the scan the atom can be reattached to the surface by increasing the separation between the tip and the surface. In this way adsorbed atoms can be rearranged on the surfaces of materials, and structures can be built on the surfaces atom by atom. The surface of the material has to be cooled to liquid helium temperatures in order to reduce thermal vibrations, which may cause the atoms to diffuse thermally, thereby disturbing the arrangement of atoms being assembled. Thermal diffusion is a problem because this method of construction can be carried

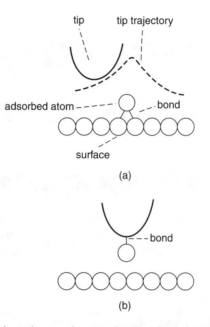

Figure 13.4. Illustration of a trajectory of a scanning tunneling microscope (STM) tip over an atom adsorbed on the surface of a material: (a) *imaging mode*, when the tip is not in contact with the surface but is close enough to obtain an image; (b) *manipulation mode*, when the tip is sufficiently close to the surface such that the adsorbed atom bonds to the tip. (Adapted from D. Eigler, in *Nanotechnology*, G. Timp, ed., AIP Press, New York, 1998, p. 428.)

out only on materials in which the lateral or in-plane interaction between the adsorbed atom and the atoms of the surface is not excessive. The manipulation also has to be done under ultra-high-vacuum conditions in order to keep the surface of the material clean.

Figure 13.5 depicts a circular array of iron atoms on a copper surface, called a "quantum corral", assembled by STM manipulation. The wavelike structure inside the corral is the surface electron density distribution inside the well corresponding to three quantum states of this two-dimensional circular potential well, in effect providing a visual affirmation of the electron density predicted by quantum theory. This image is taken using an STM with the tip at such a separation that it does not move any of the atoms. The adsorbed atoms in this structure are not bonded to each other. The atoms will have to be assembled in three-dimensional arrays and be bonded to each other to use this technique to build nanostructures. Because the building of three-dimensional structures has not yet been achieved, the slowness of the technique together with the need for liquid helium cooling and high vacuum all indicate that STM manipulation is a long way from becoming a large-scale fabrication technique for nanostructures. It is important, however, in that it demonstrates that building nanostructures atom by atom is feasible, and it can be

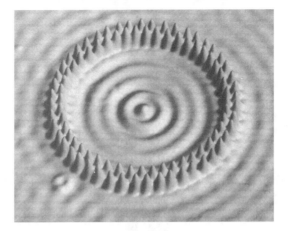

Figure 13.5. A circular array of iron atoms placed on a copper surface using an STM tip, as described in the text, forming a "quantum corral." The ripples inside the corral represent the surface distribution of electron density that arises from the quantum-mechanical energy levels of a circular two-dimensional potential well. [With permission from D. Eigler, in *Nanotechnology*, G. Timp, ed., Springer Verlag, Heidelberg, 1998, p. 433.]

used to build interesting structures such as the quantum corral in order to study their physics.

13.2.2. Nanodevices and Nanomachines

Actuators are devices that convert electrical energy to mechanical energy, or vice versa. It is known that single-walled carbon nanotubes deform when they are electrically charged. An actuator based on this property has been demonstrated using single-walled carbon nanotube paper. Nanotube paper consists of bundles of nanotubes having their long axis lying in the plane of the paper, but randomly oriented in the plane. The actuator consisted of 3×20-mm strips of nanopaper 25–50 μm thick. The two strips are bonded to each other in the manner shown in Fig. 13.6 by double-stick Scotch tape. An insulating plastic clamp at the upper end supports the paper and holds the electrical contacts in place. The sheets were placed in a one molar NaCl electrolytic solution. Application of a few volts produced a deflection of up to a centimeter, and could be reversed by changing the polarity of the voltage, as shown in Fig. 13.6. Application of an AC voltage produced an oscillation of the cantilever. This kind of actuator is called a *bimorph cantilever actuator* because the device response depends on the expansion of opposite electrodes. Strictly speaking, this actuator is neither a NEMS nor MEMS device because of the size of the electrodes. However, it works because of the effect of charging on the individual carbon nanotubes, and indicates that nanosized actuators employing three single-walled carbon nanotubes are possible. Such a device might

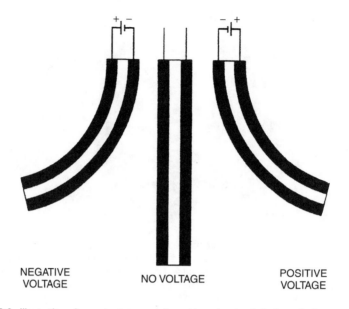

NEGATIVE
VOLTAGE

NO VOLTAGE

POSITIVE
VOLTAGE

Figure 13.6. Illustration of an actuator consisting of two sheets of single-walled nanopaper held together by insulating double-stick Scotch tape. The figure shows the positive-voltage state (right), the resting state (center), and the negative-voltage state (left). [Adapted from R. H. Baughman et al., *Science* **284**, 340 (1999).]

consist of three fibers aligned with their axes parallel and in contact. The outer two tubes would be metallic and the inner tube insulating.

Although electron-beam lithography can be used to fabricate silicon structures less than 10 nm in size, nanomachines have not been produced to any large extent. A number of difficulties must be overcome before significant progress can be made. The first is the problem of communicating with and sensing the motion of the nanoscale devices. The second obstacle is that little is known or understood about the mechanical behavior of objects, which have up to 10% of their atoms on or near the surface.

The resonant frequency f_0 of a clamped beam is given by

$$f_0 \sim \left[\frac{E}{\rho}\right]^{1/2} \frac{b}{L^2} \tag{13.1}$$

where E is the elastic modulus, ρ is the density, b is the thickness of the beam, and L is the beam length. Experimental verification of the scaling of the frequency with $1/L^2$ is shown in Fig. 13.2 for polysilicon beams of micrometer dimension. Notice that in the micrometer range the frequencies are in the hundreds of kilohertz ($>10^5$ cycles per second). Now a beam having a length of 10 nm and thickness of 1 nm will have a resonant frequency 10^5 times greater, of the order of 20–30 GHz (2–3×10^{10} cycles per second). As the frequency increases, the amplitude of vibration decreases, and in this range of frequencies the displacements of the beam can range from a picometer (10^{-12} m) to a femtometer (10^{-15} m).

These high frequencies and small displacements are very difficult, if not impossible, to detect. Optical reflection methods such as those used in the micrometer range on the cantilever tips of scanning tunneling microscopes are not applicable because of the diffraction limit. This occurs when the size of the object from which light is reflected becomes smaller than the wavelength of the light. Transducers are generally used in MEMS devices to detect motion. The MEMs accelerometer shown in Fig. 13.1 is an example of the detection of motion using a transducer. In the accelerometer mechanical motion is detected by a change in capacitance, which can be measured by an electrical circuit. It is not clear that such a transducer sensor can be built that can detect displacements as small as 10^{-15} to 10^{-12} m, and do so at frequencies up to 30 GHz. These issues present significant obstacles to the development of NEMS devices.

There are, however, some noteworthy advantages of NEMS devices that make it worthwhile to pursue their development. The small effective mass of a nanometer-sized beam renders its resonant frequency extremely sensitive to slight changes in its mass. It has been shown, for example, that the frequency can be affected by adsorption of a small number of atoms on the surface, which could be the basis for a variety of very high-sensitivity sensors.

A weight on a spring would oscillate indefinitely with the same amplitude if there were no friction. However, because of air resistance, and the internal spring friction, this does not happen. Generally the frictional or damping force is proportional to the velocity dx/dt of the oscillating mass M. The equation of motion of the spring is

$$M \frac{d^2x}{dt^2} + b \frac{dx}{dt} + Kx = 0 \tag{13.2}$$

where K is the spring of constant.

The solution $X(\omega)$ to this equation for a small damping factor b is

$$X(\omega) = A \exp\left(\frac{-bt}{2M}\right) \cos(\omega t + \delta) \tag{13.3}$$

with the frequency ω given by

$$\omega = \left[\left(\frac{K}{M}\right) - \left(\frac{b}{2M}\right)^2\right]^{-1/2} \tag{13.4}$$

Equation (13.3) describes a system oscillating at a fixed frequency ω with an amplitude exponentially decreasing in time. The displacement as a function of time is plotted in Fig. 13.7a. For a clamped vibrating millimeter-sized beam, a major

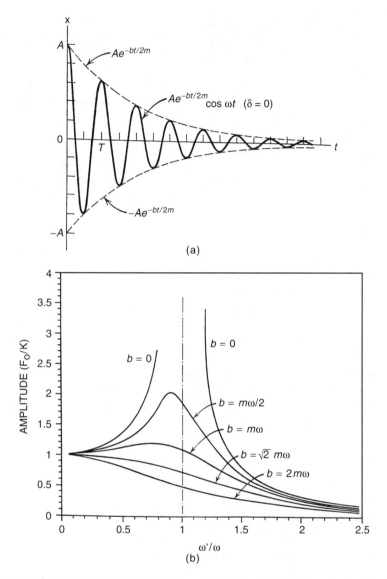

Figure 13.7. Damped harmonic oscillator showing (a) dependence of the displacement of a free-running (no driving force) oscillator on time, and (b) dependence of the amplitude of a driven oscillator versus the ratio of the driving frequency ω' to the undamped natural frequency ω for different damping constants b, where F_0 is the amplitude of the driving force. (Adapted from D. Halliday and R. Resnick, *Physics*, John Wiley & Sons, Inc., New York, 1960, Vol. 1, p. 311.)

source of damping will be air resistance, which is proportional to the area of the beam. For the case of a nanosized beam, this area is very small, so the damping factor b is small. Nanosized beams dissipate very little energy over their vibrational cycles.

If one applies an external oscillating force $F_0 \cos(\omega' t)$ to a damped harmonic oscillator, a very large increase in amplitude occurs when the frequency of the applied force ω' equals the natural resonant frequency ω of the oscillator. This is called *resonance*. The increase in amplitude depends on the magnitude of the damping term b in Eq. (13.2), which is the cause of dissipation. Figure 13.7b shows how the magnitude of the damping factor affects the amplitude at resonance for a vibrating mass on a spring. Notice that the smaller the damping factor, the narrower the resonance peak, and the greater the increase in amplitude. The quality factor Q for the resonance given by the expression $Q = \omega_0/\Delta\omega$, where $\Delta\omega$ is the width of the resonance at half height and ω_0, the resonant frequency. The quality factor is the energy stored divided by the energy dissipated per cycle, so the inverse of the quality factor $1/Q$ is a measure of the dissipation of energy. Nanosized cantilevers have very high Q values and dissipate little energy as they oscillate. Such devices will be very sensitive to external damping, which is essential to developing sensing devices. High-Q devices also have low thermomechanical noise, which means significantly less random mechanical fluctuations. The Q values of high-Q electrical devices are in the order of several hundreds, but NEMS oscillators can have Q values 1000 times higher. Another advantage of NEMS devices is that they require very little power to drive them. A picowatt (10^{-12} W) of power can drive an NEMS device with a low signal-to-noise ratio (SNR).

Computer simulation has been used to evaluate the potential of various nano-machine concepts. One example, shown in Fig. 13.8, is the idea of making gears out of nanotubes. The "teeth" of the gear would be benzene molecules bonded to the outer walls of the tube. The power gear on the left side of the figure is charged in order to make a dipole moment across the diameter of the tube. Application of an alternating electric field could induce this gear to rotate. An essential part of any

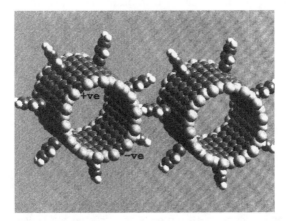

Figure 13.8. Illustration of a proposed method for making gears by attaching benzene molecules to the outside of carbon nanotubes. (With permission from D. Srivastava et al., in *Handbook of Nanostructured Materials and Nanotechnology*, H. S. Nalwa, ed., Academic Press, San Diego, 2000, Vol. 2, Chapter 14, p. 697.)

nanosized machine is the capability of moving the power gear. The idea of using an electric field, which does not require contacts with the nanostructure, to roll a C_{60} molecule over a flat surface has been proposed. The idea is illustrated in Fig. 13.9. An isolated C_{60} molecule is adsorbed on the surface of an ideally flat ionic crystal such as potassium chloride. The application of an electric field would polarize the C_{60} molecule, putting plus and minus charges on opposite sides of the sphere, as shown in the figure. Because the molecule has a large polarizability and a large diameter, a large electric dipole moment is induced. If the interaction between the dipole moment and the applied electric field is greater than the interaction between the moment and the surface of the material, rotation of the electric field should cause the C_{60} molecule to roll across the surface.

The atomic force microscope, which is described in chapter 3, employs a sharp tip mounted on a cantilever spring, which is scanned closely over the surface of a material. The deflection of the cantilever is measured. In the region of surface atoms the deflection is larger because of the larger interaction between the tip and the atoms. The cantilevers are fabricated by photolithographic methods from silicon, silicon oxide, or silicon nitride. They are typically 100 μm long and 1 μm thick, and have spring constants between 0.1 and 1.0 N/m (newton per meter). Operating in the tapping mode, where the change in the amplitude of an oscillating cantilever driven near its resonance frequency is measured as the tip taps the surface, can increase the sensitivity of the instrument. One difficulty is that too hard a tap might break the tip. A group at Rice University has demonstrated that using carbon nanotubes as tip material can provide a possible solution to this problem. A multiwalled nanotube (MWNT) was bonded to the side of a tip of a conventional silicon cantilever using a soft acrylic adhesive as illustrated in Fig. 13.10. If the nanotube crashes into the surface, generating a force greater than the Euler buckling force, the nanotube does not break, but rather bends away and then snaps back to its original position. The nanotube's tendency to buckle rather than break makes it unlikely that the tip will

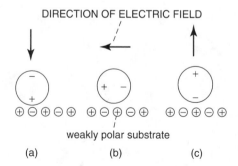

Figure 13.9. Illustration of how a C_{60} molecule (large circles) possessing an induced electric dipole moment, adsorbed on the surface of an ionic crystal with alternating charged atoms (small circles), could be rolled over the surface by a rotating external electric field (arrows). (Adapted from M. S. Dresselhaus, G. Dresselhaus, and P. C. Eklund, *Science of Fullerenes and Carbon Nanotubes*, Academic Press, San Diego, 1996, p. 902.)

CANTILEVER

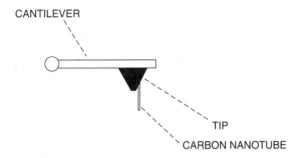

TIP

CARBON NANOTUBE

Figure 13.10. Illustration of a single-walled carbon nanotube mounted on an STM tip attached to a cantilever arm of an atomic force microscope. [Adapted from H. Dai et al., *Nature* **384**, 147 (1996).]

break. The MWNT tip can also be used in the tapping mode. When the nanotube bends on impact, there is a coherent deexcitation of the cantilever oscillation. The MWNT serves as a compliant spring, which moderates the impact of each tap on the surface. Because of the small cross section of the tip, it can reach into deep trenches on the surface that are inaccessible to normal tips. Since the MWNTs are electrically conducting, they may also be used as probes for an STM.

The azobenzene molecule, shown in Fig. 13.11a, can change from the trans isomer to the cis isomer by subjecting it to 313-nm light. Isomers are molecules having the same kind of atoms and the same number of bonds but a different equilibrium geometry. Subjecting the cis isomer to light of wavelength greater than 380 nm causes the cis form to return to the original trans form. The two forms can be distinguished by their different optical absorption spectra. Notice that the cis isomer is shorter than the trans isomer. Azobenzene can also form a polymer consisting of a chain of azobenzene molecules. In the polymer form it can also undergo the trans-to-cis transformation by exposure to 365-nm light. When this occurs, the length of the polymer chain decreases. A group at the University of Munich have constructed a molecular machine based on the photoisomerization of the azobenzene polymer. They attached the trans form of the polymer to the cantilever of an atomic force microscope as shown in Fig. 13.11b and then subjected it to light of 365-nm wavelength, causing the polymer to contract and the beam to bend. Exposure to 420-nm light causes the polymer to return to the trans form, allowing the beam to return to its original position. By alternately exposing the polymer to pulses of 420- and 365-nm light, the beam could be made to oscillate. This is the first demonstration of an artificial single-molecule machine that converts light energy to physical work.

13.3. MOLECULAR AND SUPRAMOLECULAR SWITCHES

The lithographic techniques used to make silicon chips for computers are approaching their limits in reducing the sizes of circuitry on chips. Nanosize architecture is

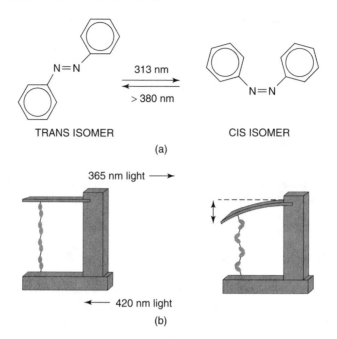

Figure 13.11. (a) Cis-to-trans UV-light-induced isomerization of azobenzene; (b) a molecular machine based on light-induced isomeric changes of the azobenzene polymer, which contracts when it is converted to the cis form, causing the cantilever beam to bend.

becoming more difficult and more expensive to make. This has motivated an effort to synthesize molecules, which display switching behavior. This behavior might form the basis for information storage and logic circuitry in computers using binary systems. A molecule A that can exist in two different states, such as two different conformations A and B, and can be converted reversibly between the two states by external stimuli, such as light or a voltage, can be used to store information. In order for the molecule to be used as a zero or one digital state, necessary for binary logic, the change between the states must be fast and reversible by external stimuli. The two states must be thermally stable and be able to switch back and forth many times. Furthermore the two states must be distinguishable by some probe, and the application of the probe R is called the *read mode*. Figure 13.12 is a schematic

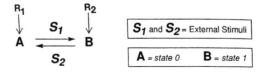

Figure 13.12. Schematic representation of the elements of a molecular switch. An external stimulus S_1 changes a molecule from state 0 to state 1, and S_2 returns the molecule to state 0.

representation of the basic elements of a molecular switch, in which stimulus S_1 brings about a conversion from state 0 to state 1, and stimulus S_2 induces the reverse conversion. There are a number of different kinds of molecular switches.

An example of a molecular switch is provided by the azobenzene molecule, which has the two isomeric forms sketched in Fig. 13.11a. Unfortunately the cis form of azobenzene is not thermally stable, and a slight warming causes it to return to the trans form, so optical methods of switching are not of practical use for applications in computing. Employing electrochemical oxidation and reduction can overcome this thermal instability of azobenzene. Figure 13.13 shows how the cis isomer is reduced to hydrazobenzene by the addition of hydrogen atoms at a more anodic (negative) potential, and then converted back to the trans isomer by oxidation, which removes the hydrogen atoms.

A chiroptical molecular switch, such as the one sketched in Fig. 13.14, uses circularly polarized light (CPL) to bring about changes between isomers. The application of left circularly polarized light (−)-CPL to the molecular conformation M on the left side of the figure causes a rotation of the four-ring group on the top from a right-handed helical structure to a left-handed helical arrangement P, as shown. Right circularly polarized light (+)-CPL brings about the reverse transformation. Linearly polarized light (LPL) can be used to read the switch by monitoring the change in the axis of the light polarizer. The system can be erased using unpolarized light (UPL).

Conformational changes involving rearrangements of the bonding in a molecule can also be the basis of molecular switching. When the colorless spiropyran, shown on the left in Fig. 13.15, is subjected to UV light, hv_1, the carbon–oxygen bond opens, forming merocyanine, shown on the right in Fig. 13.15. When the merocyanine is subjected to visible (red) light, hv_2, or heat (Δ), the spiropyran reforms.

A catenane molecule has been used to make a molecular switch that can be turned on and off with the application of a voltage. A *catenane* is a molecule with a

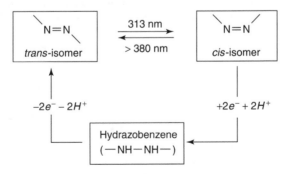

Figure 13.13. Schematic of controlling the azobenzene switching process using both photo-isomerization (top of figure), and electrochemistry (bottom of figure), making this a dual-mode switch. (With permission from M. Gómez-López and F. J. Stoddart, in *Handbook of Nano-structured Materials and Nanotechnology*, H. S. Nalwa, ed., Academic Press, San Diego, 2000, Vol. 5, Chapter 3, p. 230.)

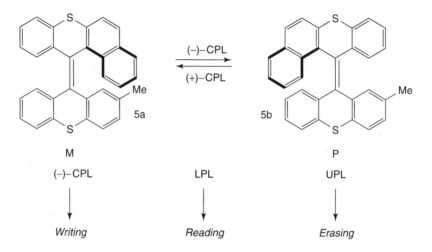

M   P

(−)−CPL   LPL   UPL

↓   ↓   ↓

Writing   *Reading*   *Erasing*

Figure 13.14. Sketch of a molecule that can be switched between two states denoted by M and P by using circularly polarized light (CPL). The molecular switch position is read using linearly polarized (LPL), and the information can be erased using unpolarized light (UPL). (With permission from M. Gómez-López and F. J. Stoddart, in *Handbook of Nanostructured Materials and Nanotechnology*, H. S. Nalwa, ed., Academic Press, San Diego, 2000, Vol. 5, Chapter 3, p. 231.)

molecular ring mechanically interlinked with another molecular ring, as shown by the example sketched in Fig. 13.16. Its two different switched states are shown in Figs. 13.16a and 13.16b. This molecule is 0.5 nm long and 1 nm wide, making it in effect a nanoswitch. For this application a monolayer of the catenane anchored with amphiphilic phospholipid counterions is sandwiched between two electrodes. The structure in Fig. 13.16a is the open switch position because this configuration does not conduct electricity as well as the structure in Fig. 13.16b. When the molecule is oxidized by applying a voltage, which removes an electron, the tetrathiafulvalene group, which contains the sulfurs, becomes positively ionized and is thus electro-

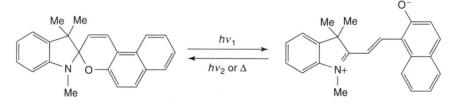

Figure 13.15. Photochemical switching of spiropyran (left) to merocyanine (right) by ultraviolet light hv_1, where red light (hv_2) or heat (Δ) induces the reverse-direction conformational change in the molecule. (With permission from M. Gómez-López and F. J. Stoddart, in *Handbook of Nanostructured Materials and Nanotechnology*, H. S. Nalwa, ed., Academic Press, San Diego, 2000, Vol. 5, Chapter 3, p. 233.)

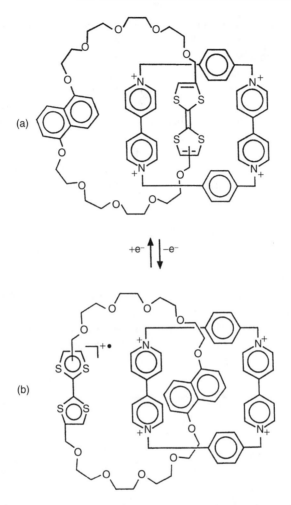

Figure 13.16. Illustration of a switchable catenane that changes its conformation when subjected to a voltage that induces oxidation $(-e^-)$ from the upper $(\cdot\cdot)$ isomeric state to the lower $(\cdot)$ state, and reduction $(+e^-)$ for the reverse transformation. [With permission from J. F. Stodart, *Chem. Eng. News* 28 (Oct. 16, 2000).]

statically repelled by the cyclophane group, the ring containing the nitrogen atoms. This causes the change in structure shown in Fig. 13.16, which essentially involves a rotation of the ring on the left side of the molecule to the right side.

An interesting aspect of this procedure is the observation that some molecules can conduct electricity, although not in large amounts. The STM has been used to measure the conductivity of long chainlike molecules. In a more recent study a monolayer of octanethiol was formed on a gold surface by self-assembly. The sulfur group at the end of the molecule bonded to the surface in the manner illustrated in Figs. 10.2–10.4 (of Chapter 10). Some of the molecules were then removed using a solvent technique and replaced with 1,8-octanedithiol, which has sulfur groups at both ends of the chain. A gold-coated STM tip was scanned over the top of the monolayer to find the 1,8-octanedithiol. The tip was then put in contact with the end of the molecule, forming an electric circuit between the tip and the flat gold surface. The octanethiol molecules, which are bound only to the bottom gold electrode, serve as molecular insulators, electrically isolating the octanedithiol wires. The voltage between the tip and the bottom gold electrode is then increased and the current measured. The results yield five distinct families of curves, each an integral multiple of the fundamental curve, which is the dashed curve in the Fig. 13.17. In the figure we only show the top and bottom curve. The fundamental curve corresponds to electrical conduction through a single dithiol molecule; the other curves correspond to conduction through two or more such molecules. It should be noted that the current is quite low, and the resistance of the molecule is estimated to be $900\,M\Omega$.

Having developed the capability to measure electrical conduction through a chain molecule, researchers began to address the question of whether a molecule could be designed to switch the conductivity on and off. They used the relatively simple molecule sketched in Fig. 13.18, which contains a thiol group (SH−) that can be attached to gold by losing a hydrogen atom. The molecule, 2-amino-4-ethylnylphenyl-4-ethylnylphenylphenyl-5-nitro-1-benzenethiolate, consists of three benzene rings linked in a row by triple bonded carbon atoms. The middle ring has an

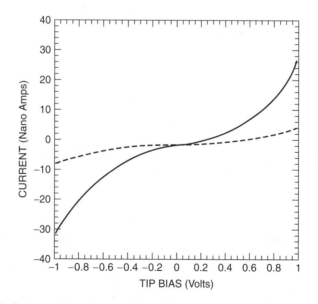

Figure 13.17. Current–voltage characteristics of an octanethiol monolayer on a gold substrate measured by STM using a gold-coated tip. Five curves are actually observed, but only two, the lowest (----) and the highest (—), are shown here. The solid curve corresponds to 4 times the current of the dashed curve. [Adapted from X. D. Cui et al., *Science* **294**, 571 (2001).]

A molecular electronic device

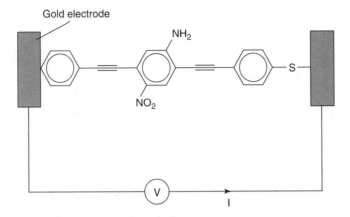

Figure 13.18. Illustration of an electronic switch made of a conducting molecule bonded at each end to gold electrodes. Initially it is nonconducting; however, when the voltage is sufficient to add an electron from the gold electrode to the molecule, it becomes conducting. A further voltage increase makes it nonconducting again with addition of a second electron. [Adapted from J. Chen, *Science* **286**, 1550 (1999).]

amino (NH_2) group, which is an electron donor, pushing electric charge toward the ring. On the other side is an electron acceptor nitro (NO_2) group, which withdraws electrons from the ring. The net result is that the center ring has a large electric dipole moment. Figure 13.19 shows the current–voltage characteristics of this molecule, which is attached to gold electrodes at each end. There is an onset of current at 1.6 V, then a pronounced increase, followed by a sudden drop at 2.1 V. The result was observed at 60 K but not at room temperature. The effect is called *negative-differential resistance*. The proposed mechanism for the effect is that the molecule is initially nonconducting, and at the voltage where a current peak is observed the molecule gains an electron, forming a radical ion, and becomes conducting. As the voltage is increased further, a second electron is added, and the molecule forms a nonconducting dianion.

Of course, demonstrating that a molecule can conduct electricity, and that the conduction can be switched on and off, is not enough to develop a computer. The molecular switches have to be connected together to form logic gates. A roxatane molecule, shown in Fig. 13.20, which can change conformation when it gains and loses an electron by rotation of the oxygen ring on the left of Fig. 13.20, similar to the changes in the catenane of Fig. 13.16, has been used to make switching devices that can be connected together. A schematic cross section of an individual switch is shown in Fig. 13.21. Each device consists of a monolayer of rotaxane molecules sandwiched between two parallel electrodes made of aluminum (Al). The upper electrode on the figure has a layer of titanium (Ti) on it, and the lower one has an alumina (Al_2O_3) layer that acts as a tunneling barrier.

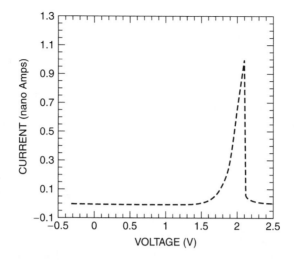

Figure 13.19. Current–voltage characteristics of the electronic switch shown in Fig. 13.18. [Adapted from J. Chen, *Science* **286**, 1550 (1999).]

To fabricate this switch, an aluminum electrode was made by lithographically patterning 0.6-μm-diameter aluminum wires on a silica substrate. This electrode was then exposed to oxygen to allow a layer of Al_2O_3 to form on it. Next a single monolayer of the roxatane molecule, such as the one shown in Fig. 13.20, was deposited as a Langmuir–Blodgett film, which was discussed in Chapter 10. Then a 5-μm layer of titanium followed by a thicker layer of aluminum were evaporated through a contact mask using electron beam deposition.

Figure 13.22 shows the current–voltage characteristics of this device. The application of −2 V, the read mode, caused a sharp increase in current. The switch could be

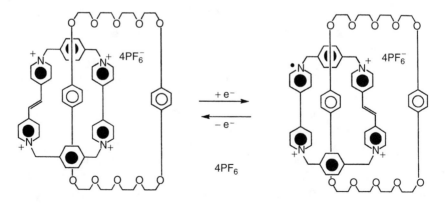

Figure 13.20. Roxatane molecule used in a switch illustrated in Fig. 13.21 to make molecule-based logic gates. [Adapted from C. P. Collier et al., *Science* **285**, 391 (1999).]

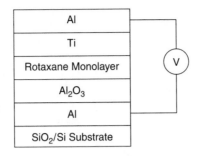

Figure 13.21. Cross section of a molecular switch prepared lithographically, incorporating roxatane molecules that have the structure sketched in Fig. 13.20, sandwiched between aluminum electrodes, and mounted on a substrate. The upper electrode has a layer of titanium, and the lower one is coated with aluminum oxide. [Adapted from C. P. Collier et al., *Science* **285**, 391 (1999).]

opened by applying 0.7 V. The difference in the current between the open and closed switch was a factor of 60–80. A number of these switches were wired together in arrays to form logic gates, essential elements of a computer. Two switches (A and B) connected as shown in the top of Fig. 13.23 can function as an AND gate. In an AND gate both switches have to be on for an output voltage to exist. There should be little or no response when both switches are off, or only one switch is on. Figure 13.23 shows the response of the gate for the different switch positions, A and B (called *address levels*). There is no current flow for both switches off ($A = B = 0$), and very little current for only one switch on ($A = 0$ and $B = 1$, or $A = 1$ and $B = 0$). Only the combination $A = 1$ and $B = 1$ for both switches on provides an appreciable output current, showing that the

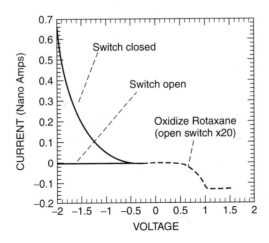

Figure 13.22. Current–voltage characteristics of the molecular switch shown in Fig. 13.21. [Adapted from C. P. Collier et al., *Science* **285**, 391 (1999).]

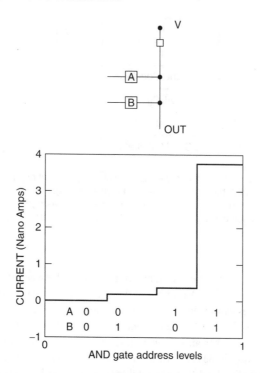

Figure 13.23. Measured truth table for an AND gate constructed from two molecular switches *A* and *B* wired as shown at the top. The lower plot shows that there is only an output (high current) when both switches are on (*A* = 1, *B* = 1). The address levels listed toward the bottom refer to the possible on (1) and off (0) arrangements of the switches *A* and *B*. [Adapted from C. P. Collier et al., *Science* **285**, 391 (1999).]

device can function as an AND gate. These results demonstrate the potential of molecular switching devices for future computer technology.

FURTHER READING

R. P. Andres et al., "The Design, Fabrication, and Electronic Properties of Self Assembled Molecular Nanostructures," in *Handbook of Nanostructured Materials and Nanotechnology*, H. S. Nalwa, ed., Academic Press, San Diego, 2000, Vol. 3, Chapter 4.

D. Bishop, P. Gammel, and C. R. Giles, "Little Machines that Are Making It Big," *Phys. Today* **54**, 38 (Oct. 2001).

R. Dagani, "Building from the Bottom Up," *Chem. Eng. News* 28 (Oct 16, 2000).

C. M. Lieber, "The Incredible Shrinking Circuit," *Sci. Am.* **39**, 59 (Sept. 2001).

M. A. Reed and J. M. Tour, "Computing with Molecules," *Sci. Am.* **38**, 86 (June 2000).

M. Roukes, "Plenty of Room Indeed," *Sci. Am.* **39**, 48 (Sept. 2001).

M. Roukes, "Nanoelectromechanical Systems Face the Future," *Phys. World* (Feb. 2001).

D. Ruger and P. Hansma, "Atomic Force Microscopy," *Phys. Today* **43**, 23 (Oct. 1990).

J. A. Stroscio and D. M. Eigler, "Atomic and Molecular Manipulation with a Scanning Tunneling Microscope," *Science* **254**, 1319 (1991).

G. M. Whiteside and J. C. Love, "The Art of Building Small," *Sci. Am.* **39**, 285 (Sept. 2001).

APPENDIX A

FORMULAS FOR DIMENSIONALITY

A.1. INTRODUCTION

We saw in Chapter 9, on quantum wells, quantum wires and quantum dotes, that the dimensionality of the system has a pronounced effect on its properties. This is one of the dramatic new factors that makes nanoscience so interesting. In this appendix we gather together some of the important formulas that summarize the role played by dimensionality.

A.2. DELOCALIZATION

The regions occupied by a system of conduction electrons delocalized in one-, two-, and three-dimensional coordinate space are the length, area, and volume given in the

Introduction to Nanotechnology, by Charles P. Poole Jr. and Frank J. Owens.
ISBN 0-471-07935-9. Copyright © 2003 John Wiley & Sons, Inc.

Table A.1. Properties of coordinate and k space in one, two, and three dimensions

Coordinate Region	k-Space Unit Cell	Fermi Region	Value of k^2	Dimensions
Length L	$2\pi/L$	$2k_F$	k_X^2	One
Area $A = L^2$	$(2\pi/L)^2$	πk_F^2	$k_X^2 + k_Y^2$	Two
Volume $V = L^3$	$(2\pi/L)^3$	$4\pi k_F^3/3$	$k_X^2 + k_Y^2 + k_Z^2$	Three

first column of Table A.1. Column 2 of the table gives the size of the unit cell in reciprocal or k space, and column 3 gives the size of the Fermi region that is occupied by the delocalized electrons, where the Fermi energy E_F has the value $E_F = \hbar^2 k_F^2/2m$, and in this region $E < E_F$. Column 4 gives expressions for k^2 in the three systems. The numbers of electrons N in the occupied regions of column 3 at the temperature of absolute zero, as well as the density of states $D(E)$ defined by the expression $D(E) = dN(E)/dE$, are given in Table A.2. We see from this table that the density of states decreases with the energy for one dimension, is constant for two dimensions, and increases with increasing energy for three dimensions. Thus the number of electrons and the density of states as functions of the energy have quite different behaviours for the three cases, as indicated by the plots of Figs. 9.9, 9.10, and 9.15.

A.3. PARTIAL CONFINEMENT

The conduction electrons in nanostructures can be partially confined and partially delocalized, depending on the shape and the dimensions of the structure. One limiting case is a quantum dot in which they are totally confined, and the other

Table A.2. Number of electrons $N(E)$ and density of states $D(E) = dN(E)/dE$ as function of energy E for electrons delocalized in one, two, and three spatial dimensions, where $A = L^2$ and $V = L^3$

Number of Electrons N	Density of States $D(E)$	Delocalization Dimensions
$N(E) = \dfrac{4k_F}{2\pi/L} = \dfrac{2L}{\pi}\left(\dfrac{2m}{\hbar^2}\right)^{1/2} E^{1/2}$	$D(E) = \dfrac{L}{\pi}\left(\dfrac{2m}{\hbar^2}\right)^{1/2} E^{-1/2}$	1
$N(E) = \dfrac{2\pi k_F^2}{(2\pi/L)^2} = \dfrac{A}{2\pi}\left(\dfrac{2m}{\hbar^2}\right) E$	$D(E) = \dfrac{A}{2\pi}\left(\dfrac{2m}{\hbar^2}\right)$	2
$N(E) = \dfrac{2(4\pi k_F^3/3)}{(2\pi/L)^3} = \dfrac{V}{3\pi^2}\left(\dfrac{2m}{\hbar^2}\right)^{3/2} E^{3/2}$	$D(E) = \dfrac{V}{2\pi^2}\left(\dfrac{2m}{\hbar^2}\right)^{3/2} E^{1/2}$	3

limiting case is a bulk material, in which they are all delocalized. The intermediate cases are a quantum wire, which is long in one dimension and very small in its transverse directions; and a quantum well, which is a flat plate nanosized in thickness and much larger in length and width. The quantum wire exhibits electron confinement in two dimensions and delocalization in one dimension, and the quantum well reverses these characteristics. Table A.3 lists the numbers of electrons $N(E)$ and the densities of states $D(E)$ for these four cases, and Figs. 9.9 and 9.10, respectively, provide plots of how they depend on the energy. The degeneracies d_i refer to potential well energy levels. For futher details, see L. Jacak, P. Hawrylak and A. Wojs, *Quantum Dots*, Springer, Berlin, 1998, Section 3.1.

The formulas presented in this appendix are for idealized cases of isotropic systems with circular Fermi limits in two dimensions, and spherical Fermi surfaces in three dimensions. The bulk case on the last row of Table A.3 assumes the presence of one conduction band. In practical cases the bands are more numerous and more complex, but these simplified expressions do serve to clarify the roles played by the effects of electron delocalization and electron confinement in nanostructures.

Table A.3. Number of electrons $N(E)$ and density of states $D(E) = dN(E)/dE$ as a function of the energy E for electrons delocalized/confined in quantum dots, quantum wells, quantum wires, and bulk material[a]

Type	Number of Electrons N	Density of States $D(E)$	Dimensions Delocalized	Dimensions Confined
Dot	$N(E) = 2\Sigma d_i \Theta(E - E_{iw})$	$D(E) = 2\Sigma d_i \delta(E - E_{iw})$	0	3
Wire	$N(E) = \dfrac{2L}{\pi}\left(\dfrac{2m}{\hbar^2}\right)^{1/2} \Sigma d_i (E - E_{iw})^{1/2}$	$D(E) = \dfrac{L}{\pi}\left(\dfrac{2m}{\hbar^2}\right)^{1/2} \Sigma d_i (E - E_{iw})^{-1/2}$	1	2
Well	$N(E) = \dfrac{A}{2\pi}\left(\dfrac{2m}{\hbar^2}\right) \Sigma d_i (E - E_{iw})$	$D(E) = \dfrac{A}{2\pi}\left(\dfrac{2m}{\hbar}\right) \Sigma d_i$	2	1
Bulk	$N(E) = \dfrac{V}{3\pi^2}\left(\dfrac{2m}{\hbar^2}\right)^{3/2} (E)^{3/2}$	$D(E) = \dfrac{V}{2\pi^2}\left(\dfrac{2m}{\hbar^2}\right)^{1/2} (E)^{1/2}$	3	0

[a]The degeneracies of the confined (square or parabolic well) energy levels are given by d_i. The Heaviside step function $\theta(x)$ is zero for $x < 0$ and one for $x > 0$; the delta function $\delta(x)$ is zero for $x \neq 0$, infinity for $x = 0$, and integrates to a unit area.

APPENDIX B

TABULATIONS OF
SEMICONDUCTING
MATERIAL PROPERTIES

In this book we have discussed various types of nanostructures, many of which are semiconductors composed of the group IV elements Si or Ge, type III–V compounds such as GaAs, or type II–VI compounds such as CdS with the sphalerite (zinc blende) structure of Fig. 2.8. The properties of these materials often become modified when they are incorporated into nanoscale structures or devices. It will be helpful to bring together in this appendix a set of tables surveying and comparing their various properties in the bulk state so that they can be referred to when needed throughout the chapters of the book.

Tables B.1–B.5 contain crystallographic and related data, and electronic information such as bandgaps, effective masses, mobilities, donor/acceptor ionization energies, and dielectric constants are presented in Tables B.6–B.11. These are the

Introduction to Nanotechnology, by Charles P. Poole Jr. and Frank J. Owens.
ISBN 0-471-07935-9. Copyright © 2003 John Wiley & Sons, Inc.

Table B.1. Lattice constant _a_ in units of nanometers for types III–V (left) and II–VI (right) semiconductors[a]

	P	As	Sb		S	Se	Te
Al	0.545	0.566	0.614	Zn	0.541	0.567	0.610
Ga	0.545	0.565	0.610	Cd	0.583	0.605	0.648
In	0.587	0.606	0.648	Hg	0.585	0.608	0.646

[a]The values for Si and Ge are 0.543 and 0.566 nm, respectively.

Source: Data from R. W. G. Wyckoff, _Crystal Structures_, Wiley, New York, 1963, Vol. 1, p. 110.

tables most often referred to throughout the book. For the convenience of the reader, we gather together in Tables B.12–B.21 data on some additional properties that may be of interest, presented in the same format as the earlier tabulated data. These include indices of refraction (Table B.12), melting points and heats of formation (Tables B.13 and B.14), and some thermal properties such as Debye temperatures θ_D, specific heats, and thermal conductivities (Tables B.15–B.17). The mechanical properties tabulated are linear expansion, volume compressibility, and microhardness (Tables B.18–B.20). Finally, we list in Table B.21 the diamagnetic susceptibilities of several III–V compounds. Some tables include more limited data entries than others because complete data sets are not readily available for every property.

Table B.2. Atomic radii from monatomic crystals and ionic radii of several elements found in semiconductors

Group	Atomic Number	Atom	Radius	Ion	Radius
II	30	Zn	0.133	Zn^{2+}	0.074
II	48	Cd	0.149	Cd^{2+}	0.097
II	80	Hg	0.151	Hg^{2+}	0.110
III	13	Al	0.143	Al^{3+}	0.051
III	31	Ga	0.122	Ga^{3+}	0.0602
III	49	In	0.163	In^{3+}	0.081
IV	14	Si	0.118		
IV	32	Ge	0.123		
V	15	P	0.110	P^{3-}	0.212
V	33	As	0.124	As^{3-}	0.222
V	51	Sb	0.145	Sb^{3-}	0.245
VI	16	S	0.101	S^{2-}	0.184
VI	34	Se	0.113	Se^{2-}	0.191
VI	52	Te	0.143	Te^{2-}	0.211

Source: Data from _Handbook of Chemistry and Physics_, CRC Press, Boca Raton, FL, 2002, pp. F189, F-191.

Table B.3. Comparison of the nearest-neighbor distance $\frac{1}{4}\sqrt{3}a$ of several semiconductor III–V and II–VI binary compounds AC with the sums of the corresponding ionic radii, and with the sums of the corresponding monatomic lattice radii[a]

Compound	$\frac{1}{4}\sqrt{3}a$ Distance	$A^{n-} + C^{n+}$ Ionic Sum	$A + B$ Monatomic Sum	Compound Type
ZnS	0.234	0.258	0.231	Semiconductor
AlP	0.236	0.263	0.253	Semiconductor
GaAs	0.245	0.284	0.246	Semiconductor
CdS	0.252	0.281	0.247	Semiconductor
HgTe	0.279	0.321	0.294	Semiconductor
InSb	0.281	0.326	0.308	Semiconductor
NaCl	0.282	0.278	0.350	Alkali halide
KBr	0.353	0.353	0.404	Alkali halide
RbI	0.367	0.367	0.419	Alkali halide
MgO	0.211	0.198	0.232	Alkaline-earth chalcogenide
CaS	0.285	0.283	0.299	Alkaline-earth chalcogenide
SrSe	0.301	0.303	0.419	Alkaline-earth chalcogenide

[a]For comparison purposes data are shown for some alkali halides and alkaline-earth chalcogenides. The distances are in nanometers.

Table B.4. Molecular mass of groups III–V and II–VI compounds[a]

	P	As	Sb		S	Se	Te
Al	57.95	101.90	148.73	Zn	97.43	144.34	192.99
Ga	100.69	144.64	191.47	Cd	144.46	191.36	240.00
In	145.79	189.74	256.57	Hg	232.65	279.55	328.19

[a]The atomic masses of Si and Ge, respectively, are $\frac{1}{2}(56.172)$ and $\frac{1}{2}(145.18)$.

Source: Data from *Handbook of Chemistry and Physics*, CRC Press, Boca Raton, FL, 2002, pp. 12–57, 12–58.

Table B.5. Density in g/cm³ of groups III–V and II–VI compounds[a]

	P	As	Sb		S	Se	Te
Al	2.42	3.81	4.22	Zn	4.08	5.42	6.34
Ga	4.13	5.32	5.62	Cd	4.82	0.567	5.86
In	4.79	5.66	5.78	Hg	7.73	8.25	8.17

[a]The densities of Si and Ge, respectively, are 2.3283 and 5.3234.

Source: Data from *Handbook of Chemistry and Physics*, CRC Press, Boca Raton, FL, 2002, pp. 12–57, 12–58.

Table B.6. Bandgaps E_g expressed in electronvolts for type III–V semiconductors (on left) and type II–VI materials (on right)[a]

	P	As	Sb		S	Se	Te
Al	(2.45)/3.62	(2.15)/3.14	(1.63)/2.22	Zn	3.68	2.7	2.26
Ga	(2.27)/2.78	1.43	0.70	Cd	2.49	1.75	1.43
In	1.35	0.36	0.18	Hg	—	−0.061	−0.30
							(4.4 K)

[a]Indirect bandgaps are given in parentheses. The values for Si and Ge are (1.11)/3.48 and (0.66)/0.81 eV, respectively.

Source: Data from P. Y. Yu and M. Cardona, *Fundamentals of Semiconductors*, Springer, Berlin, 2001, table on inside front cover.

Table B.7. Temperature dependence of bandgap dE_g/dT (in meV/°C) and pressure dependence of bandgap dE_g/dP (in meV/GPa)[a]

	P	As	Sb		S	Se	Te
		Temperature Dependence dE_g/dT					
Al	—	(−0.4)/−0.51	(−3.5)	Zn	−0.47	−0.45	−0.52
Ga	(−0.52)/−0.65	−0.395	−0.37	Cd	−0.41	−0.36	−0.54
In	−0.29	−0.35	−0.29				
	Si (−0.28)		Ge (−0.37)/−0.4				
		Pressure Dependence dE_g/dP					
Al	—	(−5.1)/102	(−15)	Zn	57	70	83
Ga	(−14)/105	115	140	Cd	45	50	80
In	108	98	157				
	Si (−14)		Ge (50)/121				

[a]The corresponding values for Si and Ge are listed below the III–V values.

Source: Data from P. Y. Yu and M. Cardona, *Fundamentals of Semiconductors*, Springer, Berlin, 2001, table on inside of front cover; see also *Handbook of Chemistry and Physics*, CRC Press, Boca Raton, FL, 2002, pp. 12–105.

Table B.8. Effective masses m^* relative to free-electron mass m_e of conduction band electrons and three types of valence band holes[a]

	Electron Effective Mass			Heavy-Hole Effective Mass			Light-Hole Effective Mass			Splitoff Hole Effective Mass			Spin–Orbit Splitting Δ_{SO} (eV)		
	P	As	Sb	P	As	Sb	P	As	Sb	P	As	Sb	P	As	Sb
Ga	—	0.067	0.047	0.57	0.53	0.8	0.18	0.08	0.05	0.25	0.15	0.12	0.08	0.34	0.75
In	0.073	0.026	0.014	0.58	0.4	0.42	0.12	0.026	0.016	0.12	0.14	0.43	0.11	0.38	0.81
		Ge 0.41			Si 0.54, Ge 0.34			Si 0.15, Ge 0.043			Si 0.23, Ge 0.09			Si 0.044, Ge 0.295	

[a]The spin–orbit splitting parameter Δ_{SO} is also listed. The splitoff hole effective mass arises from spin–orbit coupling, which was not taken into account in this book. Corresponding values for Si and Ge are listed below the III–V values.

Source: Data are from P. Y. Yu and M. Cardona, *Fundamentals of Semiconductors*, Springer, Berlin, 2001, pp. 71, 75, supplemented by G. Burns, *Solid State Physics*, Academic Press, New York, p. 312.

Table B.9. Mobilities at room temperature [in cm^2/(V·s)] for types III–V and II–VI semiconductors^a

	P	As	Sb		S	Se	Te
			Electron Mobilities				
Al	80	1,200	200–400	Zn	180	540	340
Ga	300	8,800	4,000	Cd			1,200
In	4,600	33,000	78,000	Hg	250	20,000	25,000
			Hole Mobilities				
Al	—	420	550	Zn	5 (400°C)	28	100
Ga	150	400	1,400	Cd	—	—	50
In	150	460	750	Hg	—	≈1.5	350

^{a}The electron mobilities are given first, and the hole mobilities are presented below them. The mobilities for Si and Ge for electrons are 1900 and 3800, respectively, and corresponding values for holes are 500 and 1820.

Source: Data from *Handbook of Chemistry and Physics*, CRC Press, Boca Raton, FL, 2002, p. 12–101.

Table B.10. Ionization energies (in meV) of group III acceptors and group V donors in Si and Ge

			Ionization Energy (meV)	
Ion	Group	Type	Si	Ge
B	III	Acceptor	45	10
Al	III	Acceptor	67	11
Ga	III	Acceptor	71	11
In	III	Acceptor	155	12
P	V	Donor	45	13
As	V	Donor	54	14
Sb	V	Donor	43	10
Bi	V	Donor	71	13

Source: Data from G. Bums, *Solid State Physics*, Academic Press, New York, 1985, p. 172.

Table B.11. Static relative dielectric constants $\varepsilon/\varepsilon_0$ for types III–V and II–VI semiconductors^a

	P	As	Sb		S	Se	Te
Al		10.9	11	Zn	8.9	9.2	10.4
Ga	11.1	13.2	15.7	Cd	—	—	7.2
In	12.4	14.6	17.7				

^{a}The values for Si and Ge are 11.8 and 16, respectively.

Source: Data from *Handbook of Chemistry and Physics*, CRC Press, Boca Raton, FL, 2002, p. 12–101.

Table B.12. Index of refraction *n* [optical region, $n = (\varepsilon/\varepsilon_0)^{1/2}$] for types III–V and II–VI semiconductors[a]

	P	As	Sb		S	Se	Te
Al	—	—	3.2	Zn	2.36	2.89	3.56
Ga	3.2	3.30	3.8	Cd	—	—	2.50
In	3.1	3.5	3.96	Hg	2.85		—

[a]The values for Si and Ge are 3.49 and 3.99, respectively.

Source: Data from *Handbook of Chemistry and Physics*, CRC Press, Boca Raton, FL, 2002, p. 12–101.

Table B.13. Melting point (in kelvins) for types III–V and II–VI semiconductors[a]

	P	As	Sb		S	Se	Te
Al	~2100	2013	1330	Zn	2100	1790	1568
Ga	1750	1510	980	Cd	1750	1512	1365
In	1330	1215	798	Hg	1820	1070	943

[a]The values for Si and Ge are 1685 and 1231 K, respectively.

Source: Data from *Handbook of Chemistry and Physics*, CRC Press, Boca Raton, FL, 2002, pp. 12–97, 12–98.

Table B.14. Heats of formation (in kJ/mol) at 300 K for types III–V and II–VI semiconductors[a]

	P	As	Sb		S	Se	Te
Al	—	627	585	Zn	477	422	376
Ga	635	535	493	Cd	—	—	339
In	560	477	447	Hg	—	247	242

[a]The values for Si and Ge are 324 and 291 kJ/mol, respectively.

Source: Data from *Handbook of Chemistry and Physics*, CRC Press, Boca Raton, FL, 2002, p. 12–101.

Table B.15. The Debye temperature Θ_D (in kelvins) for types III–V and II–VI semiconductors[a]

	P	As	Sb		S	Se	Te
Al	588	417	292	Zn	530	400	223
Ga	446	344	265	Cd	219	181	200
In	321	249	202	Hg	—	151	242

[a]The values for Si and Ge are 645 and 374 K, respectively.

Source: Data from *Handbook of Chemistry and Physics*, CRC Press, Boca Raton, FL, 2002, pp. 12–97, 12–98.

Table B.16. The specific heat C_P [in J/(kg·K)] at 300 K for type III–V semiconductors[a]

	P	A	Sb		S	Se	Te
Al	—	—	—	Zn	472	339	264
Ga	—	—	320	Cd	330	255	205
In	—	268	144	Hg	210	178	164

[a]The values for Si and Ge are 702 and 322 J/(kg · K), respectively.

Source: Data from *Handbook of Chemistry and Physics*, CRC Press, Boca Raton, FL, 2002, pp. 12–98.

Table B.17. Thermal conductivity [in mW/(cm·K)] at 300 K for types III–V and II–VI semiconductors[a]

	P	As	Sb		S	Se	Te
Al	920	840	600	Zn	251	140	108
Ga	752	560	270	Cd	200	90	59
In	800	290	160	Hg	—	10	20

[a]The values for Si and Ge are 1240 and 640 mW/(cm·K), respectively.

Source: Data from *Handbook of Chemistry and Physics*, CRC Press, Boca Raton, FL, 2002, pp. 12–97, 12–98.

Table B.18. Linear thermal expansion coefficient (in $10^{-6}\,K^{-1}$) at 300 K for type III–V semiconductors[a]

	P	As	Sb		S	Se	Te
Al	—	3.5	4.2	Zn	6.4	7.2	8.2
Ga	5.3	5.4	6.1	Cd	4.7	3.8	4.9
In	4.6	4.7	4.7	Hg	—	5.5	4.6

[a]The values for Si and Ge are 2.49×10^{-6} and $6.1 \times 10^{-6}\,K^{-1}$, respectively.

Source: Data from *Handbook of Chemistry and Physics*, CRC Press, Boca Raton, FL, 2002, p. 12–98.

Table B.19. Volume compressibility (in $10^{-10}\,m^2/N$) for type III–V semiconductors[a]

	P	As	Sb
Al	—	—	0.571
Ga	0.110	0.771	0.457
In	0.735	0.549	0.442

[a]The values for Si and Ge are 0.306 and $0.768 \times 10^{-10}\,m^2/N$, respectively.

Source: Data from *Handbook of Chemistry and Physics*, CRC Press, Boca Raton, FL, 2002, p. 12–101.

Table B.20. Microhardness (in N/mm^2) of types III–V and II–VI semiconductors[a]

	P	As	Sb		S	Se	Te
Al	5.5 (M)	5000	4000	Zn	1780	1350	900
Ga	9450	7500	4480	Cd	1250	1300	600
In	4100	3300	2200	Hg	3(M)	2.5(M)	300

[a]The values for Si and Ge are 11,270 and 7644 N/mm^2, respectively. M indicates microhardness values on the Mohs scale.

Source: Data from *Handbook of Chemistry and Physics*, CRC Press, Boca Raton, FL, 2002, pp. 12–97, 12–98.

Table B.21. Atomic magnetic susceptibility (in 10^{-6} cgs) for types III–V and II–VI semiconductors[a]

	P	As	Sb
Ga	−13.8	−16.2	−14.2
In	−22.8	−27.7	−32.9

[a]The values for Si and Ge are −3.9 and −0.12 × 10^{-6} cgs, respectively, and the value for ZnS is −9.9 × 10^{-6} cgs.

Source: Data from *Handbook of Chemistry and Physics*, CRC Press, Boca Raton, FL, 2002, p. 12–101.

INDEX